Recent Developments in Separation Science

Volume III
Part B

Editor

Norman N. Li, Sc.D.

Exxon Research and Engineering Company
Linden, New Jersey

Published by

CRC PRESS, Inc.
18901 Cranwood Parkway • Cleveland, Ohio 44128

Library of Congress Cataloging in Publication Data

Li, Norman N
    Recent developments in separation science.

    (CRC uniscience series)
    Includes bibliographical references.
    1. Separation (Technology) I. Title.
TP156.S45L5   660.2'842   72-88417
ISBN 0-8493-5031-X (Complete Set)
ISBN 0-87819-001-5 (Former Complete Set)

    This book represents information obtained from authentic and highly regarded sources. Reprinted material is quoted with permission, and sources are indicated. A wide variety of references are listed. Every reasonable effort has been made to give reliable data and information, but the author and the publisher cannot assume responsibility for the validity of all materials or for the consequences of their use.

    All rights reserved. This book, or any parts thereof, may not be reproduced in any form without written consent from the publisher.

© 1977 by CRC Press, Inc.

International Standard Book Number 0-8493-5031-X (Complete Set)
Former International Standard Book Number 0-87819-001-5 (Complete Set)
International Standard Book Number 0-8493-5481-1 (Volume III, Part A)
International Standard Book Number 0-8493-5482-X (Volume III, Part B)

Library of Congress Card Number 72-88417
Printed in the United States

PREFACE

It was about 5 years ago that we published our first two volumes. At that time, we did not have plans to publish additional volumes. Volumes I and II, which contain a total of 22 chapters, were therefore designed to cover practically all the important aspects of separation science and technology. Our readers' favorable responses in the past years have convinced us to make the book a multivolume series so that additional volumes can be published from time to time. This two-part volume is the first of several volumes that we plan to publish in the next few years.

We are very proud to note that in this volume (Parts A and B) we are able to present 15 chapters written by the leading authorities in separation science and technology. These chapters present discourses on unifying theories for separation processes, on the theories describing the interactions of fluid dynamics, interfacial phenomena, and mass transfer, and on a variety of separation methods and processes. The last category includes dynamically formed membranes, electrolytical purification, facilitated transport through membranes, gas absorption, ion exchange process, liquid clathrates, liquid membranes, supercritical extraction, ultrafiltration, and waste water treatment for fermentation processes.

The materials discussed in each chapter are, in general, the author's own research work and his critical review of the current state of art. The authors had complete freedom in choosing certain important areas for emphasis. As a result, some chapters treat the related chemistry or mathematics in more detail than others, and some deal more with the engineering and economics aspects of a separation process. Each chapter, consequently, possesses its own special feature and appealing points.

All the chapters were reviewed by Dr. E. W. Funk of Exxon and myself, with two chapters reviewed also by Professor R. W. Rousseau of North Carolina State University of Raleigh. I wish to thank both of them for their help. I would like to express my sincere appreciation also to the authors and the Staff of CRC Press for their effort in making Volume III possible. Special thanks are due to Ms. M. Magee and Ms. T. Weintraub of CRC Press for their editorial assistance.

Norman N. Li
Linden, New Jersey

THE EDITOR

Norman N. Li, Sc.D., heads the Separation Science Group at the Corporate Research Laboratories of Exxon Research and Engineering Company. He has numerous scientific publications and patents on topics of crystallization, extraction, blood oxygenation, enzyme membranes, water treatment, liquid membranes, and polymeric membranes. Of the many patents he holds, 21 alone deal with the basic invention and various applications of liquid surfactant membranes. He was a consultant on gas diffusion for the Apollo project and has given lectures at many universities and industrial research laboratories. Dr. Li teaches two short courses for the American Institute of Chemical Engineers, "New Separation Processes" and "Surface Chemistry and Emulsion Technology." He served as chairman of two Gordon Research Conferences, "Separation and Purification" and "Transport Phenomena in Synthetic and Biological Membranes," and of several symposia on separations for the American Chemical Society and the American Institute of Chemical Engineers.

CONTRIBUTORS

Jerry L. Atwood, Ph.D.
Department of Chemistry
University of Alabama
University, Alabama

Alan R. Berens, Ph.D.
B.F. Goodrich Co.
Research and Development Center
Brecksville, Ohio

Dibakar Bhattacharyya, Ph.D.
Department of Chemical Engineering
University of Lexington
Lexington, Kentucky

Heinz-Günter Blaschke, Ph.D.
Farbenfabrik Bayer AG
Leverkusen, West Germany

Ulf Brunke, D.Sc.
Henkel AG
Düsseldorf, West Germany

Enrico Drioli, Ph.D.
Facolta di Ingegneria
Instituto di Principi di Ingegneria Chimica
Universita degli Studi di Napoli
Naples, Italy

John W. Frankenfeld, Ph.D.
Exxon Reearch and Engineering Co.
Linden, New Jersey

Edward W. Funk, Ph.D.
Exxon Research and Engineering Co.
Linden, New Jersey

R. B. Grieves, Ph.D.
Department of Chemical Engineering
University of Kentucky
Lexington, Kentucky

Harold B. Hopfenberg, Sc.D.
Department of Chemical Engineering
North Carolina State University
Raleigh, North Carolina

Cyrus Irani, Ph.D.
Exxon Research and Engineering Co.
Linden, New Jersey

Isao Karube, D. Eng.
Research Laboratory of Resources Utilization
Tokyo Institute of Technology
Tokyo, Japan

Takashi Koike, B.S.
Kyowa Hakko Kogyo Co. Ltd.
Tokyo, Japan

Russel J. Lander, M.S.
Department of Chemical and Biochemical
    Engineering
University of Pennsylvania
Philadelphia, Pennsylvania

Ho-Lun Lee, Ph.D.
Department of Chemical Engineering
University of Wisconsin
Madison, Wisconsin

Norman N. Li, Sc.D.
Exxon Research and Engineering Co.
Linden, New Jersey

Edwin N. Lightfoot, Jr., Ph.D.
Department of Chemical Engineering
University of Wisconsin
Madison, Wisconsin

R. N. Maddox, Ph.D.
School of Chemical Engineering
Oklahoam State University
Stillwater, Oklahoma

William S. Miller, B.S.
The Permutit Company
Research and Development Center
Princeton, New Jersey

Albert B. Mindler, B.S.
The Permutit Company
Research and Development Center
Princeton, New Jersey

Minoru Nagashima, M.S.
Kyowa Hakko Kogyo Co. Ltd.
Tokyo, Japan

Sadao Noguchi, B.S.
Kyowa Hakko Kogyo Co. Ltd.
Tokyo, Japan

Robert L. Pigford, Ph.D.
Department of Chemical Engineering
University of Delaware
Newark, Delaware

John A. Quinn, Ph.D.
Department of Chemical and Biochemical
  Engineering
University of Pennsylvania
Philadelphia, Pennsylvania

Joachim F. G. Reis, Ph.D.
Department of Chemical Engineering
University of Wisconsin
Madison, Wisconsin

Hirotoshi Samejima, Ph.D.
Tokyo Research Laboratory
Kyowa Hakko Kogyo Co. Ltd.
Tokyo, Japan

Karl Schügerl, Ph.D.
Institute for Technical Chemistry
Technical University of Hanover
Hanover, West Germany

Jerome S. Schultz, Ph.D.
Department of Chemical Engineering
University of Michigan
Ann Arbor, Michigan

T. Thomas Shih, Ph.D.
Chemical Research Laboratory
Allied Chemical Company
Morristown, New Jersey

Douglas R. Smith, M.S.
Department of Chemical and Biochemical
  Engineering
University of Pennsylvania
Philadelphia, Pennsylvania

Rolf Streicher, D.Sc.
Lurgi Mineralöltechnik GmbH
Frankfurt am Main, West Germany

Shuichi Suzuki, D.Sc.
Research Laboratory of Resources Utilization
Tokyo Institute of Technology
Tokyo, Japan

Mario S. Waissbluth, Ph.D.
Department of Biotechnology and Bioengineering
Center for Advanced Studies
National Polytechnic Institute
Mexico City, Mexico

TABLE OF CONTENTS

PART A

The Systematic Description and Development of Separations Processes 1
H. L. Lee, E. N. Lightfoot, J. F. G. Reis, and M. D. Waissbluth

Interaction of Fluid Dynamics, Interfacial Phenomena, and Mass Transfer in Extraction Processes . . 71
K. Schügerl, H. G. Blaschke, U. Brunke, and R. Streicher

Removal of Salt from Water by Thermal Cycling of Ion-Exchange Resins 129
T. T. Shih and R. L. Pigford

Ion Exchange Separation of Metal Ions from Water and Waste Water 151
W. S. Miller and A. B. Mindler

Separations Using Supercritical Gases 171
C. A. Irani and E. W. Funk

PART B

Liquid Clathrates 195
J. L. Atwood

Gas Absorption Sweetening of Natural Gas 211
R. N. Maddox

Carrier-Mediated Transport in Synthetic Membranes 225
D. R. Smith, R. J. Lander, and J. A. Quinn

Carrier-Mediated Transport in Liquid-Liquid Membrane Systems 243
J. S. Schultz

Charged Membrane Ultrafiltration 261
D. Bhattacharyya and R. B. Grieves

Waste Water Treatment by Liquid Ion Exchange in Liquid Membrane Systems 285
J. W. Frankenfeld and N. N. Li

Removal of Solvent and Monomer Residuals from Glassy Polymers 293
A. R. Berens and H. B. Hopfenberg

Treatment of Fermentation Waste Waters, Its Problems and Practices 313
H. Samejima, T. Koike, S. Noguchi, and M. Nagashima

Dynamically Formed and Transient Membranes 343
E. Drioli

Electrolytic Deacidification 355
S. Suzuki and I. Karube

LIQUID CLATHRATES

J. L. Atwood

TABLE OF CONTENTS

Introduction 195
   A. Clathrate Compounds 195
   B. Initial Observation of Liquid Clathrate Behavior 196

Preparation of Parent Compounds $M[Al_2R_6X]$ 196

Anionic Structure of $M[Al_2R_6X]$ 197

Definition of Liquid Clathrate Behavior 198

Preparation of Liquid Clathrates 199

Factors Affecting the Constitution of the Liquid Clathrate 202

Solid State Clues to Liquid Clathrate Behavior 205

Origin of Liquid Clathrate Behavior 207

Stability of Liquid Clathrates 207

Application of Liquid Clathrates to Aromatic Separation Problems 207

References 209

INTRODUCTION

Clathrate Compounds

The term clathrate is used generally to define a class of substances formed from two different stable compounds without the existence of a chemical bond between them. This behavior is normally found when one compound (the host) crystallizes, because of its shape, in a fashion such that there exist cavities or holes in the lattice. The second compound (the guest) can then be trapped in these holes. When the unoccupied space in the host latttice is cage-like in nature the term conforms to its original interpretation (from the Latin word *clathratus* meaning enclosed or protected by the cross bars of a grating). These substances may, however, also exhibit other types of inclusion such as would be formed with open channels or layers in the host lattice.

The range of clathrate compounds may be subdivided into three classifications: (1) inclusion compounds with a host lattice constructed from small molecules, (2) inclusion compounds of host molecules which themselves contain holes, and (3) macromolecular inclusion compounds such as molecular sieves.

Perhaps the best known of the inclusion compounds are those of urea and thiourea. Urea crystallizes with a lattice which contains long channels of approximately 5Å diameter. Straight chain hydrocarbons, among other substances, may be accommodated in the vacancies, but branched molecules cannot fit. Any guest with a van der Waals diameter of more than 5Å would be excluded. Selectivity is a feature of great interest, and has formed the basis of many separation schemes. This aspect of clathrate compounds has been recently reviewed by Fuller.[1]

By the definition and by the general use of the term clathrate, the substances must be in the solid state. However, the properties of a novel class of liquid substances which we have investigated are such that the term "liquid clathrate" has been applied. The new use of the word should not imply order as in the solid state, but should evoke the concept of a liquid made up of guest molecules entrapped in a host. The basis of the phenomenon and potential applications will be described.

Initial Observation of Liquid Clathrate Behavior

If one considers the known compounds containing metal atoms and aromatic molecules for which a strong interaction has been demonstrated, there are clearly two extremes. On the one side are those in which there is a strong directional bond between the metal ion and the aromatic molecule such as in $Cr(C_6H_6)_2$[2] or in the $AgClO_4$·aromatic systems.[3,4] On the other are the molecules of solvation[5,6] (inclusion compounds) which are certainly held tightly in the crystal lattice, but for which there is no true interaction between the metal ion and the aromatic center.

Recently, we reported[7,8] a class of compounds which exhibit a new type of strong interaction with small aromatic molecules. The initial studies indicated that when the anion of $M[Al_2(CH_3)_6X]$ possesses an angular geometry (1), the substance will react with

$$(CH_3)_3Al \quad \overset{\Theta}{\wedge} \quad Al(CH_3)_3$$

(1)

$$(CH_3)_3Al \text{—} \Theta \text{—} Al(CH_3)_3$$

(2)

molecules such as benzene or toluene to form liquid complexes which contain 2.5 to 5 aromatic molecules per anionic unit. On the other hand, a symmetrical anionic structure (2) was not found to produce this effect.[6] The nature of the interaction was thus related to the *shape* of the anion in $M[Al_2(CH_3)_6X]$; it was proposed that a cage or layer-like structure of oriented anions is set up with counter ions and aromatic molecules trapped inside. An appropriate designation for these substances thus appeared to be liquid clathrates.

Although subsequent investigations have led to the conclusion that lattice energy, rather than an angular anionic geometry, is the overriding consideration, the layer-like liquid structure is still regarded as the best model for the behavior.

PREPARATION OF THE PARENT COMPOUNDS $M[Al_2R_6X]$

Before delving more deeply into what is presently known about liquid clathrate behavior, a survey of the structure and chemical behavior of $M[Al_2R_6X]$ will be given.

Ziegler and co-workers[9] first reported in 1960 that aluminum alkyls react with alkali metal halides to form either 1:1 or 2:1 complexes:

$$AlR_3 + MX \rightarrow M[AlR_3X]$$
$$2\,AlR_3 + MX \rightarrow M[Al_2R_6X]$$

At the time, these substances were of interest for two reasons. In the first place, the very formation of the compounds provided a neat inorganic study of the relative importance of the lattice energies of MX and $M[AlR_3X]$ or $M[Al_2R_6X]$. Lehmkuhl[10] concluded that complexation occurs most readily for the cases in which the ionic radius of the alkali metal is large, the alkyl chain is short, and the complex ion is small. The second focus of attention came as the result of the X-ray crystallographic investigation of $K[Al_2(C_2H_5)_6F]$ by Allegra and Perego[11] in 1963. They found that the anion exhibited a completely linear Al-F-Al linkage, and postulated both *sp* hybridization of the fluorine atom and *d* orbital participation by the aluminum atom in order to explain the observed bond distances and angles.

Although the logical extension of the preparative method to include alkali metal pseudohalides was hinted at in the original paper,[9] no well-defined complexes of this sort were noted until the 1971 reports by our group,[12] and by Weller et al.[13]

Of the compounds prepared thus far, for both the 1:1 and 2:1 substances, those with alkali metal cations are oxygen- and water-sensitive solids. With a tetraalkylammonium ion, the 1:1 compounds are also air-sensitive solids, but the 2:1 moieties are air-sensitive liquids. The reactivity of $M[Al_2R_6X]$ and $M[AlR_3X]$ is, however, in general much less than that of the aluminum alkyls themselves. All manipulations with the compounds should be done in an inert atmosphere, via either a dry box or on the bench top with Schlenk techniques.

ANIONIC STRUCTURE OF $M[Al_2R_6X]$

The structures of the 1:1 complexes are not subject to speculation since they are certainly quite ionic and at present show no far-reaching synthetic prospects or any indication of liquid clathrate behavior. The configurations of the 2:1 complexes are, however, of utmost importance in understanding the phenomenon, and will be dealt with now and referred to subsequently.

For the 2:1 pseudohalide complexes two clear cases may be differentiated with reference to $[Al_2R_6SCN]^-$:

(3)

(4)

For R = $-CH_3$, Structure (3) appears to be favored in the solid state from the detailed infrared and Raman studies,[14] and by analogy to the structure of $K[Al_2(CH_3)_6N_3]$.[7,8] However, synthetic work now has given an indication that Structure (4) is dominant at high temperatures[15] and for R = $-C_2H_5$ or higher alkyl chain length.[16]

A feature of the structure of $K[Al_2(CH_3)_6N_3]$ which was originally listed as an interesting sidelight is now viewed as extremely important. The methyl groups could have two possible conformations corresponding to anionic point symmetries of C_{2v} and C_s. Viewed down the nitrogen atom chain they appear as

(C_{2v})

(C_s)

Weller and Dehnicke,[14,17] have determined from spectroscopic studies that the anions in $[N(CH_3)_4][Al_2(CH_3)_6N_3]$ and $[N(CH_3)_4][Al_2(CH_3)_6SCN]$ are of C_{2v} symmetry. However, in the crystallographic asymmetric unit of $K[Al_2(CH_3)_6N_3]$ there are *two nonequivalent* $Al_2(CH_3)_6N_3^-$ ions: one of C_{2v} and one of C_s symmetry. The bond distances and angles in both forms are quite similar, and the existence of both types may be taken as an indication of the closeness in overall energy of the two configurations in solution.

As the two independent anions differ geometrically, so also do the two potassium ions differ environmentally. One is most closely associated with the terminal nitrogen atoms of two azide ions, 2.92(1) and 2.93(1)Å, with four methyl carbon atoms filling out the coordination sphere. The other potassium ion finds itself in the less favorable electrostatic position of having no portion of the azide ion within 4.90Å. Its neighborhood consists of six methyl carbon atoms at distances from 3.08(1) to 3.24(1)Å. This anomaly is viewed as extremely important.

The halide-bridged complexes are also of interest. As was mentioned previously both the angular (1) and the linear (2) structure must be considered. Structure (1) is postulated for all Cl^-, Br^-, and I^- bridged anions, while (2) has been demonstrated for $K[Al_2(C_2H_5)_6F]$[11] and $K[Al_2(CH_3)_6F] \cdot C_6H_6$.[6]

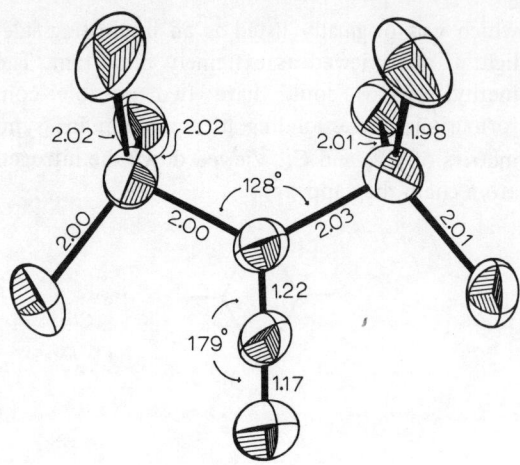

FIGURE 1A. Structure of the C_{2v} anion of $K[Al_2(CH_3)_6 N_3]$. Hydrogen atoms of the methyl groups have been omitted for clarity. (From Atwood, J. L. and Newberry, W. R., *J. Organomet. Chem.*, 65, 145, 1974. With permission.)

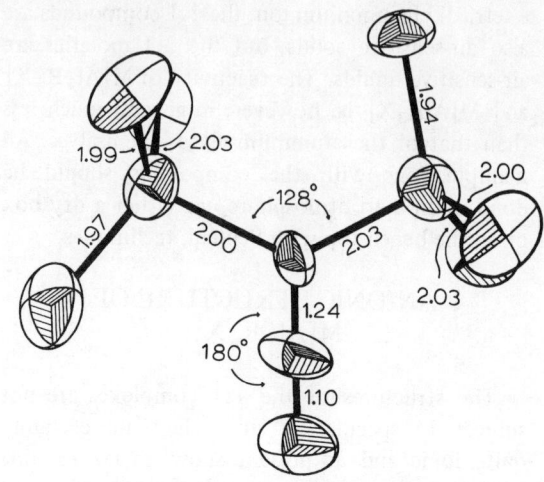

FIGURE 1B. Structure of the C_s anion of $K[Al_2(CH_3)_6 N_3]$. (From Atwood, J. L. and Newberry, W. R., *J. Organomet. Chem.*, 65, 145, 1974. With permission.)

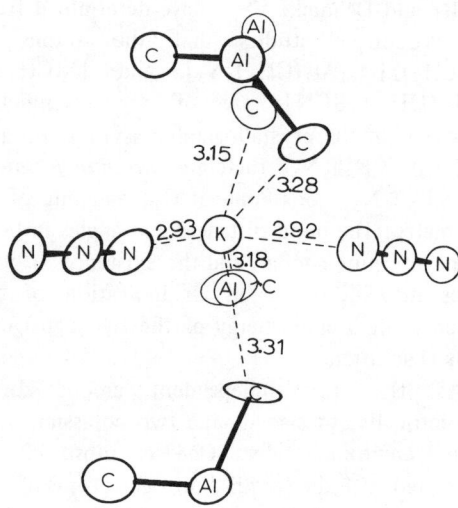

FIGURE 2A. Environment of one of the two potassium ions in the asymmetric unit of $K[Al_2(CH_3)_6 N_3]$. The ion is near the azide portion of two of the anions. (From Atwood, J. L. and Newberry, W. R., *J. Organomet. Chem.*, 65, 145, 1974. With permission.)

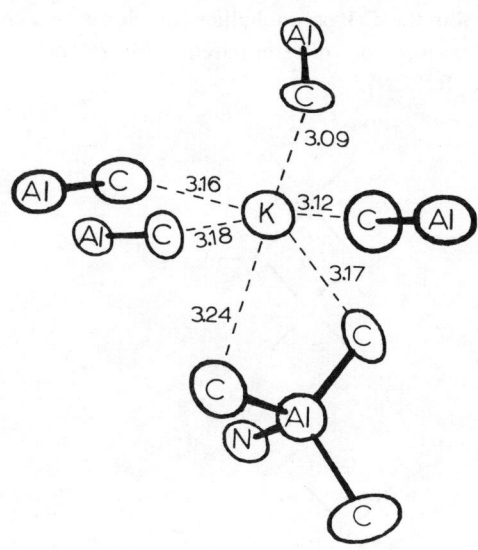

FIGURE 2B. Environment of the second potassium ion of $K[Al_2(CH_3)_6 N_3]$. The ion is surrounded by methyl groups from the anions, and is in a poorer electrostatic position than that in Figure 2A. (From Atwood, J. L. and Newberry, W. R., *J. Organomet. Chem.*, 65, 145, 1974. With permission.)

DEFINITION OF LIQUID CLATHRATE BEHAVIOR

The term liquid clathrate is used here to designate that group of nonstoichiometric liquid enclosure compounds which form upon the interaction of aromatic molecules with certain $M[Al_2 R_6 X]$ moieties. In Table 1, a compilation of different combinations of M, R, and X is given. The interaction of $M[Al_2 R_6 X]$ with an appropriate aromatic substance produces a liquid which is immiscible with excess aromatic. The visually dramatic appearance of the formation of the liquid layering is shown in Figure 3.

The aromatic molecules in the liquid clathrate

TABLE 1

Components of the Parent M[Al$_2$R$_6$X] Compounds Which Have Been Used to Form Liquid Clathrates

M	R	X
K	–CH$_3$	F$^-$
Rb	–C$_2$H$_5$	Cl$^-$
Cs	–C$_3$H$_7$	Br$^-$
NR$'_4$	–C$_4$H$_9$	I$^-$
PR$'_4$		N$_3^-$
		SCN$^-$
		SeCN$^-$
		NO$_3^-$
		HCOO$^-$
		CH$_3$COO$^-$

are truly trapped much as they would be in a solid clathrate, and can be freed by a change in temperature and reclaimed unchanged. These observations may be summarized in the form of the equilibrium

M[Al$_2$R$_6$X] + n aromatic $\underset{\text{L.T.}}{\rightleftharpoons}$ M[Al$_2$R$_6$X]·n aromatic liquid clathrate

The exact temperature at which the transformation occurs is strongly dependent on the nature of M, R, X, and the aromatic substance. For example, K[Al$_2$(CH$_3$)$_6$N$_3$]·5.8 C$_6$H$_6$ is metastable at room temperature, and will deposit crystals of K[Al$_2$(CH$_3$)$_6$N$_3$] in a few hours, while Cs[Al$_2$(CH$_3$)$_6$N$_3$]·7.4 C$_6$H$_6$ is stable for weeks; [N(C$_5$H$_{11}$)$_4$][Al$_2$(CH$_3$)$_6$I]·13.0 C$_6$H$_6$ is stable at room temperature indefinitely.

PREPARATION OF LIQUID CLATHRATES

Table 2 contains a complete compilation of all known liquid clathrates;[16] the two different ways in which they may be prepared are best illustrated by reference to the K[Al$_2$(CH$_3$)$_6$N$_3$] complex. Following the method previously described,[6] 0.010 mol Al(CH$_3$)$_3$ was added to 0.005 mol KN$_3$ in an N$_2$ atmosphere dry box. The mixture was sealed in a bomb tube, removed from the dry box, heated, returned to the dry box, and opened. Another 0.005 mol Al(CH$_3$)$_3$ was then added to the powdered contents. After three cycles of grinding, adding Al(CH$_3$)$_3$, and heating, the white crystalline product was dried under vacuum. Addition of benzene (~0.10 mol), followed by heating at 60°C for 1 hr afforded the liquid clathrate K[Al$_2$(CH$_3$)$_6$N$_3$]·5.8 C$_6$H$_6$.

A decidedly improved method for the production of the compounds involves simply the addition of 0.005 mol KN$_3$ and 0.010 mol Al(CH$_3$)$_3$ to ~0.10 mol C$_6$H$_6$ in the dry box. The liquid clathrate identical in composition to the one prepared by the previous method was obtained in 1 hr. All liquid clathrates reported here were synthesized in this fashion. As was previously noted the formation of the liquid clathrate is a visually dramatic event. As the reaction proceeds, a separation of two liquid layers (liquid clathrate and excess aromatic) becomes obvious; the layers appear upon shaking just as oil and water.

FIGURE 3. Liquid layering effect in the K[Al$_2$(CH$_3$)$_6$N$_3$]·benzene system. Lower phase contains 5.8 benzene molecules per anion.

TABLE 2

Composition of Various Liquid Clathrates

Compound	Aromatic	Maximum A/A ratio
$K[Al_2(CH_3)_6 N_3]$	benzene	5.8
$Rb[Al_2(CH_3)_6 N_3]$	benzene	6.1
$Cs[Al_2(CH_3)_6 N_3]$	benzene	7.4
$K[Al_2(CH_3)_6 NO_3]$	benzene	7.0
$Cs[Al_2(CH_3)_6 NO_3]$	benzene	12.0
$[N(C_2H_5)_4][Al_2(CH_3)_6 NO_3]$	benzene	9.8
$[N(CH_3)_4][Al_2(CH_3)_6 Cl]$	benzene	8.1
$[N(CH_3)_4][Al_2(CH_3)_6 I]$	benzene	6.5
$[N(C_2H_5)_4][Al_2(CH_3)_6 I]$	benzene	7.3
$[N(C_3H_7)_4][Al_2(CH_3)_6 I]$	benzene	9.0
$[N(C_4H_9)_4][Al_2(CH_3)_6 I]$	benzene	9.9
$[N(C_5H_{11})_4][Al_2(CH_3)_6 I]$	benzene	13.0
$[P(C_5H_5)_4][Al_2(CH_3)_6 I]$	benzene	16.1
$[N(CH_3)_4][Al_2(CH_3)_6 CH_3COO]$	benzene	6.3
$Rb[Al_2(C_2H_5)_6 N_3]$	benzene	3.8
$[N(CH_3)_4][Al_2(C_2H_5)_6 I]$	benzene	18.7
$[N(C_2H_5)_4][Al_2(C_2H_5)_6 I]$	benzene	15.9
$[N(C_3H_7)_4][Al_2(C_2H_5)_6 I]$	benzene	17.1
$[N(C_4H_9)_4][Al_2(C_2H_5)_6 I]$	benzene	18.0
$[N(C_5H_{11})_4][Al_2(C_2H_5)_6 I]$	benzene	20.4
$[N(C_6H_{13})_4][Al_2(C_2H_5)_6 I]$	benzene	34.4
$[N(C_2H_5)_4][Al_2(C_3H_7)_6 I]$	benzene	19.0
$[N(C_3H_7)_4][Al_2(C_3H_7)_6 I]$	benzene	22.6
$[N(C_5H_{11})_4][Al_2(C_3H_7)_6 I]$	benzene	42.2
$[N(CH_3)_4][Al_2(C_4H_9)_6 F]$	benzene	18.0
$K[Al_2(CH_3)_6 SCN]$	toluene	2.5
$K[Al_2(CH_3)_6 N_3]$	toluene	3.8
$Rb[Al_2(CH_3)_6 N_3]$	toluene	5.7
$Cs[Al_2(CH_3)_6 N_3]$	toluene	6.3
$[N(C_2H_5)_4][Al_2(CH_3)_6 NO_3]$	toluene	6.2
$[N(CH_3)_4][Al_2(CH_3)_6 F]$	toluene	2.9
$[N(CH_3)_4][Al_2(CH_3)_6 Cl]$	toluene	5.6
$[N(CH_3)_4][Al_2(CH_3)_6 Br]$	toluene	5.5
$[N(CH_3)_4][Al_2(CH_3)_6 I]$	toluene	5.0
$[N(C_2H_5)_4][Al_2(CH_3)_6 I]$	toluene	6.0
$[N(C_3H_7)_4][Al_2(CH_3)_6 I]$	toluene	6.4
$[N(C_4H_9)_4][Al_2(CH_3)_6 Br]$	toluene	9.3
$[N(C_4H_9)_4][Al_2(CH_3)_6 I]$	toluene	7.0
$[N(C_5H_{11})_4][Al_2(CH_3)_6 I]$	toluene	11.0
$[N(C_6H_5)(CH_3)_3][Al_2(CH_3)_6 I]$	toluene	8.4
$[N(CH_3)_4][Al_2(C_2H_5)_6 I]$	toluene	12.9
$[N(C_2H_5)_4][Al_2(C_2H_5)_6 I]$	toluene	10.6
$[N(C_3H_7)_4][Al_2(C_2H_5)_6 I]$	toluene	11.1
$[N(C_4H_9)_4][Al_2(C_2H_5)_6 I]$	toluene	13.5
$[N(C_5H_{11})_4][Al_2(C_2H_5)_6 I]$	toluene	18.4
$[N(C_6H_{13})_4][Al_2(C_2H_5)_6 I]$	toluene	30.2
$[N(C_2H_5)_4][Al_2(C_3H_7)_6 NO_3]$	toluene	14.8
$[N(C_3H_7)_4][Al_2(C_3H_7)_6 I]$	toluene	24.7

TABLE 2 (continued)

Composition of Various Liquid Clathrates

Compound	Aromatic	Maximum A/A ratio
$[N(C_2H_5)_4][Al_2(CH_3)_6I]$	ethylbenzene	4.6
$[N(C_3H_7)_4][Al_2(CH_3)_6I]$	ethylbenzene	5.0
$[N(C_4H_9)_4][Al_2(CH_3)_6I]$	ethylbenzene	5.9
$[N(C_5H_{11})_4][Al_2(CH_3)_6I]$	ethylbenzene	11.0
$[N(CH_3)_4][Al_2(CH_3)_6F]$	ethylbenzene	2.4
$[N(CH_3)_4][Al_2(C_2H_5)_6I]$	ethylbenzene	10.6
$[N(C_2H_5)_4][Al_2(C_2H_5)_6I]$	ethylbenzene	8.6
$[N(C_3H_7)_4][Al_2(C_2H_5)_6I]$	ethylbenzene	9.1
$[N(C_4H_9)_4][Al_2(C_2H_5)_6I]$	ethylbenzene	9.7
$[N(C_5H_{11})_4][Al_2(C_2H_5)_6I]$	ethylbenzene	13.2
$[N(C_6H_{13})_4][Al_2(C_2H_5)_6I]$	ethylbenzene	15.8
$[N(C_7H_{15})_4][Al_2(C_2H_5)_6I]$	ethylbenzene	17.4
$[N(C_3H_7)_4][Al_2(C_3H_7)_6I]$	ethylbenzene	17.3
$[N(CH_3)_4][Al_2(C_2H_5)_6I]$	propylbenzene	8.0
$[N(C_3H_7)_4][Al_2(C_2H_5)_6I]$	propylbenzene	7.2
$[N(C_5H_{11})_4][Al_2(C_2H_5)_6I]$	propylbenzene	10.2
$[N(CH_3)_4][Al_2(C_2H_5)_6I]$	o-xylene	12.9
$[N(C_2H_5)_4][Al_2(C_2H_5)_6I]$	o-xylene	11.8
$[N(C_3H_7)_4][Al_2(C_2H_5)_6I]$	o-xylene	11.3
$[N(C_4H_9)_4][Al_2(C_2H_5)_6I]$	o-xylene	12.5
$[N(C_5H_{11})_4][Al_2(C_2H_5)_6I]$	o-xylene	21.5
$[N(C_6H_{13})_4][Al_2(C_2H_5)_6I]$	o-xylene	39.6
$[N(C_5H_{11})_4][Al_2(CH_3)_6I]$	m-xylene	6.0
$[N(CH_3)_4][Al_2(CH_3)_6F]$	m-xylene	3.1
$[N(CH_3)_4][Al_2(C_2H_5)_6I]$	m-xylene	9.1
$[N(C_2H_5)_4][Al_2(C_2H_5)_6I]$	m-xylene	7.6
$[N(C_3H_7)_4][Al_2(C_2H_5)_6I]$	m-xylene	6.0
$[N(C_4H_9)_4][Al_2(C_2H_5)_6I]$	m-xylene	6.8
$[N(C_5H_{11})_4][Al_2(C_2H_5)_6I]$	m-xylene	9.2
$[N(C_6H_{13})_4][Al_2(C_2H_5)_6I]$	m-xylene	14.1
$[N(C_3H_7)_4][Al_2(C_3H_7)_6I]$	m-xylene	16.3
$Cs[Al_2(CH_3)_6N_3]$	p-xylene	4.3
$[N(CH_3)_4][Al_2(CH_3)_6F]$	p-xylene	2.4
$[N(C_5H_{11})_4][Al_2(CH_3)_6I]$	p-xylene	7.0
$[N(CH_3)_4][Al_2(C_2H_5)_6I]$	p-xylene	9.2
$[N(C_2H_5)_4][Al_2(C_2H_5)_6I]$	p-xylene	7.8
$[N(C_3H_7)_4][Al_2(C_2H_5)_6I]$	p-xylene	8.5
$[N(C_4H_9)_4][Al_2(C_2H_5)_6I]$	p-xylene	10.9
$[N(C_5H_{11})_4][Al_2(C_2H_5)_6I]$	p-xylene	12.8
$[N(C_6H_{13})_4][Al_2(C_2H_5)_6I]$	p-xylene	14.8
$[N(C_3H_7)_4][Al_2(C_3H_7)_6I]$	p-xylene	13.6

TABLE 2 (continued)

Composition of Various Liquid Clathrates

Compound	Aromatic	Maximum A/A ratio
$[N(CH_3)_4][Al_2(CH_3)_6 F]$	mesitylene	1.5
$[N(C_5H_{11})_4][Al_2(CH_3)_6 I]$	mesitylene	3.7
$[N(CH_3)_4][Al_2(C_2H_5)_6 I]$	mesitylene	7.5
$[N(C_2H_5)_4][Al_2(C_2H_5)_6 I]$	mesitylene	6.1
$[N(C_3H_7)_4][Al_2(C_2H_5)_6 I]$	mesitylene	5.2
$[N(C_3H_7)_4][Al_2(C_3H_7)_6 I]$	mesitylene	7.8
$[N(C_5H_{11})_4][Al_2(C_3H_7)_6 I]$	mesitylene	10.3
$[N(CH_3)_4][Al_2(CH_3)_6 F]$	1,2,3,5-tetramethylbenzene	1.3

Once a liquid clathrate has reached its maximum composition, it is not possible to cause a further uptake of aromatic molecules. The formulation of $K[Al_2(CH_3)_6 N_3] \cdot 5.8 \ C_6H_6$ in Table 2 represents a *maximum* aromatic:anion ratio.

Analysis of liquid clathrates was done by the integration of NMR spectra recorded on a Perkin-Elmer® R20-B. The aromatic stoichiometries quoted in Table 2 are in each case the average of three preparations and integrations. A realistic standard deviation for them would be ±0.2 molecules.

The liquid clathrates, although water and oxygen sensitive, are much less reactive than the pure parent organometallic compounds. None of the azides have been found to present an explosion hazard, but extreme caution should be exercised in the handling of the formates and acetates.

FACTORS AFFECTING THE CONSTITUTION OF THE LIQUID CLATHRATE

The number of aromatic molecules in the liquid clathrate is dependent upon the size and type of cation, the nature of the anion, and the type of aromatic molecule. For insight into these factors reference should be made to the data given in Table 2, and Figures 4 to 9.

Regardless of the aromatic substance employed, the effect of the size of the cation is clear-cut: for a given variety of anions, the larger the cation, *the greater the number of aromatic molecules in the clathrate*. This is true of both the alkali metal and tetraalkylammonium cation series, although at present no meaningful size comparison can be made between the two types of cations. This effect is shown graphically in Figures 4, 5, 6, and 7.

Apparent anomalies are shown with the higher aluminum alkyls, and are typified in Figures 6 and 7. The smallest numbers of aromatic molecules are taken by $N(C_2H_5)_4^+$ and $N(C_3H_7)_4^+$ complexes. No explanation has yet been offered, and the general trend is the same as seen for the trimethylaluminum derivatives in Figures 4 and 5.

The effect of the anion can be subdivided according to the nature of the alkyl group attached to the aluminum atom and the type of halide or pseudohalide ion. The former can be seen with reference to Figure 8: *the larger the aluminum alkyl, the larger the aromatics: anion (A/A) ratio*.

The second effect is illustrated most clearly in the tetraalkylammonium series (Figure 9). *The more electronegative the halide ion, the greater the number of aromatic molecules in the clathrate*. This could also be a size dependence, but the size of the halide ion probably does not affect the dimensions of the entire anion enough to cause the large effects seen in Table 2. For example, with the tetra-*n*-butylammonium complexes, the bromide takes on 9.3 toluene molecules per anion, while the iodide assumes only 7.0 toluene molecules per anion.

Just as with solid clathrates, the effect of the size of the guest is straightforward: *the larger the* aromatic molecules, the smaller the number of aromatic molecules in the liquid clathrate. It is possible that subtle electronic effects may play a role as well, but the limitations on the accuracy of the determination of the aromatics/anion (A/A) ratio doubtless legislate against the observation of their ramifications. This general trend is seen almost without exception in the data presented in Table 2 and Figures 4 to 9.

All the trends shown for the liquid clathrates are just as one would expect for solid state clathrates, and each substance has a maximum A/A ratio. These two observations represent persuasive arguments for the existence of cages or layers of anions and cations with the guest aromatic molecules trapped within. There are, however, two other sources of information which must be examined: NMR studies and solid state analogues.

For all liquid clathrates, the NMR spectra show one feature in common: the entire spectrum is shifted 0.2 to 0.5 ppm downfield relative to the pure aromatic substance. This is believed to be a bulk diamagnetic effect, but it is significant that there is no exchange between the trapped and free aromatic molecules on the NMR time scale.

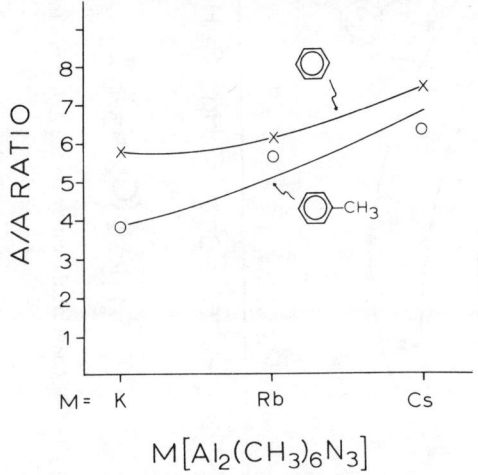

FIGURE 4. The effect of changing the cation plotted versus the aromatic/anion ratio. The standard deviations are ± 0.2 here and in Figures 5 and 9.

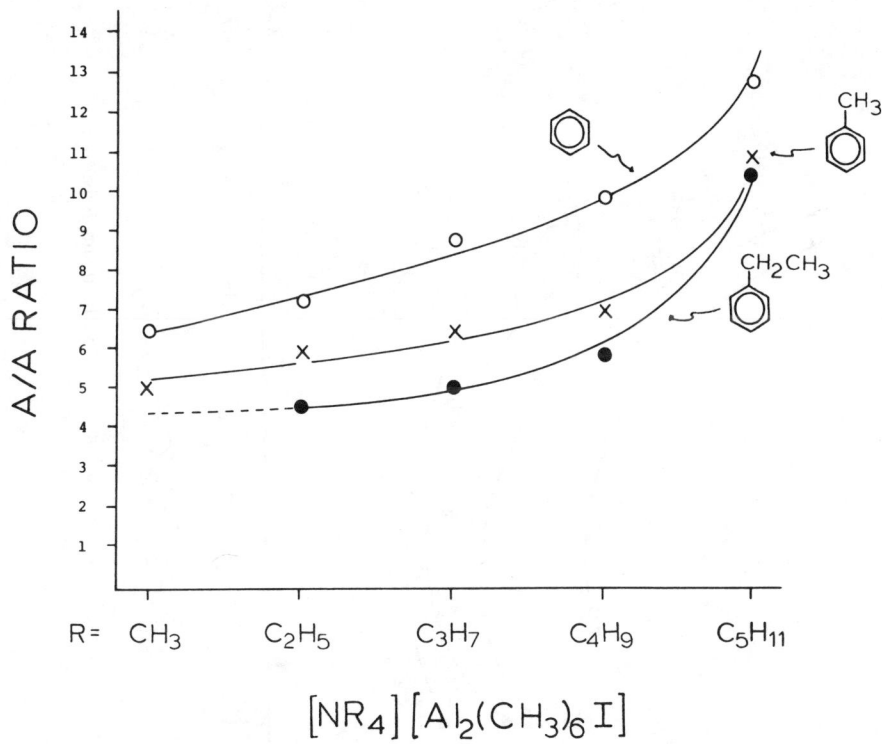

FIGURE 5. The effect of changing the tetraalkylammonium ion plotted versus the aromatic/anion ratio. [N(CH$_3$)$_4$] [Al$_2$(CH$_3$)$_6$I] disproportionates to [N(CH$_3$)$_4$] [Al(CH$_3$)$_3$I] and Al(CH$_3$)$_3$ in the presence of ethylbenzene.

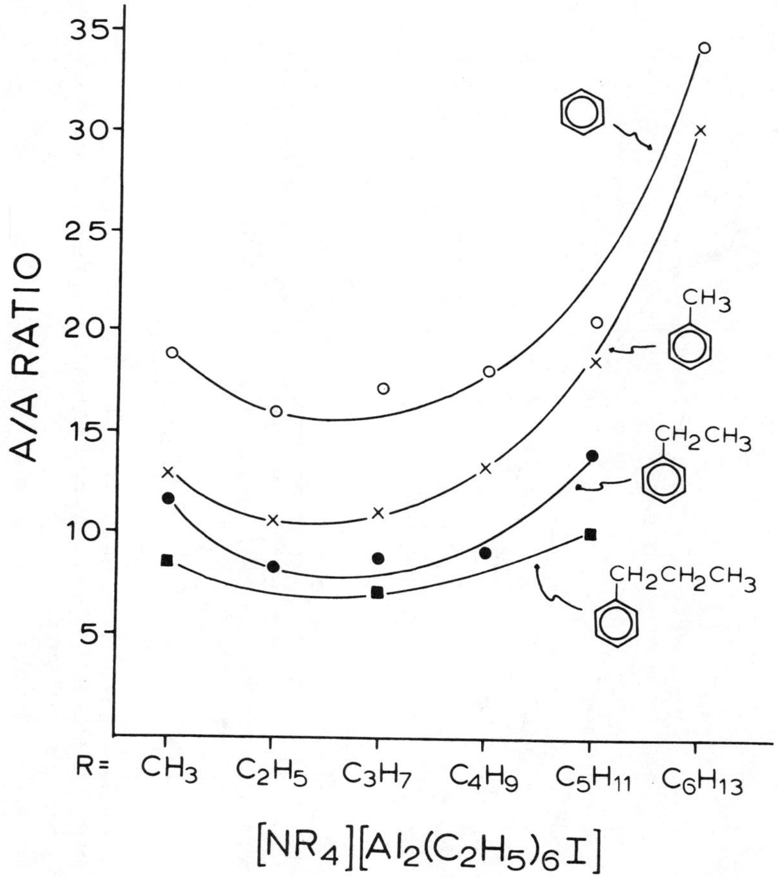

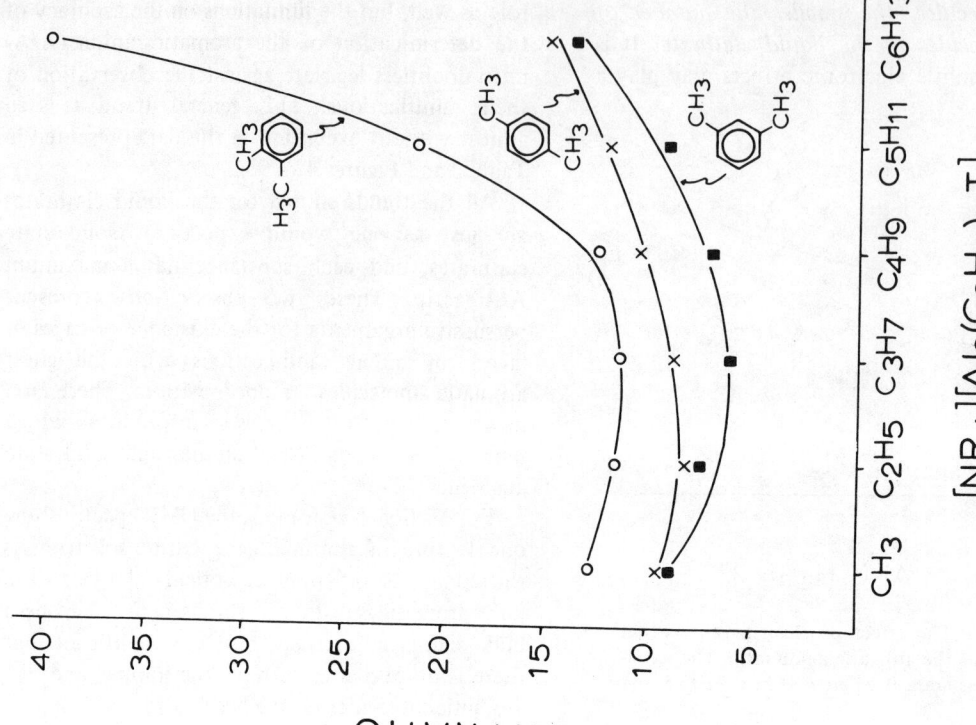

FIGURE 6. The effect of changing the cation for triethylaluminum derivatives. The standard deviations are ± 0.5 here and in Figures 7 and 8.

FIGURE 7. The effect of changing the cation on the A/A ratios for the xylenes.

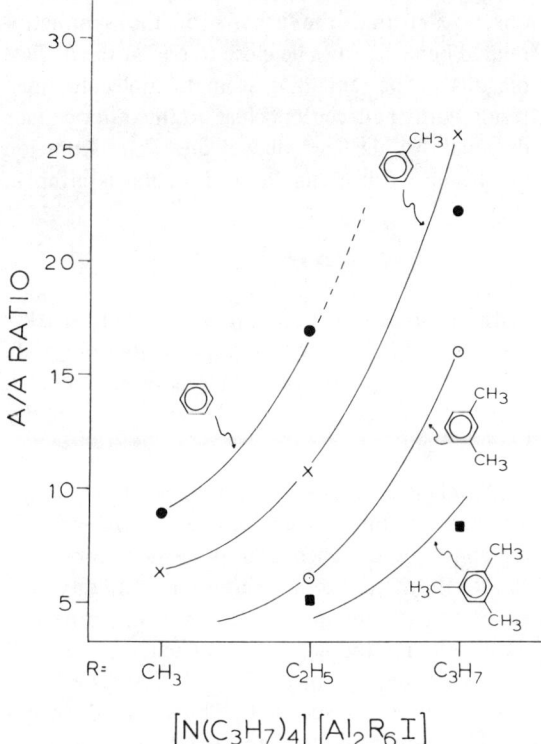

FIGURE 8. The effect of changing the alkyl group on the anion of tetrapropylammonium derivatives.

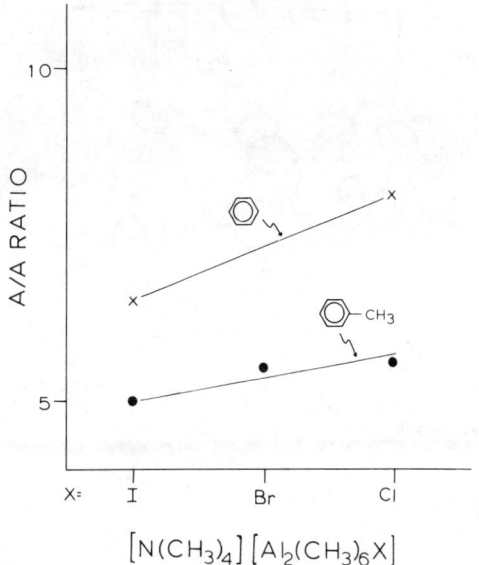

FIGURE 9. The effect of varying the halide ion on the A/A ratios with benzene and toluene as the aromatic moieties.

SOLID STATE CLUES TO LIQUID CLATHRATE BEHAVIOR

In order to understand the formation of liquid clathrates it is necessary to construct an appropriate model for the behavior. Two possibilities fit all of the observed properties: a cage-like or layer-like liquid structure. In two dimensions the models appear as in Figure 10. The layer-like structure seems much more likely, since it allows for a great reduction in the electrostatic repulsion of the cations. Although it is difficult to obtain direct experimental evidence to confirm the layer-like model, possible substantiation has been found in the form of an exciting new crystal structure.[18]

The novel selenide, $K[CH_3Se\{Al(CH_3)_3\}_3] \cdot 2 C_6H_6$, is a stable (albeit oxygen- and water-sensitive) crystalline solid which has many of the features ascribed to the liquid clathrates. For an excellent set of X-ray data, the conventional agreement index R equals 0.092 for the $K[CH_3Se\{Al(CH_3)_3\}_3]$ portion of the molecule. (The unit cell packing diagram is given as Figure 11.) Even though the atoms of the benzene molecules comprise some 32% of the total electron density,

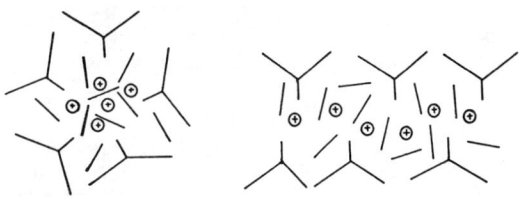

FIGURE 10. Two possible models of liquid clathrate behavior shown in two dimensions. The symbol "|" indicates an aromatic molecule, and " Y " represents an anion. The layer structure is favored from electrostatic arguments. (From Atwood, J. L. and Seale, S. K., *J. Organomet. Chem.*, 114, 107, 1976. With permission.)

the R factor indicates the structure is solved even in their absence.

In order to explain the lack of importance of the benzene molecules to the total X-ray scattering, two possibilities may be envisioned. First, extremely high thermal motion of the aromatic rings would have the effect of reducing the scattering ability of the carbon atoms. More specifically, such atoms would be expected to contribute strongly to only a few low angle reflections. The same net effect could also be achieved by a highly disordered arrangement of the benzene rings.

By a detailed analysis of electron density difference maps it was possible to locate one of the two independent benzene molecules. The positions of these molecules are shown in Figure

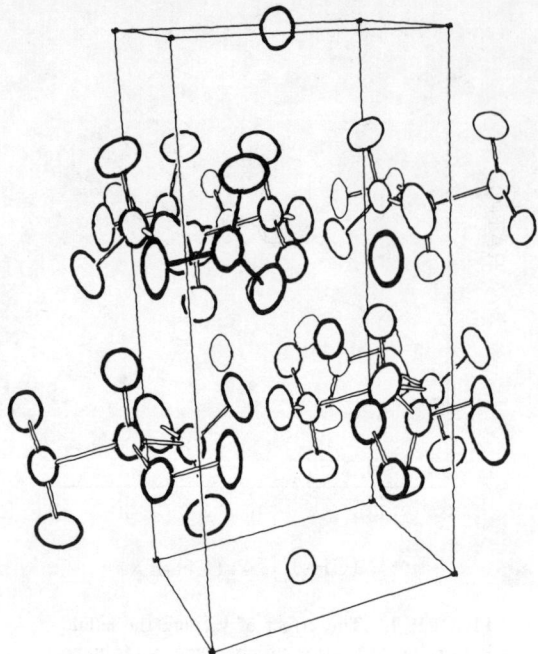

FIGURE 11. Unit cell packing of $K[CH_3Se\{Al(CH_3)_3\}_3] \cdot 2C_6H_6$ before location of the benzene molecules. The R factor was 9.2% at this point in the structure refinement. (From Atwood, J. L. and Seale, S. K., *J. Organomet. Chem.*, 114, 107, 1976. With permission.)

12. Careful inspection of Figure 12 affords the observation that the centers of the symmetry-related benzene rings lie close to one of the bc face diagonals. The remaining aromatic molecules must reside with their centers close to the other bc face diagonal. An idealized view of the "liquid" portion of the unit cell along the a direction is given in Figure 13.

The similarity between the structure given in Figures 12 and 13 and the proposed liquid clathrate structure is striking. It is possible that in the case of $K[CH_3Se\{Al(CH_3)_3\}_3]$, a solid-state mirror of the liquid ordering has been obtained. Other examples of this behavior are currently being sought.

Another interesting feature is that in the asymmetric unit there are also two crystallographically independent potassium ions. Although both reside on centers of symmetry, the environments are quite different. One is packed within the methyl groups of the anions: the K–C distances range from 3.15 to 3.22Å. The other potassium ion is associated with two carbon atoms of the anion at 3.49Å, and the remainder of its contacts must be with the benzene molecules.

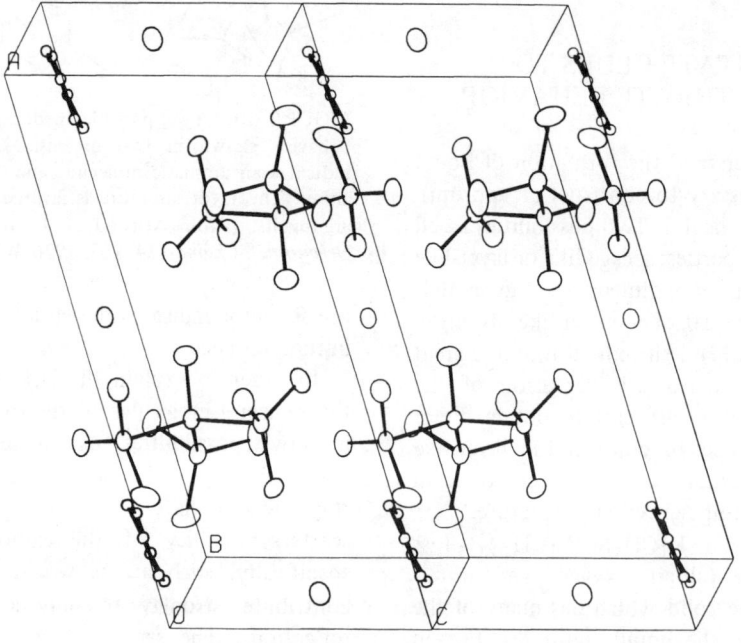

FIGURE 12. Two unit cells of $K[CH_3Se\{Al(CH_3)_3\}_3] \cdot 2C_6H_6$ with one of the two crystallographically independent benzene molecules shown. (From Atwood, J. L. and Seale, S. K., *J. Organomet. Chem.*, 114, 107, 1976. With permission.)

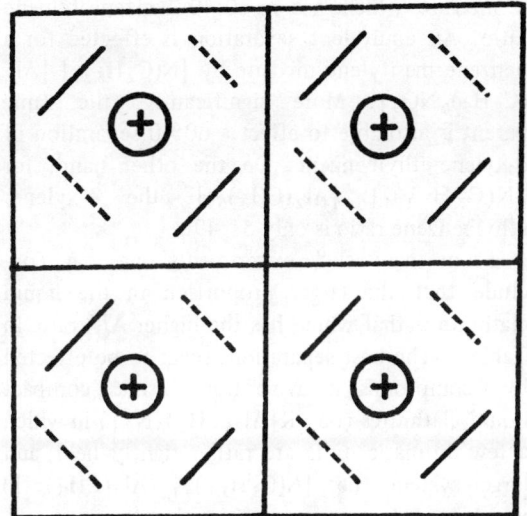

FIGURE 13. Approximate orientation of the benzene molecules with respect to the potassium ions (viewed down the *a* axis) in K[CH$_3$Se{Al(CH$_3$)$_3$}$_3$]·2C$_6$H$_6$. A solid line indicates an aromatic found on the difference Fourier; a dashed line indicates the predicted location of an aromatic. (From Atwood, J. L. and Seale, S. K., *J. Organomet. Chem.*, 114, 107, 1976. With permission.)

ORIGIN OF LIQUID CLATHRATE BEHAVIOR

In order to expand the scope of the novel solution behavior, it is necessary to originate a correct model which accounts for the experimental observations. We now believe that the parent compound must possess two properties in order to form a liquid clathrate: (1) the substance must have a relatively low lattice energy, and (2) the substance must be capable of exhibiting a very strong cation-anion interaction. These two factors seem at first glance to be contradictory, but they may be rationalized with reference to an example.

In the solid state the lattice energy of KN$_3$ is quite high. It is easy for the K$^+$ ions to pack close to the N$_3^-$ ions. The same situation also can be envisioned for K[Al(CH$_3$)$_3$N$_3$], where the Al(CH$_3$)$_3$ group has blocked a portion of the N$_3^-$ ion. However, the crystal structure of K[Al$_2$(CH$_3$)$_6$N$_3$] shows that the presence of two Al(CH$_3$)$_3$ groups causes the K$^+$ ions to pack in two different environments: one close to the N$_3^-$ ions, but one among the methyl groups. The intrinsic ability of the K$^+$ ion to interact with the N$_3^-$ group is probably not greatly diminished by the presence of the Al(CH$_3$)$_3$ groups, but the lattice energy is nonetheless relatively low.

In this circumstance the expansion of the structure by addition of the aromatic molecules can have the effect of allowing the K$^+$ ions packed among methyl groups to migrate to a more favorable electrostatic position, as in Figure 10. However, the cation-anion interaction is too strong to allow the ions to move apart as in a normal solution. Thus, the structure expands until the layers are filled, but no further.

Aromatic molecules are especially effective in promoting the effect primarily because of their ability to "insulate" the positive charges from each other. It is possible to substitute a moderate percentage of nonaromatic molecules into a liquid clathrate, but a solid parent compound, i.e., K[Al$_2$(CH$_3$)$_6$N$_3$], shows no interaction with molecules such as cyclohexane.

STABILITY OF LIQUID CLATHRATES

In the isolated liquid clathrate there are two principal modes of decomposition. The first is governed by the equilibrium

$$M[Al_2R_6X] + n \text{ aromatic} \underset{L.T.}{\rightleftharpoons} M[Al_2R_6X] \cdot n \text{ aromatic}$$

and forms the basis for the removal of the aromatic from the compound.

The second consideration is tied up in the equilibrium

$$M[Al_2R_6X] \cdot n \text{ aromatic} \rightleftharpoons M[AlR_3X] + AlR_3 + n \text{ aromatic}$$

With every constituent under discussion, M[AlR$_3$X] is a solid, and all liquid clathrates must be somewhat unstable with respect to this method of decomposition.

All liquid clathrates except those based on the acetate and formate ions are thermally stable to at least 100°C.

APPLICATIONS OF LIQUID CLATHRATES TO AROMATIC SEPARATION

From the data given in Table 2 and shown graphically in Figures 4 to 9, the potential of liquid clathrates for organic separations is apparent. Very dramatic differences for similar molecules are often seen. For example,

$[N(C_6H_{13})_4]$ $[Al_2(C_2H_5)_6I]$ will add 14 to 15 molecules of *m*- or *p*-xylene, but will take on 40 molecules of *o*-xylene. The compound, $[N(CH_3)_4]$ $[Al_2(CH_3)_6I]$, "complexes" 5.1 molecules of toluene, but does not interact with ethylbenzene.

It is also possible to work out simple, effective methods for the removal of the aromatic from the parent compound. A system based on the use of $K[Al_2(CH_3)_6N_3]$ for the separation of benzene and toluene has been worked out in principal. Solid $K[Al_2(CH_3)_6N_3]$ is warmed to 60°C in the presence of the mixture. The compound takes up proportionally more benzene than toluene, and the more dense liquid clathrate is removed by gravity. Upon cooling to 0°C, the parent compound crystallizes, and the aromatic mixture — enriched in benzene — is released. In this manner, the liquid clathrate functions as a catalyst for the separation. A single pass is not sufficient to effect a complete separation, but many passes are comparatively easy to carry out. In this respect the application of liquid clathrates is more facile than is that of solid-state clathrates.

Since there are literally thousands of liquid clathrates based on $M[Al_2R_6X]$, it will take time to evaluate the systems. If one adds in the important variables of temperature and time, then the task becomes even greater. However, our preliminary experiments have revealed several important facts about the applications.[19] As a model system, we have begun with 50:50 mixtures of aromatics. From benzene:toluene, $[N(C_2H_5)_4]$ $[Al_2(CH_3)_6NO_3]$ is able to remove in a single pass a mixture which has a 60:40 benzene:toluene ratio. An equivalent separation is effected for a benzene:mesitylene mixture by $[N(C_2H_5)_4]$ $[Al_2(C_2H_5)_6NO_3]$. More significantly, the same parent is also able to effect a 60:40 separation of *o*-xylene:ethylbenzene. On the other hand, for $[N(C_5H_{11})_4]$ $[Al_2(CH_3)_6I]$, the *o*-xylene:ethylbenzene ratio is only 51:49.

From many such experiments one may conclude that the larger proportion in the liquid clathrate is that which has the higher A/A ratio in Table 2. The best separations seem to be effected by a compromise of two extremes: small, compact liquid clathrates (i.e., $K[Al_2(CH_3)_6N_3]$) in which a few aromatic units are rather tightly held, and larger systems (i.e., $[N(C_5H_{11})_4]$ $[Al_2(CH_3)_6I]$) which contain many aromatic molecules.

Research into the use of liquid clathrates in effecting separations has only just begun. The range of compounds and variables is large enough to perhaps permit the design of specific parent compounds for specific applications.

ACKNOWLEDGMENTS

The author is grateful to the National Science Foundation for financial support of this work under NSF-GP-24852 and CHE-75-04927. Appreciation is also expressed to past students Dr. Stephen K. Seale and Dr. William R. Newberry for their assistance during the early phases of this research, and present students Karen D. Crissinger and M. Beth Humphrey for their aid in the expansion of the field.

REFERENCES

1. **Fuller, J.,** Clathrates and other inclusion compounds, in *The Encyclopedia of Chemical Processing and Design,* McKetta, J. J., Ed., Elsevier Sequoia, S. A., Lausanne, 1977.
2. **Bailey, M. F. and Dahl, L. F.,** The structure of hexamethylbenzenechromium tricarbonyl with comments on the dibenzenechromium structure, *Inorg. Chem.,* 4, 1299, 1965.
3. **Rodesiler, P. F., Griffith, E. A. H., and Amma, E. L.,** Metal ion-aromatic complexes. XIII. Trigonal planar silver(I) in the structure of indene silver perchlorate, *J. Am. Chem. Soc.,* 94, 761, 1972.
4. **Griffith, E. A. H. and Amma, E. L.,** Metal ion-aromatic complexes. XI. The crystal and molecular structure of bis(cyclohexylbenzene)silver (I) perchlorate, *J. Am. Chem. Soc.,* 93, 3167, 1971.
5. **Kobayashi, Y., Iitaka, Y., and Yamazaki, H.,** The crystal and molecular structures of a novel μ-allyl palladium complex, μ-allyl-μ-iodo-bis(triphenylphosphinepalladium) benzene solvate, *Acta Crystallogr. Sect. B.,* 28, 899, 1972.
6. **Atwood, J. L. and Newberry, W. R.,** The interaction of aromatic hydrocarbons with organometallic compounds of the main group elements. III. The crystal structure of $K[Al_2(CH_3)_6F] \cdot C_6H_6$, *J. Organomet. Chem.,* 66, 15, 1974.
7. **Atwood, J. L. and Newberry, W. R.,** Solid state structure and solution behavior of compounds of the type $M[Al_2(CH_3)_6X]$, *J. Organomet. Chem.,* 42, C77, 1972.
8. **Atwood, J. L. and Newberry, W. R.,** The interaction of aromatic hydrocarbons with organometallic compounds of the main group elements. II. Solution behavior and crystal structure of $K[Al_2(CH_3)_6N_3]$, *J. Organomet. Chem.,* 65, 145, 1974.
9. **Ziegler, K., Koster, R., Lehmkuhl, H., and Reinert, K.,** Neue Komplexverbindungen der Aluminiumalkyle, *Justus Liebigs Ann. Chem.,* 629, 33, 1960.
10. **Lehmkuhl, H.,** Complex formation with organoaluminum compounds, *Angew. Chem. Int. Ed. Engl.,* 3, 107, 1964.
11. **Allegra, G. and Perego, G.,** The crystal structure of the $KF \cdot 2Al(C_2H_5)_3$ complex, *Acta Crystallogr.,* 16, 185, 1963.
12. **Atwood, J. L., Milton, P. A., and Seale, S. K.,** Thermal behavior of anionic organoaluminum thiocyanates, *J. Organomet. Chem.,* 28, C29, 1971.
13. **Weller, F., Wilson, I. L., and Dehnicke, K.,** Halogen-und pseudohalogenkomplexe von aluminum- und galliumalkylen, *J. Organomet. Chem.,* 30, C1, 1971.
14. **Weller, F. and Dehnicke, K.,** Darstellung und Schwingungsspektren der Trimethylaluminum-pseudohalogeno-komplexe $[(Me_3Al)_2SCN]^-$, $[Me_3AlSCN]^-$, $[(Me_3Al)_2SeCN]^-$ und $[(Me_3Al)_2CN]^-$, *J. Organomet. Chem.,* 36, 23, 1972.
15. **Atwood, J. L. and Bowles, L. K.,** unpublished results.
16. **Atwood, J. L. and Atwood, J. D.,** Non-stoichiometric liquid enclosure compounds ("liquid clathrates"), *Inorganic Compounds with Unusual Properties,* Adv. Chem. Ser., Vol. 150, King, R. B., Ed., American Chemical Society, Washington, 1976, chap. 11.
17. **Weller, F. and Dehnicke, K.,** Darstellung und Schwingungsspektren der Trimethylaluminum-azido-komplexe $[(Me_3Al)_2N_3]^-$ und $[Me_3AlN_3]^-$, *J. Organomet. Chem.,* 35, 237, 1972.
18. **Atwood, J. L. and Seale, S. K.,** The interaction of aromatic hydrocarbons with organometallic compounds of the main group elements, IV. The preparation and structure of the novel selenide $K[CH_3Se\{Al(CH_3)_3\}_3] \cdot 2C_6H_6$, *J. Organomet. Chem.,* 114, 107, 1976.
19. **Atwood, J. L. and Humphrey, M. B.,** unpublished results.

GAS ABSORPTION SWEETENING OF NATURAL GAS

R. N. Maddox

TABLE OF CONTENTS

Introduction 211

Process Developments 212

Hot Carbonate Processes 212

Physical Solvent Processes 213

Combination Processes 214

Solids Processes 214

Pollution Control Processes 215

New Data 216

Calculational Procedures 220

References 223

INTRODUCTION

There have been a number of recent developments in gas sweetening. New processes have been developed, new data reported, and new calculation procedures suggested. Perhaps the most important and significant development is none of these three, but rather the emphasis by federal and state governments on control of air pollution and regulated quantities of sulfur compounds that can be released to the atmosphere. This has completely changed the sulfur production picture. Where once the decision to produce sulfur was based on economic analyses and market projection, it now is looked on as the most convenient and realistic way for disposing of waste material that cannot be otherwise consumed or released. Because of this, Buckingham and Homan[1] project a continuing surplus of produced sulfur through at least 1980.

Actually their data and projections indicate a worsening supply-demand picture for sulfur producers through the next decade. Escalations in energy prices in recent years have clouded the projection picture as far as sulfur is concerned. This is because the Frasch process for sulfur production is highly energy intensive and is also the largest supplier of sulfur for world markets. The result is an unclear sulfur supply-demand picture for the next few years.

Removal of H_2S and/or CO_2 from natural gas streams is necessary to make such gas streams suitable for consumer consumption. Specifications on the degree of hydrogen sulfide removal required are essentially standard in the United States: one quarter grain of H_2S per 100 standard cubic feet of natural gas. This standard remains unchanged as the criterion for so-called "sweet gas" in most natural gas applications. The degree of carbon dioxide removal varies. In general, carbon dioxide may be transported in natural gas streams so long as the quantity of carbon dioxide does not reach the point of seriously lowering the heat content of the gas.

PROCESS DEVELOPMENTS

The first successful processes applied to removing hydrogen sulfide from natural gas involved reaction of the acid gas constituents with monoethanolamine (MEA). The reactions proceed as follows: For H_2S

$$2RNH_2 + H_2S \rightleftharpoons (RNH_3)_2S \quad (1)$$

$$(RNH_3)_2S + H_2S \rightleftharpoons 2RNH_3HS \quad (2)$$

For CO_2

$$2RNH_2 + H_2O + CO_2 \rightleftharpoons (RNH_3)_2CO_3 \quad (3)$$

$$(RNH_3)_2CO_3 + H_2O + CO_2 \rightleftharpoons 2RNH_3HCO_3 \quad (4)$$

or

$$2RNH_2 + CO_2 \rightleftharpoons RNHCOONH_3R \quad (5)$$

These reactions proceed to the right at low temperatures and are reversed at high temperatures. The gas is sweetened in an absorber, usually at elevated pressure, with the fouled amine solution then circulating to a low pressure regenerator tower where the amine solution is regenerated.

For years, monoethanolamine was by far the most popular of the amines for use as a sweetening agent. Because MEA is the most reactive of the amines, it can readily treat gases to sweet gas specifications. MEA reacts irreversibly with carbonyl sulfide, and for this reason diethanolamine (DEA) tended to be used for refinery gases.

Recently diethanolamine has become increasingly popular as a natural gas sweetening agent. The SNPA-DEA process[2] proved that diethanolamine can be successfully used for sweetening natural gas streams. Used in higher concentrations, it can be competitive with monoethanolamine in acid gas carrying capacity per gallon of solution circulated. In addition, it appears to be capable of higher acid gas loading with lower corrosion rates than is the case for the MEA solutions.

Diisopropanolamine is also used for an acid gas removal. The Adip® process[3] is licensed by Shell and has advantages claimed for it of lower heat of reaction with acid gas constituents in comparison with DEA and MEA, and a reversible reaction with COS as compared with MEA. It is claimed to be particularly well suited for removal of COS from liquefied gas constituents such as propane and butane.

The Fluor Econamine process is a recently developed process which uses 2-(2-aminoethoxy)ethanol as a sweetening agent.[4] The reactions involved are essentially the same as those with the ethanolamines and are reversible by heating at low pressure. The process uses a high concentration of solution (60 to 65% by weight), which results in lower solution circulation with attendant lower investment and operating costs. The Econamine process was developed for higher acid gas concentrations and shows to best advantage when the sour gas contains more than 1% total acid gas.

General problems encountered in operating amine-type sweetening units include corrosion, foaming, and solution degradation. Maintaining these problems within reasonable limits is important if the sweetening unit is to operate satisfactorily. The causes of these problems are interrelated and a unit suffering from one ordinarily has one or both of the other problems also.

Experience has shown that maintaining the quality of the amine solution will serve to eliminate or minimize all operating problems. The quality of the circulating solution can be best guaranteed by a combination of reclaiming and/or filtration. A reclaimer is a batch distillation operated on a small side stream of the circulating solution. It vaporizes the amine and water, leaving behind solids and heavier materials resulting from degradation, as well as other undesirable products. Ballard[5] gave an excellent summary of the construction and operation of reclaimers for amine-type units. Reclaimers are not normally used on diethanolamine solutions. Wendt and Daily[2] report that satisfactory quality diethanolamine solution can be maintained by filtration alone.

Filters for amine-type plants normally include cloth or bag filters for removing particles down to the range of 10 to 25 μm in size. In addition, an activated charcoal filter on a side stream of the circulating solution is recommended for removal of smaller particles and degradation products.

HOT CARBONATE PROCESSES

The potassium carbonate process was developed by the U.S. Bureau of Mines for the bulk removal of carbon dioxide from synthesis gas streams. It is referred to as the "hot" carbonate process because both the absorber and the regenerator run at elevated temperatures, normally in the range of 230 to 240°F. The reactions involved are

$$K_2CO_3 + CO_2 + H_2O \rightleftharpoons 2KHCO_3 \quad (6)$$

$$K_2CO_3 + H_2S \rightleftharpoons KHS + KHCO_3 \quad (7)$$

These reactions make clear the reason for high temperature operation; whether the acid gas constituent is carbon dioxide or hydrogen sulfide, the bicarbonate salts result in 2 mol of salt being formed for each mole of acid gas reacting. The hot carbonate process does not treat sour gas to sweet-gas specifications and encounters difficulty in treating to CO_2 contents much below 1%. It is generally restricted to pressures above 300 psia in the absorber.

There are a number of compounds which will serve as "activators" or catalysts to speed the reaction of the potassium carbonate solution with acid gas constituents. The Catacarb® process,[6] the Benfield process,[7] and the Giammarco-Vetrocoke process[8] are notable activated carbonate processes. Only the Giammarco-Vetrocoke process has been reviewed in detail. The Giammarco-Vetrocoke H_2S removal process is based on the absorption of hydrogen sulfide in a solution of alkaline arsenites are arsenates. In contrast with most sweetening processes, the Giammarco-Vetrocoke process results in the production of sulfur. The chemistry of the process is complex but can be represented as:

Absorption

$$KH_2AsO_3 + 3H_2S \rightarrow KH_2AsS_3 + 3H_2O \quad (8)$$

Digestion

$$KH_2AsS_3 + 3KH_2AsO_4 \rightarrow 3KH_2AsO_3S + KH_2AsO_3 \quad (9)$$

Acidification

$$3KH_2AsO_3S \rightarrow 3KH_2AsO_3 + 3S \quad (10)$$

Oxidation

$$6KH_2AsO_3 + 3O_2 \rightarrow 6KH_2AsO_4 \quad (11)$$

The catalyzed potassium carbonate process can treat natural gas to sweet-gas specifications[9] and also reduce CO_2 levels below those of the nonactivated carbonate solutions. Parrish and Neilson[10] report on the degree of removal of CO_2 and H_2S that can be obtained in a single-stage unactivated hot carbonate process. The results were obtained in a pilot study with the absorber at 300 psig and 230°F.

Feed Concentration (mol %)		Outlet gas (ppm)	
CO_2	H_2S	CO_2	H_2S
4.8	4.7	3750	800
8.2	1.7	3650	390
9.5	0.5	3500	180

They also report on the removal of trace sulfur contaminants. Carbonyl sulfide is hydrolized to form CO_2 and H_2S.

$$COS + H_2O \rightleftharpoons CO_2 + H_2S \quad (12)$$

They report carbonyl sulfide removal from 75 to 99% in commercial installations.

Carbon disulfide hydrolizes in activated potassium carbonate solutions to form COS and H_2S. COS would then hydrolize as outlined above.

$$CS_2 + H_2O \rightleftharpoons COS + H_2S \quad (13)$$

In a commercial unit in Scotland, operating on gas from a coal gasifier, 75% of the 2.5 ppm CS_2 in the feed gas is absorbed by the carbonate solution.

Mercaptans are slightly acidic and react with activated potassium carbonates to form mercaptides:

$$K_2CO_3 + RSH \rightleftharpoons RSK + KHCO_3 \quad (14)$$

At the same unit in Scotland mentioned above, complete removal of 4 ppm mercaptan was reported.

In the carbonate processes, care must be taken to avoid "cold spots" in the process, particularly where the solution rich in acid gas is encountered. Formation of solids and scale does present some problems and solution filtration is highly recommended.

PHYSICAL SOLVENT PROCESSES

The physical solvent utilizes its attraction for the acid gas constituents for their solution in the solvent. The ideal physical solvent would be inexpensive, noncorrosive, stable under all operating conditions, have a low viscosity, be nonvolatile and nonflammable, and have no foaming characteristics. It also would have infinite solubility for the solute and, in the case of sweetening applications, would have a high degree of selectivity for hydrogen sulfide over carbon dioxide. Obviously, the ideal solvent has not yet been discovered nor is it likely to be.

A number of new physical solvents have been introduced. Notable among these are the Fluor Solvent process,[11] which utilizes propylene carbonate as a solvent; the Purisol process licensed by Lurgi Gesellschaft für Wärme- und Chemotecknik m.b.H. and Ralph M. Parsons Company, which uses n-methyl-2-pyrrolidone (M-Pyrol®) as a solvent;[12] the Selexol® process licensed by Allied Chemical Corporation, which uses the dimethyl ether of polyethylene glycol as a solvent;[13] a process described by Woertz[14] which uses methyl cyanoacetate as a solvent; and the Rectisol® process, also licensed by Lurgi, which uses methanol as a solvent.[15]

The physical solvents do have the advantage that their carrying capacity for acid gases is a function of the partial pressure of the acid gas. This means that in gases containing a high percentage of acid gas constituents, the amount of acid gas dissolved in the solvent will be higher than might be the case for a completely reacted solvent. On the other hand, this also means that lower partial pressures result in lower solution loading. Consequently, the physical solvent processes are viewed as being ideally suited for bulk removal of acid gas constituents. Recent evidence,[12,13,15] however, indicates that physical solvent processes can be used for treating natural gas to sweet gas specifications. The most popular situation for treating pipeline-specification gas with physical solvents involves sour gases containing relatively small amounts of H_2S. In these cases, the solvent processes can selectively remove the H_2S and, if properly designed, leave a significant portion of the carbon dioxide in the sweetened gas stream.

Process flow for the physical solvents is essentially that of high pressure absorption followed by regeneration by flashing at low pressure. Most of the physical solvents have a relatively high affinity for hydrocarbons so that generally flashes at intermediate pressures are necessary to prevent excessive hydrocarbon losses. The flash gases from the intermediate pressures are either compressed for recycle to the absorber or used for fuel. In the case of treatment to sweet gas specifications, something other than atmospheric flash is generally required. Air and/or inert gas stripping and possibly heating for solvent regeneration is necessary to make pipeline-specification gas. In choosing between air and inert gas as a stripping medium, consideration must be given to the possibility of oxidizing hydrogen sulfide to sulfur by the use of air.

COMBINATION PROCESSES

A number of processes which combine two solvents in order to obtain the advantages of both have been introduced. The first of these recorded was the Shell Sulfinol process.[16,17] This process utilizes diisopropanolamine in combination with the physical solvent sulfolane (tetrahydrothiophene 1,1-dioxide). By combining a chemically reacting material with a physical solvent, the distinct advantages of high-potential acid gas loading together with good acid gas cleanup characteristics are obtained.

The Amisol® process is another Lurgi development.[18] This combines methanol as a physical solvent with a satisfactory ethanolamine. Here the advantages are the same as those claimed for the Sulfinol process.

A variation on the combination process is the Hi-Pure process licensed by Benfield Corporation.[19] This process combines an activated hot-carbonate solution with a chemically reacting solution reported by Thirkell[20] to be DEA. The advantage of this over the conventional activated carbonate is the certainty of treating to sweet-gas specifications and lower utility operating costs.

SOLIDS PROCESSES

Crystalline alumino silicates can be used for removal of H_2S and other sulfur compounds from natural gas streams. Some forms of these materials occur naturally. The most common forms used in commercial adsorption are synthetically manufactured materials. They were developed in the late 1940s by the Linde Division of the Union Carbide Corporation, and are known colloquially as "molecular sieves." This reference comes from the fact that they have very accurately and evenly sized pores in their surfaces. In some applications, molecules of a given size can "enter" the pores while larger molecules cannot, thus leading to a "sieving" action based on molecular size. In gas sweetening applications, however, they act as solid adsorbents rather than molecular sieves.

In keeping with other similar methods, the molecular sieves are restricted to gas volumes containing small amounts of acid gas constituents. Large concentrations of acid gas require excessive vessel size and adsorbent quantities.

Adsorption occurs at ambient temperatures and high pressure. Regeneration is essentially by heating and may be carried out at either high or low pressure. The regeneration stream, however, should not contain heavier hydrocarbons or other materials that might be adsorbed in the bed and not removed during the regeneration cycle. The regeneration gas is contacted by water, amine, or other solvent to remove the acid gasses stripped from the regenerated adsorption bed.

A problem than can occur in molecular sieve units involves the formation of carbonyl sulfide from hydrogen sulfide and carbon dioxide. The reaction is

$$H_2S + CO_2 \rightleftharpoons COS + H_2O \qquad (15)$$

At ambient temperatures the equilibrium constant for this reaction is such that conversion to hydrogen sulfide and carbon dioxide is almost complete. At elevated temperatures, however, such as those that may occur in the regeneration cycle, conversion is much less complete and some carbonyl sulfide can exist in the gas. This can lead to problems in two ways: high solvent consumption in the sweetening part of the regeneration gas cycle or off-specification sweetened gas. The latter occurs when there is a total sulfur content restriction on the sweetened gas. Molecular sieves will treat to specification on H_2S, but the carbonyl sulfide that may be produced in the regeneration step may throw the gas off on total sulfur specification.

POLLUTION CONTROL PROCESSES

Most processes of this kind are intended for treating gases relatively low in acid gas constituents. A typical application would be for treating the tail gas from a Claus sulfur unit. Some of the processes do have the capability for treating natural gas streams per se. When applied to tail gas from Claus units, most of the processes utilize some hydrosulfurization step to convert carbonyl sulfide and disulfides in the tail gas to hydrogen sulfide before treatment.

The Stretford A.D.A. process[21-23] was developed in Great Britain. Its name is derived from the fact that anthraquinone disulfonic acid is used in the process. Basically, the process consists of solution of hydrogen sulfide in a basic solution of sodium carbonate with resulting production of sulfur and regeneration of the solution for further use. The five different steps in the Stretford A.D.A. Process reactions are

1. Absorption of hydrogen sulfide in the alkali solution

$$Na_2CO_3 + 2H_2S \rightarrow 2NaHS + 2NaHCO_3 \qquad (16)$$

2. Reduction of the anthraquinone disulfonic acid by addition of sulfur from the sodium bisulfide

$$2RC=O + 2NaHS \rightarrow 2RC\genfrac{}{}{0pt}{}{ONa}{SH} \qquad (17)$$

where $RC=O$ is the anthraquinone disulfonic acid molecule

3. Liberation of sulfur from the reduced anthraquinone disulfonic acid by interaction with oxygen dissolved in the solution

$$2RC\genfrac{}{}{0pt}{}{ONa}{SH} + \tfrac{1}{2}O_2 \rightarrow RC=O + RC\genfrac{}{}{0pt}{}{ONa}{H}$$

$$\text{dissolved oxygen} + 2S + NaOH \qquad (18)$$

The sodium hydroxide formed reacts with sodium bicarbonate to form sodium carbonate and water as follows:

$$NaHCO_3 + NaOH \rightarrow Na_2CO_3 + H_2O \qquad (19)$$

4. Reoxidation of the reduced anthraquinone disulfonic acid

$$RC\genfrac{}{}{0pt}{}{ONa}{H} + \tfrac{1}{2}O_2 \rightarrow RC=O + NaOH \text{ air oxidation} \qquad (20)$$

Again, the sodium hydroxide formed reacts with the sodium bicarbonate to form sodium carbonate and water as shown in reaction Step 3.

5. Reoxygenation of the alkaline solution to provide dissolved oxygen for Step 3 of the reaction process.

Sweetened gas produced by the process can be as low in concentration as 0.2 ppm of H_2S. The sulfur produced is of good quality and suitable for sale.

An extension of the Stretford A.D.A. Process was developed by adding a vanadium salt to the circulating liquor. This addition increases the reaction rate for the oxidation of hydrosulfide to sulfur. Also the vanadate process can work at a lower pH than the A.D.A. process. The reaction of the Stretford A.D.A./vanadate process is

$$4NaVO_3 + 2H_2S \rightleftharpoons Na_2V_2O_9 + 2S + 2NaOH + H_2O \quad (21)$$

The vanadate reacts with the hydrosulfide to form sulfur. The vanadate is reduced in the process and is then oxidized by the A.D.A. The reduced A.D.A. is then oxidized by atmospheric oxygen provided to the oxidizer (regenerator) towers. The vanadate by itself is capable of completing the reactions to form sulfur from the hydrosulfide. The vanadium salts are not readily oxidizable by atmospheric oxygen and this is the reason a two-stage process utilizing A.D.A. as an oxygen carrier is used.

The Beavon sulfur removal process[29] is licensed by Union Oil Company of California and the Ralph M. Parsons Company.[24] It is a two-step process using a cobalt-molybdate catalyst for the hydrogenation of sulfur dioxide, carbonyl sulfide, and carbon disulfide.[25] This hydrogenation reaction gives essentially complete conversion of carbon disulfide with conversions of carbonyl sulfide approaching 98%. Final purification of the gas is accomplished via a conventional Stretford A.D.A./vanadate process.

The Institut Francais du Pétrole process[30] is a combination catalyst-solvent process for conversion of low concentrations of hydrogen sulfide and SO_2 to sulfur. The process offers high conversion of these two gases but has no effect on carbonyl sulfide, carbon disulfide, or other sulfur compounds.

Shell Development Company has described in patents a process to absorb H_2S in a solution of sulfolane. The sulfolane contains a ferrous salt catalyst, a pyridine carboxylic acid, and water. The H_2S reacts with SO_2 to produce elemental sulfur.

Tokyo Gas Company has announced the Takahax process which utilizes a solution of sodium carbonate and 1,4-naphthoquinone, 2-sulfonate sodium. This process is very similar to the Stretford process. Hydrogen sulfide reacts with sodium carbonate to form sodium bisulfide which in turn reacts with the naphthoquinone to precipitate elemental sulfur. The reduced naphthoquinone is oxidized to its original chemical form by blowing with air.

The Shell-Claus Off-Gas Treating process (SCOT)[26] involves the catalytic conversion of all sulfur values in the gas to H_2S. This is followed by absorption of the H_2S into an alkanolamine. The H_2S from the generator is combined with the main feed stream to the Claus sulfur unit.

The "Lo-Cat" process is licensed by Air Resources, Inc.[27] It uses an organometallic complex (ARI-300) as a proprietary catalyst for conversion of H_2S to sulfur. Removal of 99+% of H_2S from the gas stream is possible, with elemental sulfur provided in the range of 95%. The process is said to be cheaper than many competing processes because the catalyst is regenerated by blowing with air instead of being consumed in the H_2S removal.

NEW DATA

One of the most important pieces of recent experimental work has not attracted widespread attention. Its impact, however, could have great portent in any natural gas sweetening operation. Jones et al.[28] reported experimental data showing the interacting effect of carbon dioxide and hydrogen sulfide on the acid gas carrying capacity of 15% by weight monoethanolamine solution in water. Tables 1 and 2 show a part of the data of Jones et al. Study of these data results in two conclusions: the partial pressure of H_2S over the amine solution has an effect on the degree of conversion of the amine to its sulfide salt, and the presence of CO_2 has an effect on the amount of amine that will react with hydrogen sulfide at any gas partial pressure of hydrogen sulfide. Since most sour natural gases will contain both hydrogen sulfide and carbon dioxide, the data of Jones et al. must be considered in determining the equilibrium conversion of monoethanolamine that can be obtained in the absorbent. McCoy and Maddox[31] have plotted the Jones et al. data in nomograph form for easier use. A typical nomograph for the 60°C data is shown in Figure 1.

Operation of monoethanolamine sweetening units over the years has proven that carbon dioxide is more difficult to regenerate from the amine than is hydrogen sulfide. Study of the reactions for the two compounds with monoethanolamine will show the reason for this. Reaction with carbon dioxide involves water also.

TABLE 1

Hydrogen Sulfide and Carbon Dioxide Partial Pressures over Monoethanolamine Solutions Containing Hydrogen Sulfide and Carbon Dioxide

Temp °C	Partial pressure H_2S (mm)	Mole H_2S per mole MEA						
		$R_V^* = 0.01$	$R_V = 0.05$	$R_V = 0.10$	$R_V = 0.50$	$R_V = 1.0$	$R_V = 10$	$R_V = \infty$
40	1	0.0013	0.0035	0.0050	0.0120	0.0178	0.0500	0.128
	10	0.0039	0.0100	0.0149	0.0380	0.0540	0.1450	0.374
	100	0.0107	0.0279	0.0415	0.1050	0.1510	0.3900	0.802
	1000	–	0.0625	0.0920	0.2170	0.3050	0.7300	1.00
120	1	0.0013	0.0024	0.0031	0.0058	0.0078	0.0115	0.012
	10	0.0056	0.0107	0.0140	0.0265	0.0352	0.0520	0.056
	100	–	0.0429	0.0973	0.1110	0.1380	0.1800	0.182
	1000	–	–	–	0.3000	0.3630	0.5000	0.520

$$*R_V = \frac{pp \cdot H_2S}{pp \cdot CO_2}$$

TABLE 2

Hydrogen Sulfide Partial Pressure over Monoethanolamine Solutions Containing Carbon Dioxide and Hydrogen Sulfide

Temp °C	Partial pressure H_2S (mm)	Mole H_2 per mole MEA					
		$R_L^* = 0.01$	$R_L = 0.05$	$R_L = 0.10$	$R_L = 0.50$	$R_L = 1.0$	$R_L = \infty$
40	1	0.0047	0.0190	0.0327	0.0863	0.1140	0.128
	10	0.0066	0.0263	0.0468	0.1510	0.2220	0.374
	100	0.0092	0.0351	0.0619	0.2120	0.3260	0.802
	1000	–	0.0464	0.0830	0.2700	0.4250	1.00
120	1	0.0016	0.0031	0.0040	0.0072	0.0088	0.012
	10	0.0059	0.0120	0.0163	0.0312	0.0393	0.056
	100	–	0.0424	0.0558	0.1075	0.1400	0.182
	1000	–	–	–	0.3120	0.4050	0.520

$$*R_L = \frac{\text{mole } H_2S/\text{mole MEA}}{\text{mole } CO_2/\text{mole MEA}}$$

Since regeneration is in a steam atmosphere, the carbon dioxide regeneration will be more difficult to accomplish. Fitzgerald and Richardson[32] presented data obtained from operating plants which showed the effect of steam stripping monoethanolamine solutions. Their results are shown in Figures 2 and 3. Use of Figures 2 and 3 in conjunction with the data of Jones et al. permits estimation of equilibrium conditions on the top tray of a monoethanolamine absorber. The Fitzgerald and Richardson data also point out that under some combinations of regeneration steam rate and sour gas compositon sweetened gas will not meet normal specifications.

Lee et al.[33] have determined joint solubilities for H_2S and CO_2 in 5.0 N solutions of monoethanolamine in water. Their data for the more concentrated MEA solution reflect the same mutual reaction effect as do the Jones data. In addition, Lee et al.[34,35] have determined mutual equilibrium partial pressure data for H_2S, CO_2, and aqueous diethanolamine solutions. All the data show similar interaction effects of H_2S and CO_2 on each other. The complete DEA-H_2S-CO_2 equilibrium data are available in the original publication as well as in Maddox.[40] No data for the residual H_2S and CO_2 concentration in regenerated DEA solutions have been published.

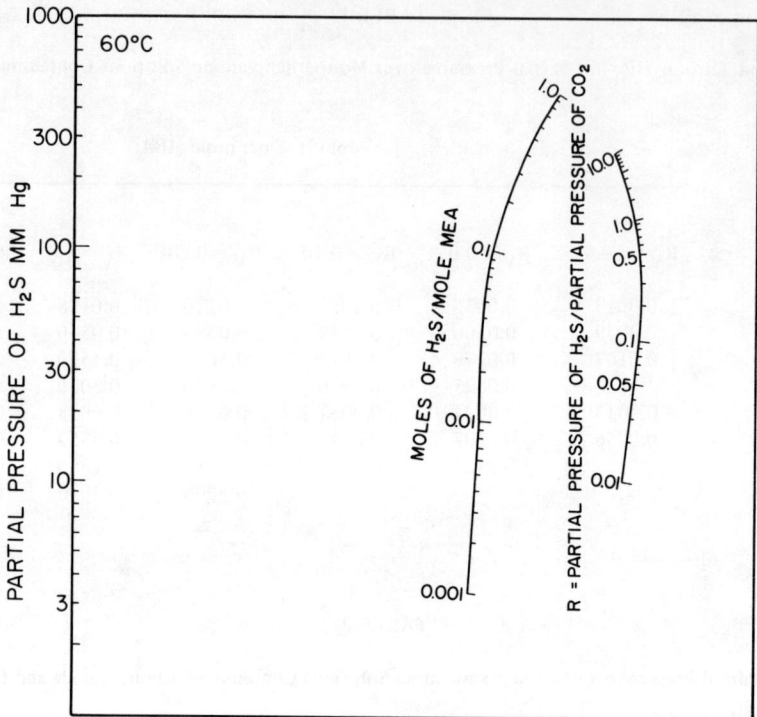

FIGURE 1. Conversion of MEA by H_2S in mixtures of H_2S and CO_2 at 60°C.

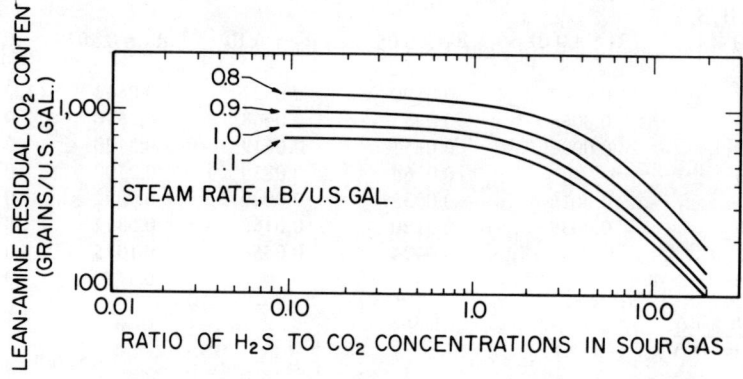

FIGURE 2. Effect of regeneration steam rate and sour gas composition on CO_2 concentration of regenerate monoethanolamine. (Adapted from Fitzgerald, K. J. and Richardson, J. A., *Oil Gas J.*, 64, 110, 1966.

Bocard and Mayland[36] reported an extensive investigation of the equilibrium between water vapor, carbon dioxide, and hot potassium carbonate solutions. Mapstone[37] plotted the data of Bocard and Mayland in nomograph form. The Mapstone nomographs make possible quick and easy estimates of solution concentrations and acid gas loading for hot potassium carbonate systems operating on carbon dioxide.

Leder[38] has studied the rate of absorption of carbon dioxide into hot carbonate solutions activated or buffered by amines. The amines studied were diethanolamine, monoethanolamine, methylaminoethanol, and ethylenediamine. Morpholine was also studied. At 2.5% by weight of amine in hot carbonate solution, Leder found the absorption rate to be enhanced by approximately 2.0 to 2.35 times. For his work, the value of

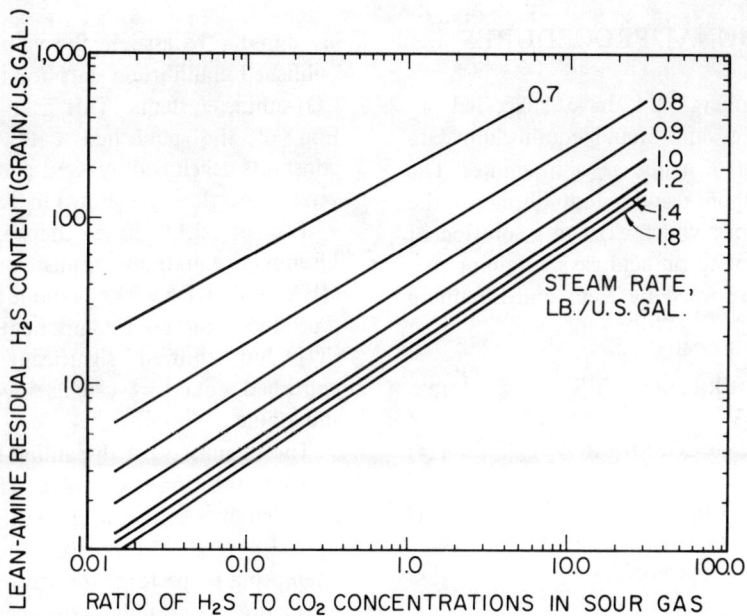

FIGURE 3. Effect of rgeneration steam rate and sour gas composition on H_2S concentration of regenerated monoethanolamine. (Adapted from Fitzgerald, K. J. and Richardson, J. A., *Oil Gas J.*, 64, 110, 1966.)

mass-transfer coefficient for the unactivated carbonate solution was

$$k_L = 0.32 \times 10^8 = \frac{mol/hr - tm^2}{mol/l}$$

Vei et al.[39] studied the removal of carbon dioxide from commercial reformer gas by propylene carbonate. They concluded that the mass-transfer coefficient does not change with change in carbon dioxide concentration from 5 to 17.5 vol% in the gas phase. They also concluded that the mass-transfer coefficient is independent of pressure over the range from approximately 5 to 30 atm. Their work was done in a 45-mm diameter column packed to a height of 1,260 mm with glass rings, 5 mm × 9 mm × 1 mm. The specific surface area of the rings was 650 m^2/m^3. For this column configuration the equation used to correlate their mass transfer rates was

$$Nu_{Liq} = 0.00202 \, Re_{Liq}^{0.76} \, Pr_{Liq}^{0.5} \qquad (22)$$

where

$$Nu_{Liq} = \frac{\beta_c \theta}{D_{Liq}}$$

$$Re_{Liq} = \frac{4W_{Liq}}{\alpha \mu_{Liq}}$$

$$Pr_{Liq} = \frac{\mu_{Liq}}{\rho_{Liq} D_{Liq}}$$

and

β_c = The mass-transfer coefficient in the liquid phase, m/sec
D_{Liq} = The diffusion coefficient of the component in the liquid, m^2/sec
W_{Liq} = The specific liquid flow rate, kg/(m^2)(sec)
α = The specific surface area of the packing, m^2/m^3
μ_{Liq} = The vicosity of the liquid
θ = The reduced thickness of the liquid film which is calculated by $\theta = (\rho^2_{Liq} g/\mu^2_{Liq})^{1/3}$
g = Acceleration due to gravity

On physical solvents, data in conventional equilibrium constant form are available for normal methyl pyrrolidone (M-Pyrol). These data for a wide variety of sulfur compounds and both saturated and unsaturated hydrocarbons can be obtained from GAF Corporation, the manufacturers of M-Pyrol. For methyl cyanoacetate, K values in natural gas systems have been reported by Woertz.[14] Little or no information is available for the other physical solvents.

CALCULATIONAL PROCEDURES

Kent and Eisenberg[41,42] have suggested a model to correlate vapor-liquid equilibrium data for the absorption of acidic gases in amines. The model is based upon reaction equilibria for the possible ionic species which exist in a solution of amine in contact with an acid gas-containing gas. The equations[3] proposed by Kent and Eisenberg are

$$RR'NH_2 + \underset{}{\overset{K_1}{\rightleftharpoons}} H^+ + RR'NH \quad (23)$$

$$RR'NCOO^- + H_2O \underset{}{\overset{K_2}{\rightleftharpoons}} RR'NH + HCO_3^- \quad (24)$$

$$H_2O + CO_2 \underset{}{\overset{K_3}{\rightleftharpoons}} H^+ + HCO_3^- \quad (25)$$

$$H_2O \underset{}{\overset{K_4}{\rightleftharpoons}} H^+ + OH^- \quad (26)$$

$$HCO_3^- \underset{}{\overset{K_5}{\rightleftharpoons}} H^+ + CO_3^= \quad (27)$$

$$H_2S \underset{}{\overset{K_6}{\rightleftharpoons}} H^+ + HS^- \quad (28)$$

$$HS^- \underset{}{\overset{K_7}{\rightleftharpoons}} H^+ + S^= \quad (29)$$

$$P_{CO_2} = H_{CO_2} [CO_2] \quad (30)$$

$$P_{H_2S} = H_{H_2S} [H_2S] \quad (31)$$

Using literature values for the equilibrium constants for the equations, Kent and Eisenberg found that they did not reproduce the H_2S or CO_2 equilibrium data with amines. Consequently, they used fitting techniques to force a fit allowing K_1 and K_2 to assume the values necessary to fit published equilibrium data for the H_2S-amine and CO_2-amine systems. This gave better representation of the published data. The equilibrium constants developed by Kent and Eisenberg for the various reactions are shown in Table 3.

Lee et al.[43] have checked the Kent and Eisenberg equations against published data for MEA and DEA. They found that the model reproduced the pure component data for H_2S and CO_2 but showed significant deviations from published data for systems involving H_2S, CO_2, and amine.

The chemistry of the amine-H_2S-CO_2 reaction in water solution is complex. The seven reactions proposed by Kent and Eisenberg are essentially equivalent to those used by other authors attempting to perform the same kind of predictive correlation. Certainly one weak point of the Kent-Eisenberg work is the use of Henry's law constants for predicting vapor-liquid equilibria for H_2S and CO_2 in the water solution. In ideal cases, these constants apply only for very dilute solutions. Neither H_2S or CO_2 are ideal in solution with water so that significant deviations could be expected. Independent checking of the Kent-Eisenberg and similar work by the author indicates that these deviations, combined with the differences in amine-H_2S-CO_2-water equilibria between individual investigators, may possibly explain most of the deviations found in the Kent and Eisenberg model.

Donnelly and Henderson[44] report apparent differences in rate of reaction in DEA-H_2S-CO_2 systems between high and low pressure operation.

TABLE 3

Constants for Kent and Eisenberg Model

($K_i = \exp A + B/T + C/T^2 + D/T^3 + E/T^4$ for T in °R)

Equilibrium constant	Units	A	B × 10⁻⁴	C × 10⁻⁸	D × 10⁻¹¹	E × 10⁻¹³
K_1	g mol/l	−3.36360	−1.053200	0.000000	0.000000	0.00000
K_1	g mol/l	−2.55100	−1.017400	0.000000	0.000000	0.00000
K_2	g mol/l	6.69425	−0.556349	0.000000	0.000000	0.00000
K_2	g mol/l	4.82550	−0.339260	0.000000	0.000000	0.00000
K_3	g ions/l	−241.81800	53.685500	−4.812300	1.940000	−2.96445
K_4	(g ions/l)²	39.55540	−17.782200	1.843000	−0.854100	−1.42920
K_5	g ions/l	−294.74000	65.589300	−5.966700	2.424900	−3.71920
K_6	g ions/l	−304.68900	69.697900	6.310070	2.555100	−3.91757
K_7	g ions/l	−657.96500	164.936000	−15.896400	6.724720	−10.60430
H_{H_2S}	mmHg/(g mol/)	104.51800	−24.625400	2.390290	−1.018980	1.59734
H_{CO_2}	mmHg/(g mol/l	22.28190	−2.489510	0.223996	−0.090918	0.12601

Their data on an operating absorber indicate that, in some cases at least, the slower rate of reaction between DEA and H_2S may lead to incomplete H_2S removal at low pressures.

Treybal[4,5] presented a calculational procedure for gas absorption and/or stripping in adiabatic packed towers. His procedure makes allowance for absorption of single or multiple solutes and evaporation or condensation of the solvent. The procedure is step wise beginning at the bottom of the absorber and proceeding upward in small increments of height of packing. Treybal chose to write his equations in terms of the Coleman-Drew mass-transfer coefficients rather than the more generally used k mass-transfer coefficient. The two can be interchanged by:

$$F_G a = k_G aP = G/H_{tG} \tag{32}$$

$$F_L a = k_L a \rho_L / M_L = L/H_{tL} \tag{33}$$

where

F_G = Gas phase mass-transfer coefficient lb mol/hr (ft^2)
F_L = Liquid phase mass-transfer coefficient, lb mol/hr (ft^2)
a = Specific interfacial surface, ft^2/ft^3
k_G = Gas mass-transfer coefficient, lb mol/hr (ft^2) atm
k_L = Liquid mass-transfer coefficient, lb mol/hr (ft^2) (lb mol/ft^3)
P = Total pressure, atm (at which k was measured)
G = Gas mass velocity, lb mol/hr (ft^2)
H_{tG} = Height of a transfer unit for the gas, ft
ρ_L = Liquid density, lb/ft^3 (for liquid in which k was measured)
M_L = Average molecular weight of liquid, lb/lb mol (for liquid in which k was measured)
L = Liquid mass velocity, lb mol/hr (ft^2)
H_{tL} = Height of a transfer unit for the liquid, ft

To use Treybal's procedure an adequately specified absorption problem is required. In addition, physical and thermodyanamic properties such as enthalpies, heat capacities, viscosity, etc., are required. The binary diffusivities are required for the solute-inert gas, the solvent-inert gas, and the solvent-solute binaries. Mass-transfer coefficient data for the individual phases must be available. Heat-transfer coefficients can be estimated from the mass-transfer data through the heat-mass-transfer analogy. A value is then assumed for the fraction of the total mass transfer in the incremental height of column that is due to solute transfer. If this fraction is R_A, then

$$R_A + R_S = 1 \tag{34}$$

$$R_A = \frac{N_A}{N_R + N_S} \tag{35}$$

where

R = Fraction of mass flux due to component indicated by subscript
N = Mass flux of component indicated by subscript, lb mol/hr (ft^2 interfacial surface), positive for absorption, negative for stripping
A = Indicates solute
S = Indicates solvent

The bulk gas and liquid concentrations are then used to calculate concentrations at the interface by:

$$y_{Ai} = R_A - (R_A - y_A)\left(\frac{R_A - x_A}{R_A - x_{Ai}}\right)^{F_{LA}/F_{GA}} \tag{36}$$

$$y_{Si} = R_S - (R_S - y_S)\left(\frac{R_S - x_S}{R_S - x_{Si}}\right)^{F_{LS}/F_{GS}} \tag{37}$$

where

y = Gas concentration, mol fraction
x = Liquid concentration, mol fraction
i = Interface

This procedure is repeated for several values of fractional flux. An important note here is that in the case of solvent vaporization (transfer of solvent from liquid phase to gas phase) the fractional flux for the solute will be greater than 1.0.

A value for the interface temperature (t_i) is assumed and the equilibrium liquid vapor liquid concentrations are then calculated. The values of fractional flux for solvent and solute are then selected which provide that $x_{Ai} + x_{Si} = 1.0$.

A heat balance is then run around the differential height of column, and the interface temperature (t_i) is computed based on this heat balance. If the calculated value does not agree with the previously assumed value, calculations must be repeated until the two values do agree. The rate of mass transfer in the section is then calculated for the short increment of packing. If the assumed temperature and concentration of the vapor leaving this section do not agree with the calculated value, new assumptions must be made and the entire procedure outlined above repeated. This sequence of calculations is repeated until the desired outlet gas composition is reached.

Treybal checked the above procedure against data for the absorption of ammonia from air into water. His calculations indicated an increase tower packing height of approximately 20% because of temperature effects considered in his calculation procedure.

Owens and Maddox[46] presented a calculation method for plate columns that incorporates several advantages. Their procedure is based on analysis of a large number of tray-to-tray absorber solutions, using a program which previously had been checked against the performance of existing absorbers.[47] In their development, Owens and Maddox used the absorption and stripping factor functions in the form presented by Edmister.[48] Using these factors the moles of one component in the vapor leaving the top tray of an absorber can be written as

$$v_1 = v_{n+1} \phi_A + 1_0 \psi_A \quad (38)$$

where

ϕ_A = $1/\Sigma_A + 1_0 \psi_A$ = Fraction of a solute component not recovered in solvent

ψ_A = $1 - [(\pi_A + 1)]$ = Fraction of solvent vaporized

π_A = $A_1 A_2 A_3 \ldots A_n$

Σ_A = $A_1 A_2 \ldots A_n + A_2 A_3 \ldots A_n + \ldots + A_n$

A = Absorption factor = L/KV

1,2,3,...n = Plate number (top plate is number 1)

From analysis of the tray-by-tray solutions, Owens and Maddox concluded that on the order of 80% of the total absorption (on a molar basis) occurring in a multicomponent absorber took place on either the top or bottom theoretical plate. With this as a basis, they developed a model for the column which requires the evaluation of three absorption factors for each component: the absorption factor for the top tray, the absorption factor for the bottom tray, and the absorption factor representing the remaining n – 2 trays in the column. In terms of these three absorption factors the product terms then become:

$$\pi_A = A_{Top} A_{Avg}^{n-2} A_{Btm} \quad (39)$$

$$\Sigma_A = A_{Top} A_{Avg}^{n-2} A_{Btm} + A_{Avg}^{n-2} A_{Btm}$$
$$+ A_{Avg}^{n-3} A_{Btm} + \ldots + A_{Btm} \quad (40)$$

To evaluate the three absorption factors, temperatures of the top and bottom trays and an average temperature for the tower must be known, together with flow rates for the top and bottom trays and the average flow rates for the column.

The temperature of the top tray of the column can be estimated from a heat balance around the top tray:

$$T_{DG} - T_{LO} = \frac{\Delta H_{Abs,Top} - C_{P_2}(T_{DG} - T_2)}{C_{P,LO}} \quad (41)$$

T	=	Temperature
$\Delta H_{Abs,Top}$	=	Heat of absorption on top tray
C_P	=	Heat capacity of stream specified by subscript
DG	=	Discharge gas stream
LO	=	Lean solvent stream

The average temperature for the n – 2 trays in the middle section of the column is calculated from a heat balance around the middle section of the column:

$$T_{Avg} = \frac{T_{DG} + [C_{P,DG} T_{RO} - T_{DG} + (1-2)\Delta H_{Abs}]}{C_{P,DG} + C_{P,OIL}} \quad (42)$$

where w is the fraction of total absorption occurring on terminal trays. The temperature of the second tray down from the top of the tower is then calculated from the empirical equation:

$$T_2 = T_{Top} + \frac{(T_{Btm} - T_{Top})}{n} + \left(T_{Avg} - \frac{T_{Top} + T_{Btm}}{2}\right) \quad (43)$$

Ratios of $\frac{L}{V}$ for the top and bottom trays result from the above calculations. The average $\frac{L}{V}$ ratio for the tower is calculated from the empirical equation:

$$\left(\frac{L}{V}\right)_{Avg} = \frac{L_n - (V_{n+1} - V_n)}{V_n - 0.05S} \quad (44)$$

where S is the total gas shrinkage in the absorber in mol.

Using this procedure, Owens and Maddox found excellent agreement with calculated results for component recoveries, solvent rates required, and temperature profiles.

Kim and Molstad[49] present an interesting analysis of the minimum cost of absorption. The costs included in their analysis are

1. The fixed cost of the absorber
2. The value of the unrecovered solute
3. The cost of recovery of the solvent used in the stripping operation.

They minimize the cost of absorption by evaluating the various terms in the dimensionless quantity where

$$\frac{C_1 \Theta r m G_m}{C_2 H_{OG} (m' - 1)}$$

where

C_1 = Total cost of stripping operation, $/lb mol of vapor produced at the base at the stripper; includes fixed charges, cost of cooling water, and cost of steam

C_2 = Annual cost of power and fixed charges per volume of absorption colum $/(ft^3) year

Θ = Hours of operation per year

r = Ratio of actual to minimum reflux ratios in the stripping column

m = Slope of the equilibrium curve (y = mx) in the absorption column; assumed constant

G_m = Molal gas velocity through absorber, lb mol/hr (ft^2)

H_{OG} = Height per transfer unit, ft; assumed constant

m' = Slope of the equilibrium curve (y = m'x) in the stripping column; assumed constant

Kim and Molstad present a graphical solution for this term. The graphical solution yields the necessary absorption factor to accomplish the specified recovery of solute from the gas stream.

REFERENCES

1. Buckingham, P. A. and Homan, H. R., *Hydrocarbon Process.*, 51, 121, 1971.
2. Wendt, C. J., Jr. and Dailey, L. W., *Hydrocarbon Process.*, 47, 155, 1967.
3. Klein, J. P., *Oil and Gas Int.*, p. 72, September 1970.
4. Dingman, J. C. and Moore, T. F., Proc. Gas Conditioning Conference, Norman, Okla., April 1968.
5. Ballard, D., *Hydrocarbon Process.*, 46, 137, 1966.
6. Eickmeyer, A. G., *Chem. Eng. Prog.*, 58, 89, 1962.
7. Anon., *Hydrocarbon Process.*, 51, 96, 1971.
8. Anon., *Hydrocarbon Process.*, 51, 105, 1971.
9. Eickmeyer, A. G., *Oil Gas J.*, 68, 74, 1971.
10. Parrish, R. W. and Neilson, H. B., Synthesis Gas Purification Including Removal of Trace Contaminants, paper presented at 167th National Meeting American Chemical Society, Division of Industrial and Engineering Chemistry, Los Angeles, California, March 31–April 5, 1974.
11. Anon., *Hydrocarbon Process.*, 51, 103, 1971.
12. Anon., *Hydrocarbon Process.*, 51, 114, 1971.
13. Anon., *Hydrocarbon Process.*, 51, 117, 1971.
14. Woertz, B. B., *J. Pet. Technol.*, 28, 483, 1971.
15. Herrin, J. P., personal communication.
16. Dunn, C. L., Freitas, E. R., Hill, E. S., and Sheeler, J. E. R., *Hydrocarbon Process.*, 44, 137, 1965.
17. Dunn, C. L., Freitas, E. R., Goodenbour, J. W., Henderson, H. T., and Papadopoulos, M. N., *Hydrocarbon Process.*, 43, 150, 1964.
18. Bratzler, K. and Doerges, A., *Hydrocarbon Process.*, p. 78, April 1974.
19. Benson, H. E. and Parrish, R. W., *Hydrocarbon Process.*, p. 81, April 1974.
20. Thirkell, H., *Hydrocarbon Process.*, 51, 115, 1972.
21. Ross, L. A. and Bennett, L. C., *Oil Gas J.*, 61, 119, 1963.
22. Anon., *Gas Coke*, 158, 822, 1963.

23. Anon., *Oil Gas J.,* 69, 68, 1971.
24. Anon., *Oil Gas J.,* 70, 66, 1972.
25. Beavon, D. K. and King, F. W., Prevention of Air Pollution by Sulfur Plants, paper presented to Canadian Natural Gas Processing Assoc., September 18, 1970.
26. Beavon, D. K. and King, F. W., *Hydrocarbon Process.,* p. 114, April 1973.
27. Beavon, D. K. and King, F. W., *Chem. Eng.,* p. 83, March 1, 1976.
28. Jones, J. H., Froning, H. R., and Claytor, E. E., Jr., *J. Chem. Eng. Data,* 4, 85, 1959.
29. Beavon, D. K., *Chem. Eng.,* 68, 71, 1971.
30. Barthel, Y. et al., *Hydrocarbon Process.,* 51, 89, 1971.
31. McCoy, D. D. and Maddox, R. N., personal communication.
32. Fitzgerald, K. J. and Richardson, J. A., *Oil Gas J.,* 64, 110, 1966.
33. Lee, J. I., Otto, F. D., and Mather, A. E., *J. Chem. Eng. Data,* 20, No. 2, 161, 1975.
34. Lee, J. I., Otto, F. D., and Mather, A. E., *J. Chem. Eng. Data,* 17, No. 4, 465, 1972.
35. Lee, J. I., Otto, F. D., and Mather, A. E., Design Data for Diethanolamine Acid Gas Treating Systems, paper presented at CNGPA Research Seminar, Calgary, Alberta, November 24, 1972.
36. Bocard, J. P. and Mayland, B. J., *Hydrocarbon Process.,* 42, 128, 1962.
37. Mapstone, G. E., *Hydrocarbon Process.,* 46, 145, 1966.
38. Leder, F., The absorption of CO_2 into chemically reactive solutions at high temperatures, accepted for publication in *Chem. Eng. Sci.*
39. Vei, D. et al., *Int. Chem. Eng.,* 10, 639, 1970.
40. Maddox, R. N., *Gas and Liquid Sweetening,* Campbell Petroleum Series, 2nd ed., John M. Campbell, Norman, Okla., 1974.
41. Kent, R. L. and Eisenberg, B., *Hydrocarbon Process.,* p. 87, February 1976.
42. Kent, R. L. and Eisenberg, B., Equilibrium of H_2S and CO_2 with MEA and DEA Solutions, paper presented at Gas Conditioning Conference, Norman, Okla., March 1976.
43. Lee, J. I., Otto, F. D., and Mather, A. E., The Solubility of Mixtures of Carbon Dioxide in Aqueous Monoethanolamine Solutions, paper presented at 25th Canadian Chemical Engineering Conference, Montreal November 5, 1975.
44. Donnelly, S. T. and Henderson, D. R., *Energy Process.* (Canada), p. 22, January–February 1975.
45. Treybal, R. E., *Ind. Eng. Chem.,* 61, 36, 1969.
46. Owens, W. R. and Maddox, R. N., *Ind. Eng. Chem.,* 60, 14, 1968.
47. Spear, R. R., Erbar, J. H., and Maddox, R. N., An evaluation of the Sujata Method for Predicting Absorber Performance, paper presented at National Meeting of American Institute for Chemical Engineers, Dallas, Texas, February 1966.
48. Edmister, W. C., *J. Am. Inst. Chem. Eng.,* 3, 165, 1957.
49. Kim, J. C. and Molstad, M. C., *Hydrocarbon Process.,* 46, 107, 1966.

CARRIER-MEDIATED TRANSPORT IN SYNTHETIC MEMBRANES

D. R. Smith, R. J. Lander, and J. A. Quinn

TABLE OF CONTENTS

Introduction 225

Recent Literature 226
    Facilitated Oxygen Transport 226
    Facilitated Carbon Dioxide Transport 226
    Nitric Oxide-Ferrous Chloride System 227

Mathematical Models 227
    Distributed versus Lumped Parameter Description 227
    $A+B \rightleftharpoons AB$ System 228
    Ionic Considerations in Facilitated Transport 228

Dimensional Analysis 229
    Dimensionless Groupings 229
    Reaction Equilibrium Limit 229

Methods of Solution 231
    Approximate Analytical Solutions 231
    Limiting Solution 232
    Absolute Permeant Flux 235

Tracer Diffusion 235

Coupled Gas Counterdiffusion 237

Future Directions 239

References 240

INTRODUCTION

Viewed as separation processes, one can only marvel at the economy and specificity which characterize biological processes at the cellular level — truly optimal design. Many of the unique features of biological transport are accomplished through reversible chemical reaction; the permeating solute reacts with a specific component, either within or attached to some membranous structure, which facilitates its transport along the conducting path. Some of the first attempts at analytical models of these reversible reaction/diffusion mechanisms grew out of Scholander's studies on oxygen transport through thin films containing hemoglobin.[1] These studies in turn stimulated research on the transport of other dissolved gases, principally CO_2, in biological systems and led to the development of synthetic reactive membranes which incorporate features of the biological process. The purpose of this review is to provide a summary of current research on synthetic reactive membranes as well as an engineering evaluation of their application to separation and purification processes.

The term "facilitated transport" has been used

to describe the process whereby a permeant flux across a thin liquid film or membrane is augmented by reaction with a mobile carrier species constrained to remain within the membrane: the flux of the permeant species from a region of higher to lower concentration is increased by the presence of a parallel flux of carrier complex. Interest has been focused on these transport systems for a number of reasons. The role of myoglobin and hemoglobin as carriers for oxygen and carbon monoxide has been demonstrated to have physiological significance.[2-7] There is incentive to apply carrier-mediated transport to separation problems since permeant flux rates can be increased several-fold over those based on simple molecular diffusion. In addition, because augmentation is based on a specific chemical interaction within the membrane, high selectivity is possible. As a side benefit, it has recently been demonstrated that steady-state facilitated flux measurements can be used as a powerful tool in the study of fast reaction kinetics.[8]

The scope of this paper is limited to systems wherein a homogeneous liquid membrane separates two gas phases. These systems are modeled by simple reaction-diffusion conservation equations. Gas permeant experiments are free of many of the complications inherent in the liquid phase counterparts: appreciable boundary layer resistance, mutual miscibility, osmotic and ionic effects, etc.

A simple, widely studied system involves the reversible bimolecular reaction:

$$A + B \underset{k_{-1}}{\overset{k_1}{\rightleftharpoons}} AB \qquad (A)$$

where A is the gaseous permeant and B and AB are the free and complexed forms of the carrier, respectively. The carrier is confined to the membrane and the permeant concentrations at the membrane boundaries are fixed by the adjoining gas phase partial pressures. Consequently, the membrane is "globally nonreactive," i.e., no net reaction occurs within the membrane.[9] The majority of gaseous permeant systems which have been studied are of the above form or have been modeled in this manner: O_2, CO, CO_2-hemoglobin;[3,5,10-16] O_2-myoglobin;[4,6,17] and NO-Fe^{+2}.[18,19]

More complicated schemes can occur. In the CO_2/HCO_3^- system, simultaneous reactions are present (see below). As another example the four-step Adair reaction sequence has been employed by Meldon[15] in analyzing the O_2-hemoglobin system.

RECENT LITERATURE

In a comprehensive review, Shultz, Goddard, and Suchdeo[20] formulate the theoretical framework for considering carrier-mediated transport systems. In addition, they present a summary of the literature on facilitation for both gas and liquid phase systems and they evaluate various methods of solution. Ward[21] discusses the potential industrial applications of carrier-containing liquid membranes. The extensive biological literature concerning the role of facilitated transport of respiratory gases in physiological systems has been summarized in reviews by Kreuzer,[22] Wittenberg,[17] and Stein.[23] Cussler[24] and Cussler and co-workers[25,26] have published a series of papers on facilitated transport in analogous liquid phase systems with emphasis on the special cases of counter- and cotransport, wherein a carrier complexes with more than one solute, competitively and noncompetitively, respectively.

Facilitated Oxygen Transport

The number of gas phase experimental systems which have been studied is relatively small. Scholander[1] measured the steady-state flux of oxygen through thin layers of hemoglobin solution supported by cellulose acetate filters. From the observed saturable transport characteristics, he postulated a "bucket brigade mechanism" whereby hemoglobin augments the diffusion of oxygen. The study of hemoglobin facilitation was expanded by Mochizuki and Forster[3] to the case of carbon monoxide.

Wittenberg[5] properly recognized the need for a mobile carrier and focused attention on the importance of the rates of complex formation and breakdown to the overall process. The mobile carrier explanation was later verified by Colton et al.[7] by comparing transport of oxygen through films of immobilized and free hemoglobin.

Facilitated Carbon Dioxide Transport

The CO_2/HCO_3^- permeant-carrier system has been studied in both a physiological and engineering context. The reaction scheme is more involved:

$$CO_2 + H_2O \rightleftharpoons H^+ + HCO_3^- \quad (B)$$

$$CO_2 + OH^- \rightleftharpoons HCO_3^- \quad (C)$$

$$HCO_3^- \rightleftharpoons H^+ + CO_3^{-2} \quad (D)$$

Reaction D can be considered sufficiently rapid that HCO_3^- and CO_3^{-2} are at equilibrium throughout the membrane. Enns[27] established the carrier role of bicarbonate and showed that facilitation could be enhanced considerably by catalyzing the hydration reaction with the enzyme which participates in CO_2 hydration in the red cell, carbonic anhydrase. Ward and Robb[28] exploited the carbon dioxide hydration catalysts — principally sodium arsenite — in the design of a novel membrane system for the removal of carbon dioxide from a mixture of carbon dioxide and oxygen. They achieved a separation factor of 4100, although a major contribution was due to the preferential salting-out of O_2 in the 6 N $CsHCO_3$ membrane solution. Otto,[29,30] Suchdeo,[31,32] and Meldon[15,33] developed various numerical and approximate analytical solution methods to interpret their experimental results. Donaldson and Quinn[34] used a tracer technique similar to that of Enns and obtained a closed form analytical solution which permitted the direct determination of diffusional and kinetic parameters from measured permeant fluxes.

It can be seen from Reactions B and C that hydrogen ion plays a role in the carrier formation reaction. In fact, carbon dioxide facilitation is increased by the addition of a weak acid or protein buffer.[15,35,36] In a series of experiments, Meldon[15] added weak acid buffers of a wide range in pK to bicarbonate solution. For the same CO_2 boundary concentrations, the facilitation is maximized at a buffer pK intermediate between the boundary pH values. The added facilitation is due to the coupling to the CO_2 hydration reaction, and the effect is an increased HCO_3^- gradient in the presence of the buffering action, hence an increased CO_2 membrane flux. A quantitative discussion of the effect of buffering species on carbon dioxide transport is given by Meldon.[15] Using a tracer diffusion technique, Lander[51] isolated the buffering contribution to CO_2 facilitation from those of the average carrier capacity and catalysis.

The buffering action of the bicarbonate-carbonate system has been exploited in facilitating the transport of hydrogen sulfide for the development of permselective membranes for scrubbing coal gas.[21] These membranes, designed to operate at elevated temperature and pressure, exhibit a high degree of selectivity for H_2S over CO_2, due to their large difference in reaction rates.[54-56]

As an extension of this concept, the possibility exists of creating multilayer asymmetric membranes which would consist of layers of varying buffering capacity chosen to augment the permeant flux via the secondary coupling to H^+. Asymmetric membranes of this type are currently being studied in our laboratory.

Nitric Oxide-Ferrous Chloride System

As a final example of dissolved-gas facilitation, Ward[18] measured the facilitated flux of nitric oxide across a thin liquid layer of ferrous chloride solution immobilized between two silicone support films. The NO reacts reversibly with Fe^{+2} by[18]

$$NO + Fe^{+2} \rightleftharpoons Fe(NO)^{+2} \quad (E)$$

Also, he later used this system to demonstrate the possibility of electrically induced carrier transport. The permeant, NO, is moved against its concentration gradient by the application of an electric potential which interacts with the charged carrier, Fe^{+2}.[19,37]

MATHEMATICAL MODELS

Distributed versus Lumped Parameter Description

Two mechanistic approaches are used in modeling the transport of a permeant through thin, carrier-containing films: "strict" carrier transport and carrier-mediated transport models. Excellent discussions of the distinguishing features of each are found in Shultz et al.[9] and Blumenthal and Katchalsky.[38] In carrier-mediated transport, the diffusing permeant species reacts with the carrier at all points within the membrane. This distributed parameter model has been successfully used to interpret transport results through artificial membranes of the order of 100 μm or greater in thickness. In the biological literature it is more common to employ a compartmental (one or more) model wherein carrier-permeant reactions are confined to the membrane boundaries — a lumped parameter model. Transport of permeant in the membrane is due solely to the diffusion of the complex and the governing equations are algebraic. The choice of this simple, discontinuous model for interpreting data on natural membranes is justifiable since these thin membranes (of order 10^2 Å) may not exhibit ordinary, macroscopic transport properties.

A + B ⇌ AB System

The carrier-mediated model is based on steady-state conservation relations valid at any point within the liquid film. For the reaction,

$$A + B \underset{k_{-1}}{\overset{k_1}{\rightleftharpoons}} AB$$

there are three independent steady-state equations of the form:

$$\nabla \cdot N_i = R_i \tag{1}$$

where N_i is the component flux given by Fick's law and R_i is the local net rate of formation of the ith species. Considering one-dimensional transport through a homogeneous medium and constant diffusivities, the differential balances are

$$D_A \frac{d^2 C_A}{dx^2} = k_1 C_A C_B - k_{-1} C_{AB} \tag{2}$$

$$D_B \frac{d^2 C_B}{dx^2} = k_1 C_A C_B - k_{-1} C_{AB} \tag{3}$$

$$D_{AB} \frac{d^2 C_{AB}}{dx^2} = -k_1 C_A C_B + k_{-1} C_{AB} \tag{4}$$

The analogous description for facilitation in a heterogeneous system is presented in Stroeve,[39] DeSimone,[40] and Donaldson and Quinn.[34]

Six constraints are required to solve these three second-order equations. The concentration of A at the boundaries is fixed by the bulk phase partial pressure of the adjoining gas phase via Henry's law:

At $x = 0$,
$$C_A = C_A^0 \tag{5}$$

At $x = L$
$$C_A = C_A^L \tag{6}$$

Neither the carrier nor its complex, B and AB, can penetrate the membrane-gas boundary:

At $x = 0$,
$$\frac{dC_{AB}}{dx} = \frac{dC_B}{dx} = 0 \tag{7}$$

At $x = L$,
$$\frac{dC_{AB}}{dx} = \frac{dC_B}{dx} = 0 \tag{8}$$

Goddard et al.[41] have shown that only three of the four flux conditions (Equations 7 and 8) are independent. This can be seen by integrating Equations 3 and 4 across the membrane and equating:

$$D_B \left(\frac{dC_B}{dx}\bigg|_L - \frac{dC_B}{dx}\bigg|_0 \right) = D_{AB} \left(\frac{dC_{AB}}{dx}\bigg|_0 - \frac{dC_{AB}}{dx}\bigg|_L \right) \tag{9}$$

The final constraint involves the total amount of carrier added to the liquid phase:

$$\int_0^L (C_B + C_{AB}) dx = C_T L \tag{10}$$

The assumption $D_B = D_{AB}$ (usually quite accurate) is often made and allows significant simplification of the analysis. Equations 3 and 4 can be summed and integrated twice. Application of the boundary conditions and the constraint, Equation 10, yields the useful relation:

$$C_B + C_{AB} = C_T \tag{11}$$

Ionic Considerations in Facilitated Transport

Another simplification pertains to facilitated transport systems involving charged species. To incorporate the effects arising from the diffusion of charged species, a second term (Nernst-Planck term[42]) is added to the Fickian diffusion relation:

$$N_i = -D_i \frac{dC}{dx} + D_i Z_i C_i \frac{FE}{RT} \tag{12}$$

where N_i is the flux of species i (Equation 1) and Z_i is its charge. The electric field is not externally imposed, but rather results from the fact that the ionic species diffuse at different rates due to their differing mobilities, and, as a result, produce small local deviations from electrical neutrality. The resulting electric field acts to speed up the less mobile ions, while retarding those that are more mobile. Thus, this term represents a restoring force and is identically zero when all charged species have equal diffusivities.[15,37]

DIMENSIONAL ANALYSIS

If the equations are expressed in dimensionless form, comparisons of dissimilar experimental systems as well as various approximate analytical solutions can be made on the basis of the resulting dimensionless parameters.

On rearranging the material-balance equations and the boundary conditions in dimensionless form,[43] C_A^* becomes a function of six dimensionless parameters, the minimum number of independent groupings:

$$C_A^* = \frac{C_A^0 - C_A}{C_A^0 - C_A^L} = C_A^* \{x^*, \gamma_A, \gamma_{AB}, \alpha, \beta, \delta\} \quad (13)$$

where

$$\gamma_A = \left[\frac{k_{-1} L^2}{D_A}\right], \gamma_{AB} = \left[\frac{k_{-1} L^2}{D_{AB}}\right], \alpha = \left[\frac{C_A^0}{C_A^0 - C_A^L}\right]$$

$$\beta = \left[\frac{C_T}{C_A^0 - C_A^L}\right], \delta = \left[\frac{k_1 C_T}{k_{-1}}\right], x^* = \left[\frac{x}{L}\right]$$

The first derivative of C_A^* evaluated at the membrane boundary is the "facilitation ratio," the ratio of the permeant flux in the presence and absence of carrier:

$$\psi \equiv \frac{N_A^T}{N_0} = \frac{-D_A \frac{dC_A}{dx}\Big|_0}{\frac{D_A}{L}(C_A^0 - C_A^L)} = \frac{dC_A^*}{dx^*}\Big|_0$$

$$= \psi \{\gamma_A, \gamma_{AB}, \alpha, \beta, \delta\} \quad (14)$$

Yung and Probstein[44] obtained four dimensionless groups for the case of zero downstream permeant concentration, in agreement with these results (let $C_A^L = 0$ in Equation 14).

Dimensionless Groupings

Equation 14 contains five dimensionless groupings. Although it is not apparent that these groups are those having most readily identifiable physical significance, experimental work does suggest that there are three important criteria for facilitation related to the quantities appearing within α, β, and δ. These criteria are related to the observed saturation behavior of reactive membranes. The role of β is illustrated by the data of Hemmingsen[4] (Figure 1). The flux of oxygen is calculated from the observed O_2^{18} flux for transport through thin layers of myoglobin and metmyoglobin solution (equal oxygen pressures are maintained on both sides of the membrane). The latter protein does not bind oxygen. As indicated by the shape of the two curves, the carrying capacity of the membrane has been reached ($\beta \to 0$) corresponding to saturation of the carrier. The facilitation is not increased by further increases in driving force, only as increases in molecular diffusion.

The other two saturation limits which preclude facilitation are displayed by α and δ. As the downstream permeant partial pressure increases, the extent of facilitation decreases markedly as the carrier becomes saturated at the downstream side ($\alpha \to \infty$). Figure 2 indicates the rapid fall-off in the extent of facilitation for the CO_2/bicarbonate system as a function of the downstream CO_2 pressure.[51] An even more dramatic decrease is shown in the results of Hemmingsen for O_2 transport across thin films of myoglobin solution.[4] Secondly, if the binding constant is sufficiently large ($\delta \to \infty$) and/or the carrier concentration small ($\delta \to 0$), the carrier is again saturated throughout the membrane, preventing any significant facilitation.

The last two dimensionless groupings, $\gamma_A = \frac{k_{-1} L^2}{D_A}$ and $\gamma_{AB} = \frac{k_{-1} L^2}{D_{AB}}$, represent the ratio of diffusion time to reaction time. This ratio is often referred to as a Damkohler number. In the limit of $\gamma \to \infty$, chemical equilibrium exists at all points within the membrane (equilibrium limit). Conversely, as γ approaches zero, one approaches the limit of nonreactive, molecular diffusion.

Reaction Equilibrium Limit

A useful relationship among the parameters C_T, C_A^0, C_A^L, k_1, and k_{-1} appearing in α, β, and δ can be seen by considering the equilibrium limit (fast reaction or very thick membrane). In this case, Equation 14 has a simple algebraic limit. This expression is derived here for the simple case of A + B ⇌ AB, but is easily extended to more complex reaction stoichiometry as well as simultaneous equilibrium (see Table 1 for other cases). The total flux of A at any point in the membrane can be expressed as the sum of the individual fluxes of A and AB by Fick's law. Addition and integration of Equations 2 and 4 yields

$$-D_A \frac{dC_A}{dx} - D_{AB} \frac{dC_{AB}}{dx} = \text{constant} \quad (15)$$

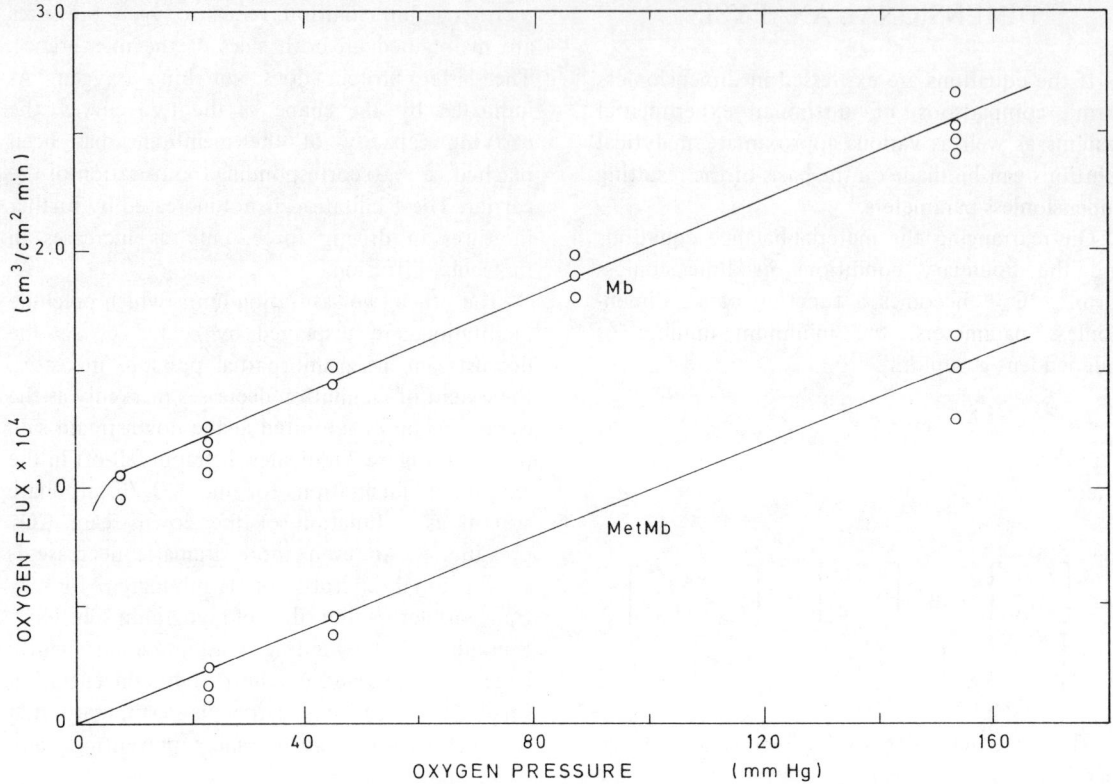

FIGURE 1. Saturation behavior exhibited by the oxygen-myoglobin system. Flux of O_2^{18} measured with equal oxygen pressure on the two sides of the membrane. (From Hemmingson, E. A., *Comp. Biochem. Physiol.*, 10, 239, 1963. With permission.)

The total flux of A, N_A^T, is thus:

$$N_A^T = -D_A \frac{dC_A}{dx} - D_{AB} \frac{dC_{AB}}{dx} \qquad (16)$$

Integration of Equation 16 across the membrane and evaluation at the boundaries yields

$$N_A^T = \frac{D_A(C_A^0 - C_A^L)}{L} + \frac{D_{AB}(C_{AB}^0 - C_{AB}^L)}{L} \qquad (17)$$

The relationship among the species at equilibrium is

$$K = \frac{C_{AB}}{C_A C_B} \qquad (18)$$

Evaluating Equations 11 and 18 at the boundaries, then substituting into Equation 17 and normalizing by the molecular flux in the absence of reaction of the permeant, yields[18]

$$\psi_{eq} = N_A^T/N_0 = 1 + \frac{D_{AB} K C_T}{D_A(1 + KC_A^0)(1 + KC_A^L)} \qquad (19)$$

$$= 1 + \frac{\gamma_A \delta}{\gamma_{AB}(1 + \frac{\alpha\delta}{\beta})(1 + \frac{\alpha\delta}{\beta}(1 - \frac{1}{\alpha}))}$$

where $N_0 = \frac{D_A(C_A^0 - C_A^L)}{L}$. Goddard et al.[41] have shown rigorously that Equation 19 is the proper asymptotic limit. Thus, at reaction equilibrium, one can obtain an upper limit of facilitation for specified values of C_T, C_A^0, C_A^L, and K.

The extent of facilitation may also be influenced by altering the rates of reaction across the membrane, for example, by means of a gradient in temperature or by the photosensitivity of the permeant-carrier binding (e.g., carbon monoxide-heme protein complexes). If the system were at reaction equilibrium, ψ_{eq} may be expressed in terms of the equilibrium constants at the membrane boundaries:

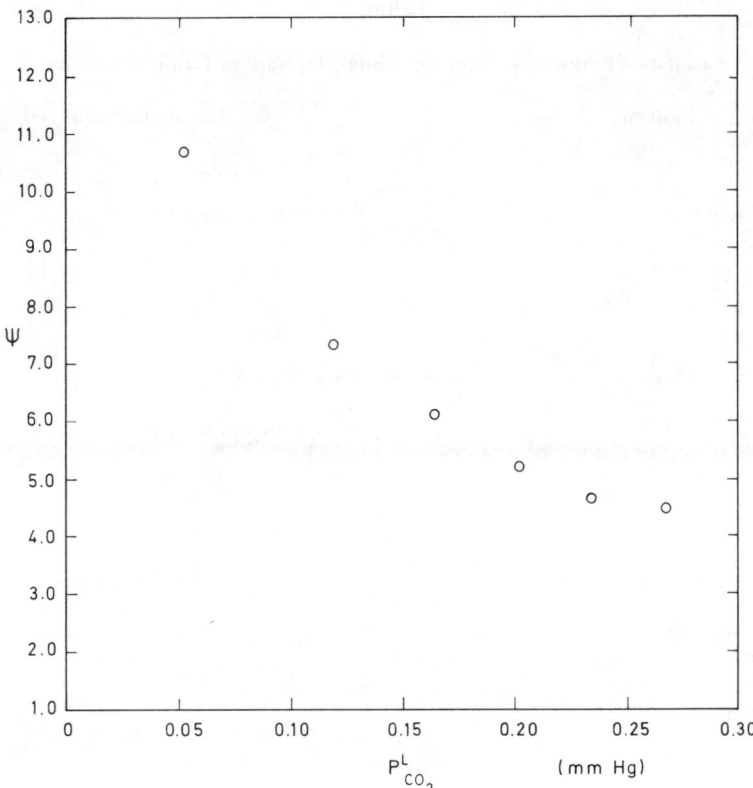

FIGURE 2. Effect of downstream pressure on the extent of facilitation for CO_2 transport across bicarbonate solution. (From Lander, R. J., CO_2 Transport Through Bicarbonate Membranes: Catalysts, Buffers, and the Role of pH Gradients, Ph.D dissertation, University of Pennsylvania, 1976.)

$$\psi_{eq} = 1 + \frac{D_{AB}C_T(K^L C_A^L - K^0 C_A^0)}{D_A(C_A^0 - C_A^L)(1 + K^0 C_A^0)(1 + K^L C_A^L)} \quad (19a)$$

Altering ψ by way of an externally activated light-sensitive membrane reaction is currently under investigation in our laboratory. The prototype reaction being studied is that of CO with myoglobin.[43]

For finite values of γ, the facilitation ratio ψ will, of course, be less than the equilibrium limit. A general solution (Equation 14) cannot be written in closed form because of the nonlinear reaction rate term in the differential equations, Equations 2 to 4. A great deal of effort has been invested in the development of numerical and approximate solutions for these equations over the last 10 years.

METHODS OF SOLUTION

A variety of numerical techniques have been developed to solve Equation 1, most notably by LaForce,[45] Kutchai et al.,[46] Ward,[18] Meldon,[15] Suchdeo,[31] and Spaan.[47] However, the computations required are extensive and some numerical schemes exhibit convergence problems.

Approximate Analytical Solutions

More expedient, though approximate, solutions are valid in asymptotic regions near the equilibrium ($\gamma \to \infty$) and diffusion ($\gamma \to 0$) limits. Smith et al.[33] have shown that in the "near-diffusion" regime the proper solution can be developed as a perturbation from the pure diffusion state ("thin film theory"):

$$\psi = 1 + \frac{\gamma_A C_B K}{12} + O(\gamma_A^2) \quad (20)$$

where $C_B = \frac{C_T}{1 + KC_{\overline{A}}}$. Other analytical solutions are of this form, including the constant carrier

TABLE 1
Facilitation Ratios at Reaction Equilibrium for Various Transport Systems

Reference	System	Facilitation ratio
18	$NO + Fe^{+2} \rightleftharpoons Fe(NO)^{+2}$ $(A + B \rightleftharpoons AB)$	$\psi_{eq} = 1 + \dfrac{D_B}{D_A} \dfrac{KC_T}{(1 + KC_A^0)(1 + KC_A^L)}$
9	$CO + Mb \rightleftharpoons CO-Mb$ $O_2 + Mb \rightleftharpoons O_2-Mb$ $\begin{pmatrix} A + B \rightleftharpoons AB \\ C + B \rightleftharpoons CB \end{pmatrix}$	$\psi_{Aeq} = 1 + \dfrac{D_B}{D_A} \dfrac{C_T K_1}{(C_A^0 - C_A^L)} \cdot \left\{ \dfrac{C_A^0}{R^0} - \dfrac{C_A^L}{R^L} \right\}$ where $R \equiv 1 + K_1 C_A + K_2 C_c$
30	$CO_2 + H_2O \rightleftharpoons H^+ + HCO_3^-$ $CO_2 + OH^- \rightleftharpoons HCO_3^-$ $K_3 = \dfrac{(H^+)(CO_3^{-2})}{(HCO_3^-)}$	$\psi_{eq} \simeq 1 + \left(\dfrac{K_3}{K_1}\right) \dfrac{D_{HCO_3^-} (C_M^+)^2}{D_{CO_2} C_{CO_2}^0 C_{CO_2}^L}$ Footnote a where C_M^+ = metal ion concentration
15	$O_2 + Hb \rightleftharpoons Hb(O_2)$ $O_2 + Hb(O_2) \rightleftharpoons Hb(O_2)_2$ $O_2 + Hb(O_2)_2 \rightleftharpoons Hb(O_2)_3$ $O_2 + Hb(O_2)_3 \rightleftharpoons Hb(O_2)_4$	$\psi_{eq} = 1 + \dfrac{D_{Hb} C_T}{D_{O_2}(C_{O_2}^0 - C_{O_2}^L)} \{B^0 - B^L\}$ where $B = \dfrac{K_1 C_{O_2} + 2K_1 K_2 C_{O_2}^2 + 3K_1 K_2 K_3 C_{O_2}^3 + 4K_1 K_2 K_3 K_4 C_{O_2}^4}{1 + K_1 C_{O_2} + K_1 K_2 C_{O_2}^2 + K_1 K_2 K_3 C_{O_2}^3 + K_1 K_2 K_3 K_4 C_{O_2}^4}$

[a] Approximate form.

solution[18] and single-point linearization,[48] applicable in the limit of small driving forces.

At the other asymptotic extreme of "near equilibrium," the use of matched asymptotic analysis ("thick film theory") improved the results of Kreuzer and Hoofd.[49,50] Thick film theory has been shown to compare very favorably with numerical solutions for significant ranges of ψ and γ_A (see Figure 3). However, the use of a matched asymptotic analysis requires the solution of a series of implicit algebraic equations.

For a system involving more complex kinetics, Suchdeo and Shultz[31,32] obtained an expansion similar to Equation 20 for the near-diffusion regime; for near-equilibrium, Smith et al.[33] suggest that thick film analysis is readily adapted and superior to the results of power series expansions.[41] Thus, an excellent estimation of facilitation can be obtained over the entire range of γ_A by the use of thick and thin film theories.

Limiting Solution

A simple solution to Equations 2 to 4 for prediction of the facilitation ratio is possible in the limiting cases of "excess carrier" or negligible downstream permeant partial pressure. Many gas phase permeation experiments have been of this latter type, where the downstream concentration is reduced to zero by drawing a vacuum or by purging with an inert gas. If the carrier concentration, C_B, is assumed constant, the solution for the facilitation ratio to the now linear differential equations is

$$\psi = \dfrac{N_{fac}}{N_0} = \dfrac{1 + F}{1 + \dfrac{F}{\phi}\tanh(\phi)} \quad (21)$$

where

$$F = \dfrac{D_{AB} K C_B}{D_A}, \text{ and } \phi = \tfrac{1}{2}\left[\dfrac{k_1 L^2 C_B}{D_A}\left(\dfrac{1+F}{F}\right)\right]^{1/2}$$

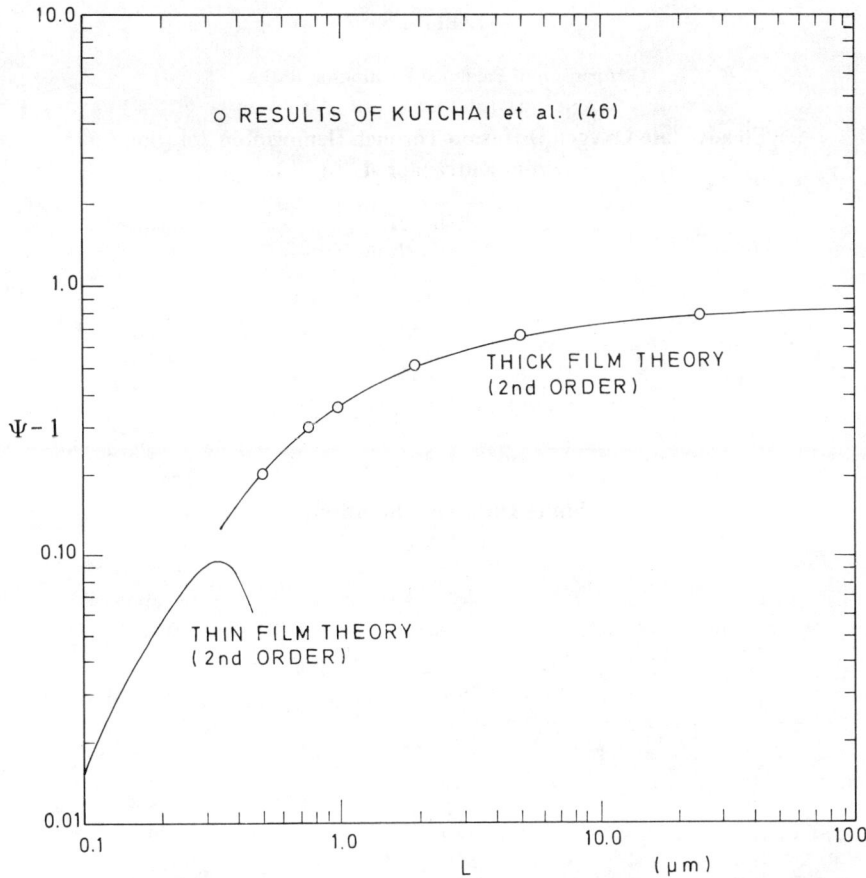

FIGURE 3. Comparison of the numerical solutions of Kutchai et al.[46] for oxygen diffusion in hemoglobin solution to the results of thin and thick film theory. (From Smith, K. A., Meldon, J. H., and Colton, C. K., *AIChE J.*, 19, 102, 1973.)

Equation 21 was first derived by Donaldson and Quinn[34] for tracer diffusion.

For the case of "excess carrier," this criterion is met if $C_B \cong C_T$, where C_T is the total amount of carrier added to the membrane. The carrier concentration may be expressed in terms of the permeant concentration through the approximate relationship:

$$C_B = \frac{C_T}{1 + KC_A} \qquad (22)$$

Equation 22 is exact at the limit of chemical equilibrium, and Smith et al.[33] have shown that it is also the proper first term in the power series expansion of C_B for the regime near molecular diffusion. Since $C_A \leq C_A^0$ at all points within the film, $C_B \cong C_T$ when the product $KC_A^0 \ll 1$. This case might be applied to the purification of a very dilute gas (small KC_A^0).

Equation 21 can also be used to predict facilitation ratios for negligible downstream partial pressure ($C_A^L = 0$) by choosing $C_A = C_A^0$ in Equation 22. The choice of a boundary permeant concentration, as opposed to, for example, a mean concentration, follows naturally from considering multilayer models[34] which show that the extent of facilitation is principally determined by reaction rates in the reaction boundary layers. Table 2 compares the facilitation ratios calculated from Equation 21 to the numerical results (for C_A^L equal to zero) of Kutchai et al.[46] and Meldon.[15] Thus, Equation 21 can be used to directly predict the facilitation ratio with modest accuracy in the useful limiting case of negligible downstream permeant concentration.

The NO/Fe^{+2} system of Ward[18] can be used as an example of the application of the preceding analysis to an $A + B \rightleftharpoons AB$ system. Table 3 lists the relevant physical and system parameters. From

TABLE 2

Comparison of Predicted Facilitation Ratios

Steady-state Oxygen Diffusion Through Hemoglobin Solution
(From Kutchai et al.[46])

$L(\mu m)$	Table VI N_{fac}/N_o	Equation 21 N_{fac}/N_o
0.5	1.196	1.160
0.75	1.286	1.275
1.0	1.351	1.364
2.0	1.500	1.550
5.0	1.646	1.692
25.0	1.769	1.780
∞	1.803	1.803

Finite Difference Solutions
(From Meldon[15])

$(k_{-1}L^2/D_B)^{1/2}$	Figure A3-1 N_{fac}/N_o	Equation 21 N_{fac}/N_o

$$\left[\frac{D_B C_T}{D_A C_A^0}, \frac{k_1 C_A^0}{k_{-1}} \right] = (1000, 10)$$

10^{-1}	2.2	1.7
10^0	16	14.9
10^1	140	129
10^2	500	568
10^3	840	858

$$\left[\frac{D_B C_T}{D_A C_A^0}, \frac{k_1 C_A^0}{k_{-1}} \right] = (10, 1)$$

10^0	1.3	1.35
10^1	4.2	4.3
10^2	5.8	5.8
10^3	6.0	6.0

$$\left[\frac{D_B C_T}{D_A C_A^0}, \frac{k_1 C_A^0}{k_{-1}} \right] = (10, 10)$$

10^0	1.6	1.6
10^1	5.5	6.4
10^2	9.0	9.5
10^3	10.0	10.0

TABLE 3

Physical and System Parameters for the NO/Fe^{+2} System

$$NO + Fe^{+2} \underset{k_{-1}}{\overset{k_1}{\rightleftharpoons}} Fe(NO)^{+2}$$

C_{NO}^0 = 2.62 × 10^{-6} mol/cm^3
C_{NO}^L = 0
L = 0.103 cm
k_1 = 7.15 × 10^{-3} cm^3/mol/sec
k_{-1} = 9 × 10^{-3}/sec
$D_{Fe^{+2}}$ = $D_{Fe(NO)^{+2}}$ = 2 × 10^{-6} cm^2/sec
D_{NO} = 1.5 × 10^{-5} cm^2/sec
C_T = 5 × 10^{-5} mol/cm^3

Data from Ward, W. J., *AIChE J.*, 16, 405, 1970.

Equation 19, substituting these values, the maximum possible facilitation at reaction equilibrium is 2.72. The expected facilitation will be less, however, and can be estimated from Equation 21. For a membrane thickness of 0.103 cm, the facilitation ratio is calculated to be 2.10. This is in good agreement with the experimental value of 2.17 (80% of the equilibrium value). Closer agreement could be expected from thick film theory.

Absolute Permeant Flux

In performing this analysis for the design of, for example, a separation system, two criteria are generally important, the degree of separation and the absolute rate of permeant transfer. The absolute permeant flux is the product:

$$N_{total} = \psi N_0 = \frac{\psi D_A (C_A^0 - C_A^L)}{L} \quad (23)$$

If Equation 21 is substituted into the above, N_{total} is seen to be a monotonically decreasing function of membrane thickness. Thus, even though a high degree of separation (large ψ) may be obtained at large L, the corresponding absolute flux may be insufficient for practical operation. Also, in regard to the degree of separation, facilitation may be increased by "loading" the membrane with carrier. However, one must also consider the possible changes in physical and thermodynamic properties resulting from gross changes in solution concentrations.

TRACER DIFFUSION

In addition to its obvious potential as a separation device, the usefulness of facilitated transport in the study of fast reaction kinetics has been demonstrated.[34,51] In the latter instance, a small quantity of isotopic tracer diffuses through a carrier-containing liquid layer equilibrated with equal partial pressures of untagged gas. The technique provides an opportunity to study the facilitated transport of (tracer) permeant under essentially equilibrium conditions; the concentration profiles of untagged species across the membrane are constant. As a result, the governing reaction-diffusion equations for the tagged permeant are then linear with respect to the permeant, and therefore a closed form solution can be obtained (Equation 21). In addition, by flushing the downstream side of the diffusion cell with an inert gas, the tracer species can be used to monitor the facilitated transport of untagged species — the specific activity of the tracer will be constant throughout.[51]

Donaldson and Quinn[34,52] have investigated the facilitated transport of CO_2 through liquid layers supported by polymer films, Millipore® filter membranes, and cross-linked protein membranes using the tracer $C^{14}O_2$, generated from $BaC^{14}O_3$. The uncatalyzed CO_2 reactions and the enzymatic reactions catalyzed by carbonic anhydrase (in solution and immobilized by cross-linking to the support filter) were studied.

The inverted form of Equation 21 is the more useful form for kinetic determinations:

$$\frac{1}{\psi} = \frac{1}{1+F} + \frac{F}{1+F} \frac{\tanh(\phi)}{\phi} \quad (24)$$

where ψ refers to the measured facilitation ratio for $C^{14}O_2$. Thus, for large ϕ ($\tanh \phi \to 1$), a Lineweaver-Burke type plot of ψ^{-1} versus (catalyst concentration)$^{-1/2}$ yielded values of the first-order carbonic anhydrase rate constant in good agreement with stopped flow techniques and afforded a means of studying the reaction at enzyme concentrations higher than previously attainable (Figure 4). Similarly, Lander[51] measured the rate constant of arsenite, another CO_2 hydration catalyst. By plotting ϕ^2 versus catalyst concentration (Figure 5), the much smaller first order catalytic rate constant of arsenite was obtained.

In another series of experiments, transport was studied in the regime of reaction equilibrium ($\frac{F}{\phi} \to 0$) by adding sufficient catalyst to the liquid layer. In Figure 6, the CO_2 facilitation ratio is plotted as

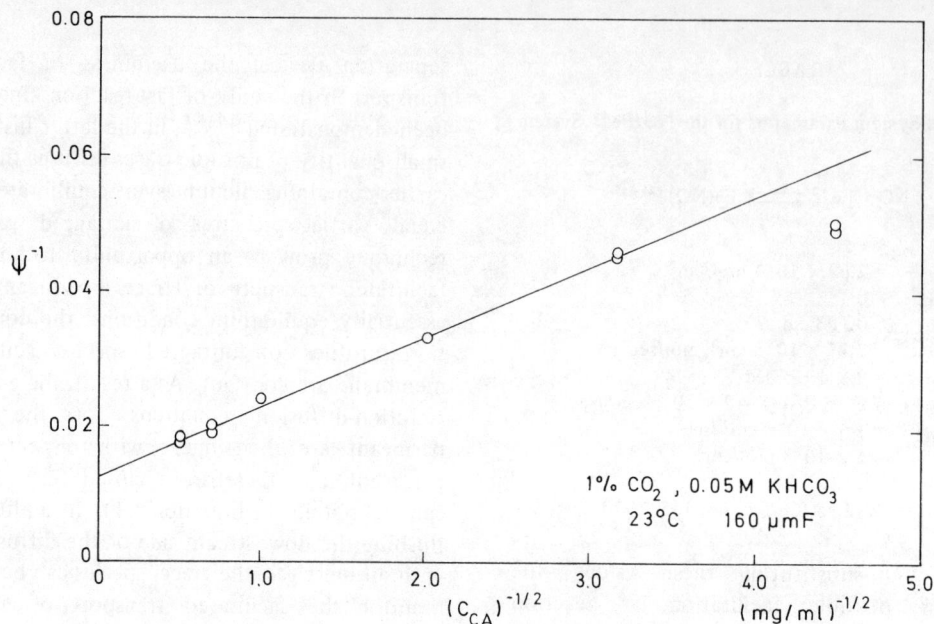

FIGURE 4. Reciprocal plot of CO_2 facilitation ratio versus square root of concentration of the catalytic enzyme, carbonic anhydrase. The solid line represents Equation 24 for tanh $\phi = 1$. (From Donaldson, T. L. and Quinn, J. A., *Chem. Eng. Sci.*, 30, 103, 1975.)

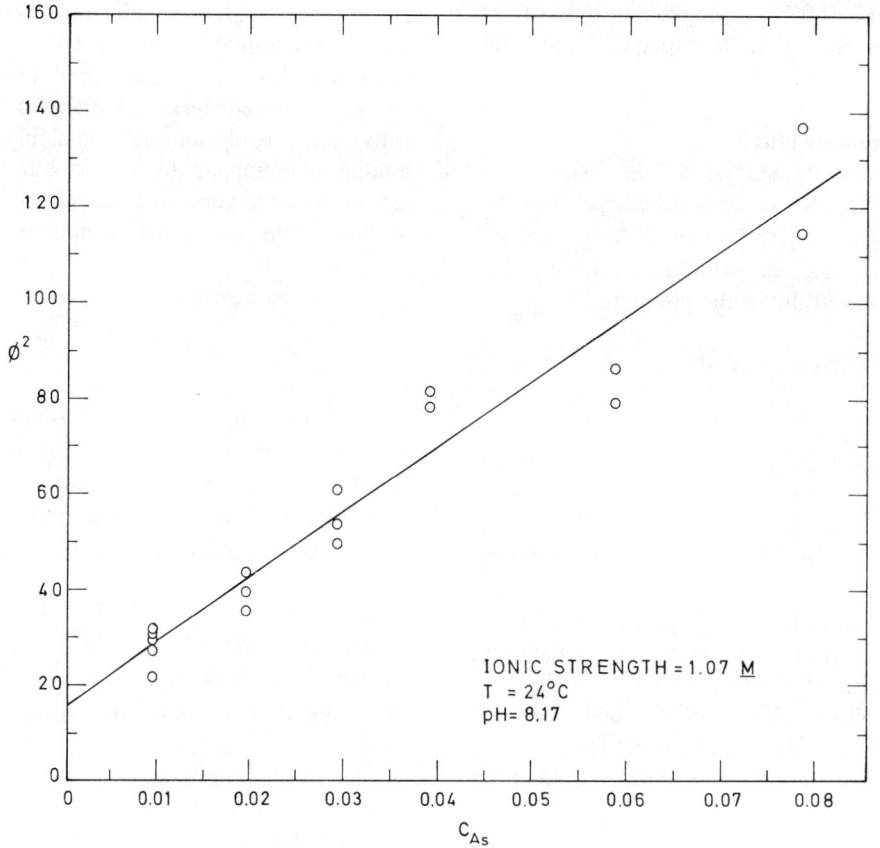

FIGURE 5. Catalytic behavior of sodium arsenite for the CO_2 hydration reaction. (From Lander, R. J., CO_2 Transport Through Bicarbonate Membranes: Catalysts, Buffers, and the Role of pH Gradients, Ph.D dissertation, University of Pennsylvania, 1977.)

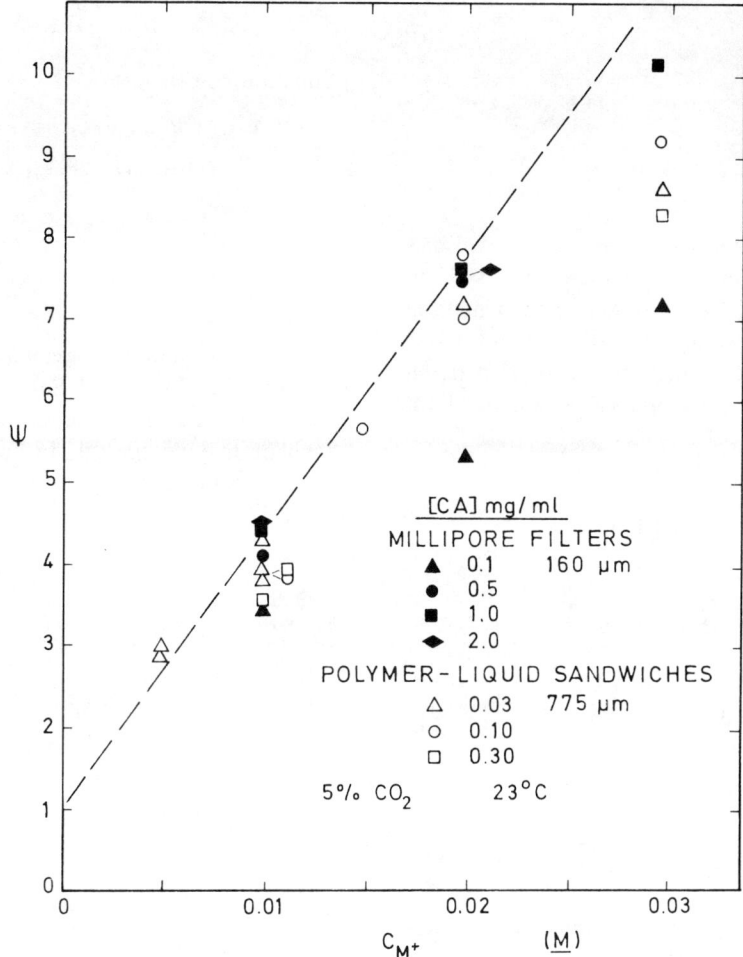

FIGURE 6. Facilitation ratio of CO_2 measured in thin liquid layer membranes and Millipore® filter membranes at several enzyme concentrations. The broken line represents the upper limit of reaction equilibrium (Equation 25). (From Donaldson, T. L. and Quinn, J. A., *Chem. Eng. Sci.*, 30, 103, 1975.)

a function of the metal ion concentration (equivalent to the bicarbonate concentration).[34] For sufficiently high carbonic anhydrase concentration, the upper limit of reaction equilibrium is reached (broken line). This line corresponds to the limiting form of Equation 21:

$$\psi_{eq} = 1 + F \tag{25}$$

The value of $D_{HCO_3^-}/D_{CO_2}$ deduced from the slope of the line is 0.60, in excellent agreement with the limited data available in the literature.[15]

COUPLED GAS COUNTERDIFFUSION

The limiting solution of equilibrium transport can provide a means of assessing the value of novel carrier systems. In principal, for example, multiple equilibria can be used as a means of coupling the transport of two permeant gases. CO_2 and H_2S are commonly occurring flue and smelter gases and the removal of H_2S has been extensively investigated.[53-56] Both are acid gases which react in aqueous solutions according to:

$$CO_2 + H_2O \underset{k_{-1}}{\overset{k_1}{\rightleftharpoons}} HCO_3^- + H^+ \tag{F}$$

where

$$K_1 = k_1/k_{-1}$$

$$H_2S \underset{k_{-2}}{\overset{k_2}{\rightleftharpoons}} HS^- + H^+ \qquad (G)$$

where

$$K_2 = k_2/k_{-2}$$

Assuming reaction equilibrium is attained in a liquid membrane exposed to these gases at both faces, the flux of H_2S is obtained by a method similar to the procedure outlined for the $A + B \rightleftharpoons AB$ system (except that the carrier material balance is replaced by the expression of local electroneutrality):

$$N_{H_2S} = \frac{D_{H_2S}}{L}\left(C^0_{H_2S} - C^L_{H_2S}\right)$$

$$+ \frac{D_{HS^-} R \bar{C}_{CO_2}}{L}\left(C^0_{H_2S} - C^L_{H_2S}\right)$$

$$+ \frac{D_{HS^-} R \bar{C}_{H_2S}}{L}\left(C^L_{CO_2} - C^0_{CO_2}\right) \qquad (26)$$

where

$$R = \frac{K_1 K_2 C_{M^+}}{(K_2 C^0_{H_2S} + K_1 C^0_{CO_2})(K_2 C^L_{H_2S} + K_1 C^L_{CO_2})}$$

and C_{M^+} = total cation concentration. This equation is similar in form to those derived by Cussler[26] for co- and countertransport of two permeants by a common carrier. The first term represents the simple molecular diffusion of H_2S. The second is the facilitation of H_2S by the carrier, HS^-, and the final term is the enhancement of flux due to the counter gradient of the second gas, CO_2 (note the opposite direction of the driving force). The coupling mechanism by which the gases interact involves the common product, H^+, of Reactions F and G and the common charge of the carrier species, HCO_3^- and HS^-. Note, to obtain this simple illustrative result, Equation 26, we have neglected the second ionization products — bisulfide and carbonate ions — as well as the hydroxyl ions.

Because of the complexity of the algebra in Equation 26, the system of acid gas countertransport is best described by several limiting cases. That "pumping" of H_2S does occur is evident from Equation 26 by assuming that no gradient in H_2S exists across the membrane:

$$C^0_{H_2S} = C^L_{H_2S} = \bar{C}_{H_2S}$$

Equation 26 becomes

$$N_{H_2S} = \frac{D_{HS^-}}{L} \frac{C_M + K_1 K_2 \bar{C}_{H_2S}(C^L_{CO_2} - C^0_{CO_2})}{(K_1 C^0_{CO_2} + K_2 \bar{C}_{H_2S})(K_1 C^L_{CO_2} + K_2 \bar{C}_{H_2S})} \qquad (27)$$

N_{H_2S} is a positive quantity in the presence of a counter-gradient in CO_2.

Two other limiting cases may be formed which describe the ratio of H_2S to CO_2 flux and the maximum attainable concentration for a given gradient in pumping gas. In both of these cases, the passive diffusion of permeants is assumed to be small:

$$\frac{D_{HS^-}}{D_{H_2S}} R \bar{C}_{CO_2} \gg 1$$

$$\frac{D_{HCO_3^-}}{D_{CO_2}} R \bar{C}_{H_2S} \gg 1 \qquad (28)$$

The ratio of N_{H_2S}/N_{CO_2} is formed by deriving an expression for N_{CO_2} analogous to Equation 26. The ratio reduces to:

$$\frac{N_{H_2S}}{N_{CO_2}} = -\frac{D_{HS^-}}{D_{HCO_3^-}} \qquad (29)$$

The negative sign reflects counter-pumping of H_2S by CO_2.

The maximum attainable concentration of H_2S is found by setting $N_{H_2S} = 0$ in Equation 26 and solving for the corresponding concentration ratio. The result is simply

$$\frac{C^L_{H_2S}}{C^0_{H_2S}} = \frac{C^L_{CO_2}}{C^0_{CO_2}} \qquad (30)$$

The simplicity of Equations 29 and 30 is a consequence of direct coupling between carriers of identical charge, namely HS^- and HCO_3^-.

The enhanced flux due to the coupling of two counter-diffusing gases can be illustrated by comparing the flux of $C^{14}O_2$ across a thin film of bicarbonate solution in the presence of untagged CO_2 concentration in both the upstream and downstream gas phases (see Table 4). The facilita-

TABLE 4

Facilitation Ratios for Tracer $C^{14}O_2$

CO$_2$ partial pressures		
Lower chamber	Upper chamber	ψ ($C^{14}O_2$)
0.05 atm	0.01 atm	15
0.05	0.05	27
0.01	0.01	48
0.01	0.05	75

Note: $C^{14}O_2$ movement from lower to upper chamber.

From Lander, R. J., CO$_2$ Transport Through Bicarbonate Membranes: Catalysts, Buffers, and the Role of pH Gradients, Ph.D. dissertation, University of Pennsylvania, 1976.

tion ratios obtained by Lander[51] for the transport of $C^{14}O_2$ from the lower to upper chamber are listed in Table 4 for various system configurations. When the ratio of the upstream to downstream CO$_2$ concentration is changed from 5:1 to 1:5 (reversing the pH profile within the membrane), the facilitation ratio increases by a factor of five. (Although the membrane contained cross-linked carbonic anhydrase, the CO$_2$ hydration catalyst, the system was not at reaction equilibrium and the foregoing analysis is not applicable.)

The chemical equilibrium limit is a convenient means for estimating the possible magnitude of counter-transport and gives useful physical insight for screening possible gas systems.

FUTURE DIRECTIONS

From an engineering viewpoint, the most immediate application of carrier-mediated transport is in the development of permselective membranes of the type developed by General Electric[54-56] for separation of gas mixtures. In addition to having low energy requirements, membrane systems can incorporate rather unique features as a separation system, such as the binding of catalytic species to the membrane support structure or accelerating the permeant flux by the counterdiffusion of another permeant. Possible permeant-carrier pairs may be suggested from considering industrial gas absorption systems, since the requirements for both are often similar.

In addition, carrier-mediated transport is important in many liquid phase systems, as in transport across biological membranes or, more recently, in terms of separation systems (see, for example, References 24 to 26). The mathematical analysis developed here for gas phase systems may be useful in describing transport across liquid phase carrier-mediated systems also.

Carrier-mediated transport that is "globally nonreactive,"[9] i.e., no net chemical conversion occurs under steady-state conditions, has been extensively investigated. However, transport studies involving globally reactive systems have been limited generally to biological transport systems. Exciting possibilities exist for studies involving facilitated transport with net chemical interconversion of the permeating species.

Finally, recent results on steady-state tracer flux experiments[34,51] indicate that this technique can be usefully exploited in determining kinetic parameters, evaluating potential facilitated transport systems, and investigating the behavior of catalytic species. The last can involve the study of enzymatic behavior — in both a bound and unbound state — in a well-described environment.

ACKNOWLEDGMENT

This investigation was supported in part by grants from the National Science Foundation and the General Electric Foundation.

REFERENCES

1. Scholander, P. F., Oxygen transport through hemoglobin solutions, *Science,* 131, 585, 1960.
2. Collins, R. E., Transport of gases through hemoglobin solutions, *Science,* 133, 1593, 1961.
3. Mochizuki, M. and Forster, R. E., Diffusion of carbon monoxide through thin layers of hemoglobin solution, *Science,* 138, 897, 1962.
4. Hemmingsen, E. A., Enhancement of oxygen transport by myoglobin, *Comp. Biochem. Physiol.,* 10, 239, 1963.
5. Wittenberg, J. B., The molecular mechanism of hemoglobin-facilitated oxygen diffusion, *J. Biol. Chem.,* 241, 104, 1966.
6. Wyman, J., Facilitated diffusion and the possible role of myoglobin as a transport mechanism, *J. Biol. Chem.,* 241, 115, 1966.
7. Colton, C. K., Stroeve, P., and Zahka, J. G., Mechanism of oxygen transport augmentation by hemoglobin, *J. Appl. Physiol.,* 35, 307, 1973.
8. Donaldson, T. L. and Quinn, J. A., Kinetic constants determined from membrane transport measurements: Carbonic anhydrase activity at high concentrations, *Proc. Nat. Acad. Sci. USA,* 71, 4995, 1974.
9. Shultz, J. S., Goddard, J. D., and Suchdeo, S. R., Facilitated transport via carrier-mediated diffusion in membranes (Part I), *AIChE J.,* 20, 417, 1974.
10. Ulanowicz, R. E. and Frazier, G. C., The transport of oxygen and carbon monoxide in hemoglobin solutions, *Math. Biosci.,* 7, 111, 1970.
11. Murray, J. D., On the molecular mechanism of facilitated oxygen diffusion by hemoglobin and myoglobin, *Proc. Roy. Soc. London Ser. B,* 178, 95, 1971.
12. Murray, J. D. and Wyman, J., Facilitated diffusion: The case of carbon monoxide, *J. Biol. Chem.,* 246, 5903, 1971.
13. Suchdeo, S. R., Goddard, J. D., and Shultz, J. S., An Analysis of the Competitive Diffusion of O_2 and CO Through Hemoglobin Solutions, in *Oxygen Transport to Tissue,* Vol. 37B, Biuley, D. F. and Bicher, H. I., Eds., Plenum Press, 1973, 951.
14. Frazier, G. C. and Shumate, S. E., II, Gaseous Transport in Hemoglobin Solutions, in *Oxygen Transport to Tissue,* Vol. 37B, Builey, D. F. and Bicher, H. I., Eds., Plenum Press, 1973, 937.
15. Meldon, J. H., Reaction-Enhanced Mass Transfer in Thin Liquid Films, Sc.D. dissertation, Massachusetts Institute of Technology, 1973.
16. Bright, P. B., The basic flow equations of electrophysiology in the presence of chemical reactions: A practical application concerning the pH and voltage effects accompanying the diffusion of O_2 through hemoglobin solutions, *Bull. Math. Biosci.,* 29, 123, 1967.
17. Wittenberg, J. B., Myoglobin-facilitated oxygen diffusion: Role of myoglobin in oxygen entry into muscle, *Physiol. Rev.,* 50, 559, 1970.
18. Ward, W. J., Analytical and experimental studies of facilitated transport, *AIChE J.,* 16, 405, 1970.
19. Ward, W. J., Electrically induced carrier transport, *Nature,* 227, 162, 1970.
20. Shultz, J. S., Goddard, J. D., and Suchdeo, S. R., Facilitated transport via carrier-mediated diffusion in membranes, *AIChE J.,* 20, 417 (Part I), 625 (Part II), 1974.
21. Ward, W. J., Immobilized Liquid Membranes, in *Recent Developments in Separation Science,* Vol. I, Li, N. N., Ed., CRC Press, 1972, 153.
22. Kreuzer, F., Facilitated diffusion of oxygen and its possible significance: A review, *Respir. Physiol.,* 9, 1, 1970.
23. Stein, W. D., *The Movement of Molecules Across Cell Membranes,* Academic Press, New York, 1967.
24. Cussler, E. L., Membranes which pump, *AIChE J.,* 17, 1300, 1971.
25. Cussler, E. L., Evans, D. F., and Matesich, M., Theoretical and experimental basis for a specific countertransport system in membranes, *Science,* 172, 377, 1971.
26. Cussler, E. L., *Multicomponent Diffusion,* Vol. 3, Elsevier, New York, 1975.
27. Enns, T., Facilitation by carbonic anhydrase of carbon dioxide transport, *Science,* 6, 44, 1967.
28. Ward, W. J. and Robb, W. L., Carbon dioxide-oxygen separation: Facilitated transport of carbon dioxide across a liquid film, *Science,* 156, 1481, 1967.
29. Otto, N. C., The Transport of Carbon Dioxide in Bicarbonate Solutions: Studies of Flux Augmentation and the Properties of Carbonic Anhydrase, Ph.D. dissertation, University of Illinois, 1971.
30. Otto, N. C. and Quinn, J. A., The facilitated transport of carbon dioxide through bicarbonate solutions, *Chem. Eng. Sci.,* 26, 949, 1971.
31. Suchdeo, S. R., Facilitated Transport of Carbon Dioxide Across Liquid Membranes Containing Bicarbonate Ion and Catalyst Carbonic Anhydrase, Ph.D. dissertation, University of Michigan, 1973.
32. Suchdeo, S. R. and Shultz, J. S., The permeability of gases through reacting solutions: The carbon dioxide-bicarbonate membrane system, *Chem. Eng. Sci.,* 28, 13, 1974.
33. Smith, K. A., Meldon, J. H., and Colton, C. K., An analysis of carrier-facilitated transport, *AIChE J.,* 19, 102, 1973.
34. Donaldson, T. L. and Quinn, J. A., Carbon dioxide transport through enzymatically active synthetic membranes, *Chem. Eng. Sci.,* 30, 103, 1975.
35. Gutknecht, J. and Tosteson, D. C., Diffusion of weak acids across lipid bilayer membranes: Effects of chemical reactions in the unstirred layers, *Science,* 182, 1258, 1973.

36. Gros, G. and Moll, W., Facilitated diffusion of CO_2 across albumin solutions, *J. Gen. Physiol.*, 64, 356, 1974.
37. Bdzil, J., Carlier, C. C., Fresch, H. L., Ward, W. J., and Breiter, M. W., Analysis of potential difference in electrically induced carrier transport systems, *J. Phys. Chem.*, 77, 846, 1973.
38. Blumenthal, R. and Katchalsky, A., The effect of the carrier association-dissociation rate on membrane permeation, *Biochim. Biophys. Acta*, 173, 357, 1969.
39. Stroeve, P., Diffusion with Reversible Chemical Reaction in Heterogeneous Media, Sc.D. dissertation, Massachusetts Institute of Technology, 1973.
40. DeSimone, J. A., Transport and Reaction in Enzymatically Active Artificial Membranes: A Theoretical and Experimental Study, Ph.D. dissertation, Harvard University, 1970.
41. Goddard, J. D., Schultz, J. S., and Bassett, R. J., On membrane diffusion with near-equilibrium reaction, *Chem. Eng. Sci.*, 25, 665, 1970.
42. Vinograd, J. R. and McBain, J. W., Diffusion of electrolytes and of the ions in their mixtures, *J. Am. Chem. Soc.*, 63, 2008, 1941.
43. Smith, D. R., Carrier-mediated Transport in Synthetic Membranes: Characterization of Transport and Evaluation of Several Potential Transport Systems, M.S. thesis, University of Pennsylvania, 1976.
44. Yung, D. and Probstein, R. F., Similarity considerations in facilitated transport, *J. Phys. Chem.*, 77, 2201, 1973.
45. LaForce, R. C., Steady-state diffusion in the carbon monoxide + oxygen + hemoglobin system, *Trans. Faraday Soc.*, 62, 1458, 1966.
46. Kutchai, H., Jacquez, J. A., and Mather, F. J., Nonequilibrium facilitated oxygen transport in hemoglobin solutions, *Biophys. J.*, 10, 38, 1970.
47. Spaan, J. A., Transfer of oxygen into hemoglobin solution, *Pflügers Arch.*, 342, 289, 1973.
48. Friedlander, S. K. and Keller, K. H., Mass transfer in reacting systems near equilibrium, Use of the affinity function, *Chem. Eng. Sci.*, 20, 121, 1965.
49. Kreuzer, F. and Hoofd, L. J., Facilitated diffusion of oxygen in the presence of hemoglobin, *Respir. Physiol.*, 8, 380, 1970.
50. Kreuzer, F., Factors influencing facilitated diffusion of oxygen in the presence of hemoglobin and myoglobin, *Respir. Physiol.*, 15, 104, 1972.
51. Lander, R. J., CO_2 Transport Through Bicarbonate Membranes: Catalysts, Buffers, and the Role of pH Gradients, Ph.D. dissertation, University of Pennsylvania, 1977.
52. Donaldson, T. L., Carbon Dioxide Transport and Carbonic Anhydrase Kinetics: Reaction and Diffusion Studies with Enzyme-Active Membranes, Ph.D. dissertation, University of Pennsylvania, 1974.
53. Dibbs, H. P., Methods for the Removal of Sulphur Dioxide From Waste Gases, Mines Branch Information Circular, IC272, Dept. of Energy, Mines, and Resources, Ottawa, Canada, 1971.
54. Matson, S. L. and Kimura, S. G., Permselective Membranes for H_2S/CO_2 Scrubbing of Coal Gas, paper presented at American Institute of Chemical Engineers National Meeting, Kansas City, Missouri, April 1976.
55. Kimura, S. G., Ward, W. J., III, and Matson, S. L., Application of Facilitated Transport Membranes to Industrial Gas Separations, paper presented at American Chemical Society Centennial Meeting, San Francisco, California, August 1976.
56. Matson, S. L. and Kimura, S. G., Permselective Membranes for the Removal of H_2S from Coal Gas, paper presented at American Chemical Society Centennial Meeting, San Francisco, California, August 1976.

CARRIER-MEDIATED TRANSPORT IN LIQUID-LIQUID MEMBRANE SYSTEMS*

J. S. Schultz

TABLE OF CONTENTS

Introduction 243

Mathematical Models 245

Carrier-mediated Separation Systems 247

$I' = 0$ Systems 247

$I' > 0$ Systems 251

Complex Forming Systems 252

Summary 258

Nomenclature 259

References 259

INTRODUCTION

Recent developments in membrane technology, e.g., hollow fibers and ultrathin membranes, have stimulated renewed interest in mechanisms of transport in membranes. Also, in the area of pharmaceutics, methodologies are being developed for more specific and local administration of drug through "controlled release" devices. Membranes with specific transport properties would be of value in this application.

Previous publications have indicated the potential selectivity and increased flux capacities of carrier-mediated facilitated transport membranes, including those by Ciani,[1] Ward,[2] Basset and Schultz,[3] Otto and Quinn,[4] and Cussler et al.[5] In this series, Ward[6] and Smith et al.[7] have reviewed the concepts and analytical methods for predicting the behavior of aqueous membranes that interact specifically with gaseous substances.

Apart from the work with lipid bilayers, experimental studies of liquid-liquid membrane separation systems have been few, possibly due to the lack of information on candidate systems. However, there is an extensive literature on extraction systems which provides a resource of information that can be used for the development of carrier-mediated transport membranes. It is the purpose of this report to point out some types of liquid extraction systems that fall into this category and also how the data can be utilized to design membrane separation systems.

A primary requirement for a carrier-mediated transport membrane is a stable separation between the membrane phase and the bulk fluid. For gaseous permeants and liquid membranes this requirement is easily met; however, for liquid-liquid systems, emulsification and surface tension effects can become practical problems as pointed out by Moore and Schecter.[8] However, some

*This work was supported in part by a NIH Research Grant GM-15152 and a Research Career Development Award, No. IKO GM08271.

recent novel methods for the stabilization of liquid membranes, e.g., such as reported by Li et al.,[16] indicate that the problems are not insurmountable.

Schultz et al.[9] defined carried-medicated membranes as that class of reactive membranes where no net consumption of the permeant or carrier species occurs in the steady state, i.e., globally nonreactive membranes. Similarly, this discussion will be limited to passive carrier-mediated transport systems.

The characterization of whether or not a given membrane system is passive can be formally determined by some simple rules, provided all species and reactions in the membrane can be identified. The system is globally nonreactive if the number of system composition invariants, I', as defined in the following equation fulfills, the condition $I' \geq 0$.

$$I' = S_p - (F + R_e) \qquad (1)$$

where

S_p = The total number of chemical species in the membrane that are involved in the transport process

F = The number of species that are free to transport into and out of the membrane phase

R_e = The number of independent chemical reactions in the membrane that are involved in the transport process

Consider, for example, a series of reactions which result in ion-pair formation only in the aqueous phase.

$$H^+ + A \rightleftharpoons HA^+, K_{HA} = \frac{[HA^+]}{[H^+][A]} \qquad (2)$$

$$HA^+ + X^- \rightleftharpoons HAX, K_{HAX} = \frac{[HAX]}{[HA^+][X^-]} \qquad (3)$$

where the neutral compounds A and HAX can distribute into immiscible organic liquids (the overbar indicates species in the organic phase).

$$A \rightleftharpoons \overline{A}, P_A = [\overline{A}]/[A] \qquad (4)$$

$$HAX \rightleftharpoons \overline{HAX}, P_{HAX} = [\overline{HAX}]/[HAX] \qquad (5)$$

A conceptual sketch of an organic liquid membrane separating two aqueous compartments is shown in Figure 1.

S_p = 2 ($\overline{HAX}$ and $\overline{A}$ are the only species involved in transport that are in the membrane phase)

F = 2 ($\overline{HAX}$ and $\overline{A}$ are the only species that are free to move across the membrane-solution interface)

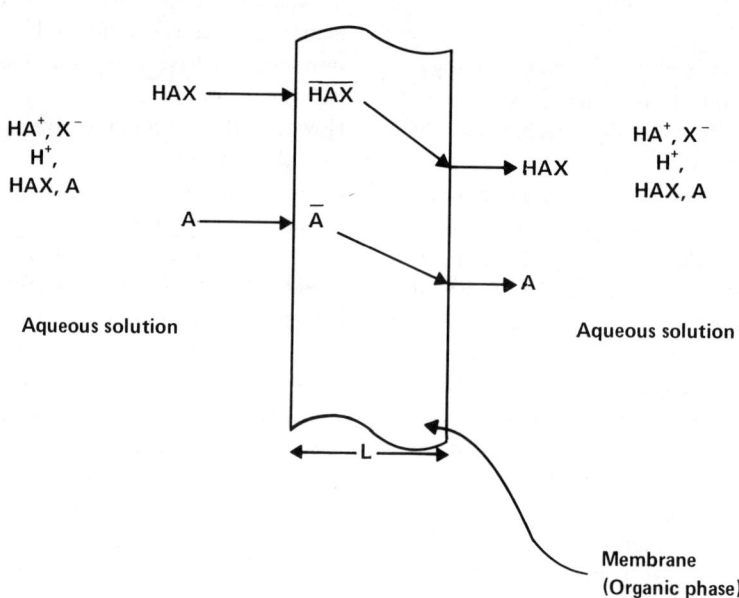

FIGURE 1. Schematic diagram of carrier-mediated transport through a membrane (organic phase) by ion-pair formation. Only the neutral ion pair, HAX, and neutral solute A are soluble in the organic membrane.

$R_e = 0$ (There are no reactions between the mobile species *within* the membrane phase)

$I' = 2 - (2 + 0) = 0$

The condition $I' = 0$, i.e., no system invariants, is a degenerate case in the sense that no "carriers" as such are confined to the membrane. The steady-state content of $\overline{HAX}$ and $\overline{A}$ in the membrane and their transport rates are determined by the boundary concentrations in the aqueous phase alone and are not at all related to any preadditions of these components to the organic liquid used in preparation of the membrane.

If an adduct reaction between HAX and S occurs in the membrane, such as

$$\overline{HAX} + 2\overline{S} \rightleftharpoons \overline{HAX \cdot S_2}$$

then

$$K_{\overline{HAX \cdot S_2}} = \frac{[\overline{HAX \cdot S_2}]}{[\overline{HAX}] \cdot [\overline{S}]^2} \quad (6)$$

Where S is a compound which is confined to the membrane phase, I' has a positive value

$S_p = 4\ (\overline{HAX}, \overline{A}, \overline{S}, \overline{HAX \cdot S_2})$
$F = 2\ (\overline{HAX}, \overline{A})$
$R_e = 1$
$I' = S_p - (F + R_e) = 4 - (2 + 1) = 1$

In this case, I' is positive with a value of 1, and therefore the membrane is still globally nonreactive, $I' > 0$. Now however, the preaddition of one component, the trapped species S, will affect steady-state transport rates.

The types of liquid-liquid extraction systems that may form the basis of carrier-mediated transport membranes generally fall into the two categories exemplified above, i.e., ion-pair forming systems and adduct-forming systems.

MATHEMATICAL MODELS

The complete description of transport in reactive membranes requires a knowledge of all transport properties (e.g., diffusion coefficients) and the kinetic mechanism and kinetic rate constants. Fortunately, one can obtain reasonable predictions of behavior without this detailed mechanistic information if the equilibrium properties of the system are known.

A detailed example of the method of calculating fluxes across a liquid membrane will be given. Consider the transport of the ion X^- according to the adduct-forming reactions (Equations 2 through 6); in the steady state three differential equations describe the diffusion-reaction process in the *membrane**

$$\mathcal{D}_{\overline{HAX}} \frac{d^2[\overline{HAX}]}{dx^2} = r_1 \quad (7)$$

$$\mathcal{D}_{\overline{S}} \frac{d^2[\overline{S}]}{dx^2} = 2r_1 \quad (8)$$

$$\mathcal{D}_{\overline{HAX \cdot S_2}} \frac{d^2(\overline{HAX \cdot S_2})}{dx^2} = -r_1 \quad (9)$$

where

$$r_1 = k_1[\overline{HAX}] \cdot [\overline{S}]^2 - k_2[\overline{HAX \cdot S_2}] \quad (10)$$

with boundary conditions

$$@\ x = 0, L\ [\overline{HAX}] = [\overline{HAX}]_0;\ [\overline{HAX}]_L \quad (11)$$

$$@\ x = 0, L\ \left.\frac{d[\overline{HAX \cdot S_2}]}{dx}\right|_{0,L} = 0;\ \left.\frac{d[\overline{S}]}{dx}\right|_{0,L} = 0 \quad (12)$$

$$\frac{1}{L} \int_0^L (2[\overline{HAX \cdot S_2}] + [\overline{S}])dx = \overline{C}^\circ_{tot}\{S\} \quad (13)$$

$\overline{C}^\circ_{tot}\{S\}$ is the initial uniform concentration of the adduct-forming substance in the membrane. The $C_{tot}\{i\}$ notation will be used to denote the total concentration of radical i in all its chemical forms.

Equations 8 and 9 can be combined with boundary conditions Equation 12 to give

$$\mathcal{D}_{\overline{S}}[\overline{S}] + 2\mathcal{D}_{\overline{HAX \cdot S_2}}[\overline{HAX \cdot S_2}] = \alpha \quad (14)$$

and if $\mathcal{D}_{\overline{S}} = \mathcal{D}_{\overline{HAX \cdot S_2}}$, Equations 13 and 14 show that

$$2[\overline{HAX \cdot S_2}] + [\overline{S}] = \overline{C}^\circ_{tot}\{S\} \quad (15)$$

That is, at all points in the membrane the total concentration of the membrane-trapped radical {S} is the same, and equal to the original "loading" concentration $\overline{C}^\circ_{tot}\{S\}$. This property of trapped radicals is general, provided that the diffusivities for all species containing the radical

---

*Since we are only concerned with transport within the membrane phase in this paper, the symbol for diffusivity, $\mathcal{D}_i$, is for species i in the membrane.

are approximately equal, e.g., $\mathscr{D}_S = \mathscr{D}_{HAX \cdot S_2}$. Differences in diffusion coefficients between species of ±20% will not have a great effect on the calculated fluxes.

The simplification represented by Equation 15 makes the calculation of transmembrane fluxes very much easier. The flux of X across the membrane is given by the flux of HAX

$$N_X = \frac{\mathscr{D}_{HAX}}{L} \Delta[\overline{HAX}] + \frac{\mathscr{D}_{HAX \cdot S_2}}{L} \Delta[\overline{HAX \cdot S_2}] \quad (16)$$

and if diffusivities are about the same ($\mathscr{D}_{HAX} = \mathscr{D}_{HAX \cdot S_2}$) then

$$N_X = \frac{\mathscr{D}_{HAX}}{L} (\Delta[\overline{HAX}] + \Delta[\overline{HAX \cdot S_2}]) \quad (17)$$

or

$$N_X = \frac{\mathscr{D}_{HAX}}{L} \Delta([\overline{HAX}] + [\overline{HAX \cdot S_2}]) \quad (18)$$

where $\Delta([\overline{HAX}] + [\overline{HAX \cdot S_2}])$ is the difference in the *total* concentration of premeable solutes evaluated at the boundaries of the membrane, i.e.,

$$\Delta([\overline{HAX}] + [\overline{HAX \cdot S_2}]) = ([\overline{HAX}] + [\overline{HAX \cdot S_2}])_0 - ([\overline{HAX}] + [\overline{HAX \cdot S_2}])_L \quad (19)$$

In general the evaluation of species concentrations at the boundaries of the membrane requires the solution of nonlinear differential equations of the form

$$\mathscr{D}_{HAX} \frac{d^2[\overline{HAX}]}{dx^2} = k_1 [\overline{HAX}] \cdot [\overline{S}]^2 - k_2 [\overline{HAX \cdot S_2}] \quad (20)$$

However, if reaction rates in the membrane are very much larger than diffusion rates, that is, if the Damkohler number for the membrane is large $\frac{k_2 L^2}{HAX} >>> 1$, then instead of Equation 10 one may write[10]

$$0 = k_1 [\overline{HAX}][\overline{S}]^2 - k_2 [\overline{HAX \cdot S_2}] \quad (21)$$

or

$$K_{\overline{HAX \cdot S_2}} = \frac{[\overline{HAX \cdot S_2}]}{[\overline{HAX}][\overline{S}]^2}$$

which is equivalent to assuming that the reaction between permeants and adduct is at equilibrium everywhere within the membranes, including the boundaries.

Methods for evaluating the validity of the equilibrium assumption for particular cases have been discussed previously,[9] but in the context of fast ionic and coordination reactions involved in extraction systems this approximation would seem to be a valid presumption. Usually, extraction reactions do not show chemical kinetic limitations, but some exceptions are known.[11]

The equilibrium assumption usually permits one to calculate the membrane boundary concentrations directly, provided the appropriate equilibrium constants are known. However, in some cases experimental distribution or partition data might be used directly without a precise understanding of the reactions in the system.

Usually, one does not specify or have knowledge of the concentration of permeants in the bulk phases adjacent to the membrane. Ignoring for the moment boundary layer (film) resistances in the aqueous phases, the interfacial membrane concentrations can be related to bulk aqueous concentrations by means of phase distribution equilibria. If one defines a practical or experimental distribution coefficient of the radical X by

$$D_X = \frac{\overline{c}_{tot}\{X\}}{c_{tot}\{X\}} = \frac{[\overline{HAX}] + [\overline{HAX \cdot S_2}]}{[HAX] + [X^-]} \quad (22)$$

Then by substituting Equation 22 into Equation 18, the flux of X across the membrane will be given by

$$N_X = \frac{\mathscr{D}_{HAX}}{L} ((D_X c_{tot}\{X\})_0 - (D_X c_{tot}\{X\})_L) \quad (23)$$

This rather simple result arises from the assumptions of equal diffusivities and chemical equilibrium in the organic phase.

The distribution coefficient, D_X, is an easily determined experimental quantity, but will be a function of many parameters, such as pH, adduct concentration, and the presence of competing anions. For these reasons, it may be desirable to relate the distribution coefficient to the more fundamental equilibrium and partition constants, as will be demonstrated in the example below. However, in order to develop the necessary relationships many other factors need to be known, which may be more difficult to measure or to find in the literature.[12] At least for the transport of one species, using experimentally determined distribution coefficients in Equation

19 may be the most direct way of making flux estimates.

In this case of an adduct-forming reaction, the total radical X content in the membrane can be related to the concentration of the ion pair HAX by the equilibrium assumption, Equation 21.

$$\overline{C}_{tot}\{X\} = [\overline{HAX}] + [\overline{HAX \cdot S_2}] = [\overline{HAX}](1 + K_{\overline{HAX \cdot S_2}}[\overline{S}]^2) \quad (24)$$

The total boundary concentration of $\overline{C}_{tot}\{X\}$ in the membrane phase is related to the aqueous concentration of $[X^-]$ by partition coefficients (Equation 5), and the aqueous phase equilibrium constant (Equation 3).

$$\overline{C}_{tot}\{X\}$$
$$= P_{HAX} K_{HAX}[HA^+][X^-](1 + K_{\overline{HAX \cdot S_2}}[\overline{S}]^2) \quad (25)$$

$$2K_{HAX}[HA^+][X^-]P_{HAX}K_{\overline{HAX \cdot S_2}}[\overline{S}]^2 + [\overline{S}] = \overline{C}^\circ_{tot}\{S\} \quad (27)$$

Now, because of conservation relations of the type given in Equation 15, i.e., system invariants, the membrane boundary concentration of free $[\overline{S}]$ can be given in terms of the initial total concentration of the adduct in the membrane, $\overline{C}^\circ_{tot}\{S\}$.

$$2[\overline{HAX \cdot S_2}] + [\overline{S}] = \overline{C}^\circ_{tot}\{S\} \quad (26)$$

Using the partition and aqueous equilibrium expressions again gives

Equation 27 can be solved for $[\overline{S}]$ (at least numerically) in terms of the (known) aqueous concentrations, and the result substituted into Equation 25 to give the membrane concentrations of $\overline{C}_{tot}\{X\}$ in terms of the aqueous concentrations of HA^+, X^-, at each boundary. The results for $\overline{C}_{tot}\{X\}_o$ and $\overline{C}_{tot}\{X\}_L$ are then substituted into Equation 23 to determine the flux of X across the membrane.

It should be pointed out that in this development, all activity coefficients were assumed to be unity. Marcus[11] discusses a more complete theory for predicting extraction phase equilibria when information on activity coefficients is available.

CARRIER-MEDIATED SEPARATION SYSTEMS

With this introduction into the analysis of carrier-mediated transport systems, we turn to some examples of organic liquid-membrane separations.

There are at least two well-recognized mechanisms for the extraction of ionic species into organic phases: complex formation and ion-pair formation.[11] Systems are said to display ion-pair formation when the interaction of the combining species are mainly "electrostatic in nature without any changes in electronic configuration of the ions or molecules, except polarization effects."[11] In complex formation, species are formed in solution from interactions between ions and molecules involving chemical bonding and capable of independent existence.

However, from the point of view of carrier-mediated transport membranes, a different distinction of systems is more relevant, and that is whether the system composition invarient (I') for the membrane phase is characterized by $I' = 0$ or $I' > 0$. In the former case, the steady-state transport rates of various species across the membrane are independent of the initial composition of the membrane with respect to the reactive species, whereas in the latter case, where some of the reactive species are "trapped" in the membrane phase, the overall transport rate responds to changes in the amounts of these species in the membrane.

However, even this may be a difficult distinction to make because the characterization of a species as trapped, or free to exchange between the membrane and aqueous phase, depends on the magnitude of its partition coefficient between the phases. Designating a species as trapped implies an organic/aqueous partition coefficient of ∞, which is not realistic for liquid-liquid systems. Therefore, a pragmatic definition of a trapped species is required, such as: a species whose total amount in the membrane does not change by more than some small amount, say 1.0%, over the range of conditions anticipated in a given application. Partition coefficients for some compounds commonly used in extraction procedures are given in Table 1.

$I' = 0$ SYSTEMS

An extensive amount of work in analytical pharmaceutical chemistry has been directed towards the development of ion-pair-forming

TABLE 1

Partition Coefficients of Complex-forming and Ion-pairing Species Between Aqueous Solutions and Organic Solvents[11]

Extractants	Partition coefficient $(mol/l)_{aq}/(mol/l)_{org}$	
	Chloroform	Other solvents
β-Diketones		
Acetylacetone	2.5×10^1	6 (Benzene)
Benzoylacetone	1.8×10^3	1.4×10^3 (Benzene)
Dibenzoyl methane	2.5×10^5	2.2×10^5 (Benzene)
Thenoyltrifluoroacetone		4.9 (Benzene)
Oxines		
8-Hydroxyquinoline	4.3×10^2	1.6×10^1 (Toluene)
2-Methyl-8-hydroxyquinoline	1.7×10^3	4.5×10^1 (1-Butane)
5,7-Dichloro-8-hydroxyquinoline	7.2×10^3	
5,7-Dibromo-8-hydroxyquinoline	1.4×10^4	
Alkylphosphoric acids		
Di-n-butylphosphoric acid vs. 0.1 M aqueous acid	2	2.3×10^1 (isobutyl methyl carbinol) 2.0×10^{-1} (toluene)
Quaternary ammonium halides		
Hyamine 1622	4.0×10^3	
Arquad 2HT-75	2.5×10^4	
$(C_7H_{15})_4NI$	4.0×10^4	
$(C_4H_9)_4NBr$	6.3×10^{-5}	
$C_{18}H_{37}(CH_3)_3NCl$	3.2×10^{-5}	

TABLE 2

Influence of Anion Species on Extraction Constants of Some Ion-pair-forming Amines in Chloroform[14]

Anion	Extraction constant* of amines			
	Codeine	Lidocaine	Promazine	Strychnine
Cl^-	7.6×10^{-2}	3.4×10^{-2}	9.1×10^1	2.3
Br^-	2.7×10^{-1}	9.5×10^{-2}	4.4×10^2	8.7
I^-	9.3×10^{-1}	4.2×10^{-1}	3.0×10^3	7.7×10^1
ClO_4^-	1.6×10^{-1}	2.0×10^{-1}	3.3×10^2	6.0
NO_3^-	5.7×10^{-2}	8.9×10^{-2}	5.0×10^2	8.3
SCN^-	1.02	5.2×10^{-1}	3.2×10^3	1.9×10^1

*Extraction constant $\equiv E^x_{HAX} = \dfrac{\overline{C}_{HAX}}{C_{HA^+} C_{X^-}}$

systems for the isolation and separation of drugs with an amine structure.[13] These systems may be viewed as examples of facilitated transport systems for the separation of drugs and/or the use of these amines to separate inorganic or organic anions.

In Table 2 some results are listed for the extraction of large organic amines into a nonpolar solvent, chloroform. The reactions involved in formation of these ion pairs are given in Equations 2 through 5 above, where A represents the amine and X^- is the ion-pairing anion. The flux of the amine (e.g., codeine) across a chloroform "membrane" is given by

$$N_A = \frac{\mathscr{D}_A}{L} \Delta[\bar{A}] + \frac{\mathscr{D}_{HAX}}{L} \Delta[\overline{HAX}] \qquad (28)$$

Note that no reactions take place within the membrane phase ($I' = [Sp - (F + R_e)] = [2 - (2 + 0)] = 0$, so that there are no composition invarients and no "carriers" to be loaded into the membrane phase.

From the equilibrium expressions, and assuming $\mathcal{P}_A = \mathcal{P}_{HAX}$, Equation 28 becomes

$$N_A = \frac{\mathcal{P}_{HAX}}{L} \Delta\left(\frac{[HA^+]}{[H^+]} P_A K_{HA^+} + [HA^+][X^-]\right) \quad (29)$$

which indicates how the flux depends upon aqueous solution conditions, pH, anion concentration, and amine concentration HA^+. However, in the reported extraction studies,[13] aqueous solution was acid, pH < 5, resulting in very little free amine (A) in the aqueous phase; therefore the flux would be given by the second term of Equation 28

$$N_A = \frac{\mathcal{P}_{HAX}}{L} \Delta[\overline{HAX}] \quad (30)$$

The individual reaction equilibrium constants were not determined, but rather a practical extraction coefficient E^x_{HAX} was calculated which was constant over the ranges of concentrations studied.

$$E^x_{HAX} \equiv \frac{\bar{C}_{tot}\{A\}}{C_{tot}\{A\}C_{tot}\{X\}} \cong \frac{[\overline{HAX}]}{[HAX][X^-]} \quad (31)$$

The flux equation becomes

$$N_A = \frac{\mathcal{P}_{HAX}}{L} E^x_{HAX} \Delta([HAX][X^-]) \quad (32)$$

For a given amine, it is obviously better to choose an anion X^- with the largest value of the extraction coefficient to get the largest flux (see Table 2). The transport of codeine across a chloroform film is maximal with thiocyanate as the counterion. Also, it can be seen from Equation 32 that with an equal mixture of the four amines in a solution with excess ion-pairing anion $[X^-]$, the driving forces $\Delta([HAX][X])$ will be the same, and the separation rates will be approximately proportional to the extraction coefficients; e.g., promazine thiocyanate will diffuse across a chloroform layer about $3.2 \times 10^3/1.02$ or 3200 times faster than codeine thiocyanate.

Some appreciation of the range in extraction constants as a function of anion structure is shown in Table 3. For a given cation, a tetrabutylammonium quaternary amine, the extraction constants vary over 10 orders of magnitude. Also, the distribution of an ion pair into an organic phase is dependent on the relative hydrophobic nature of the amine. Increasing the alkyl chain length directly affects the distribution according to the equation

$$\log E^x_{HAX} = \log b + an \quad (33)$$

where n is the number of alkyl carbons. Again, the extraction constant increases 3 to 4 times per each addition of a carbon in the chain (see Figure 2). Schill et al.[14] have shown that the type of amine also is important.

These systems might also be used to increase the concentration of a given ion-pair-forming species. In Equation 32 it is the difference in the product [HAX] [X$^-$] across the membrane that drives the amine (or the anion) across the organic layer, so that if $([HAX][X^-])_{left} >> ([HAX][X^-])_{right}$, then the flux of A will be from left to right even though $[HAX]_{left} > [HAX]_{right}$. The maximum concentration ratio of $[HAX]_{left}/[HAX]_{right}$ will be achieved when the flux is zero, $N_A = 0$, or $[HAX]_{left}/[HAX]_{right} = [X^-]_{right}/[X^-]_{left}$. The concentration of ion-pair-forming anions can be independently adjusted by using salts of cations that do not form ion pairs, e.g., Na$^+$ X$^-$. In the case illustrated in Figure 3, the quaternary amine will diffuse from the left to right side of the membrane, since $[HAX][X^-]_{left} > [HAX][X^-]_{right}$. The use of this concept to separate amino acids and peptides was illustrated by Behr and Lehn.[19]

If both the anionic and cationic species of an ion-pair combination can protonize, then the net distribution of the ion pair in the organic phase may show an optimum with respect to pH in the aqueous phase. Schill[13] has given some examples of organic amines reacting with bromothymol blue (H_2B) that demonstrate this effect. The equilibria involved are

$$H^+ + A \rightleftharpoons HA^+ \quad K_{HA^+} = [HA^+]/[H^+][A] \quad (34)$$

$$HB^- + HA^+ \rightleftharpoons HAHB \quad K_{HAHB}$$
$$= [HAHB]/[HA^+][HB^-] \quad (35)$$

$$HAHB \rightleftharpoons \overline{HAHB} \quad P_{HAHB} = \overline{HAHB}/HAHB \quad (36)$$

$$HB^- + H^+ \rightleftharpoons H_2B \quad K_{H_2B} = [H_2B]/[HB^-][H^+] \quad (37)$$

The flux of the amine (A) across a methylene

TABLE 3

Extraction Constants of Tetrabutylammonium Ion-pairs from Aqueous Solutions into Chloroform[1,3]

Type of anion	Anionic component (X⁻)	K_{HAX}*
Inorganic ion	Cl⁻	0.7
	Br⁻	2.0×10^1
	NO₃⁻	2.6×10^1
	I⁻	1.0×10^3
	ClO₄⁻	3.0×10^3
Sulfonic acid	Toluene-4-sulfonic acid	2.1×10^2
	Napthalene-2-sulfonic acid	2.8×10^3
	Anthracene-2-sulfonic acid	1.3×10^5
	Trinitrobenzene-sulfonic acid	3.0×10^4
Phenol	Phenol	0.9
	Picric acid	8.1×10^5
	2,4-Dinitro-α-naphthol	2.8×10^6
Carboxylic acid	Acetic acid	7.6×10^{-3}
	Phenylacetic acid	1.9
	Benzoic acid	2.5
	Salicylic acid	2.6×10^2
	3-Hydroxybenzoic acid	2.9×10^2
	Benzylpenicillin	1.1×10^2
	Phenoxymethylpenicillin	6.9×10^2
	Trinitrobenzoic acid	4.0×10^2
Barbituric acid derivative	Phenobarbital	7.4×10
	Hexobarbital	5.8×10^2
	Amobarbital	3.2×10^2
Acid dye	Methyl orange	2.95×10^5
	Bromothymol blue	1.0×10^8
	Dipicrylamine	4.0×10^9

$$*K_{HAX} = \frac{[\overline{HAX}]}{[\overline{HA^+}][X^-]}$$

chloride film (assuming the free amine has a negligible partition coefficient) is given by

$$N_{HAHB} = \frac{\mathscr{D}_{HAHB} \Delta[\overline{HAHB}]}{L} \quad (38)$$

or in terms of aqueous concentrations

$$N_{HAHB} = \frac{\mathscr{D}_{HAHB} \Delta(D_{HAHB} C_{tot} \{A\})}{L} \quad (39)$$

By inspection of the various reactions one can see that at low pH values, bromothymol blue will be neutral, lowering the concentration of ion pairs. At high pH values the amine will be uncharged, also interfering with ion-pair formation, and therefore in some intermediate pH range ion-pair formation and partition into the organic membrane will be favored.

This behavior is demonstrated in Figure 4 where the distribution pattern shows a maximum in the pH range of 4 to 7 for these compounds. Some rather interesting transport patterns can be expected from systems of this type. For example, say an equal-molar mixture of piribenzil (I) and dextropropoxyphane (II) were placed on opposite sides of a dichloromethane film with the left side buffered at pH 4 and the right side at pH 7.5. Compound I would diffuse from the right to the left compartment and compound II would move in the opposite direction because of the differences in distribution coefficients D_x, as a function of pH.

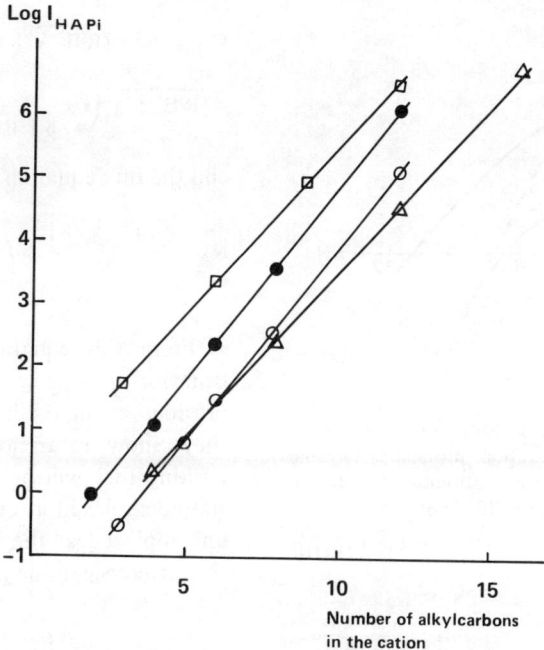

FIGURE 2. Effect of alkylamine length on the extraction constants of alkylammonium ion pairs with picrate Organic phase: CH_2Cl_2 (○ primary amine; ● secondary amine; □ tertiary amine; △ quaternary ammonium ion). (From Schill, G., Isolation of drugs and related compounds by ion-pair extraction, in *Ion Exchange and Solvent Extraction,* Vol. 6, Marinsky, J. A. and Marcus, Y., Eds., Marcel Dekker, New York, 1974. With permission.)

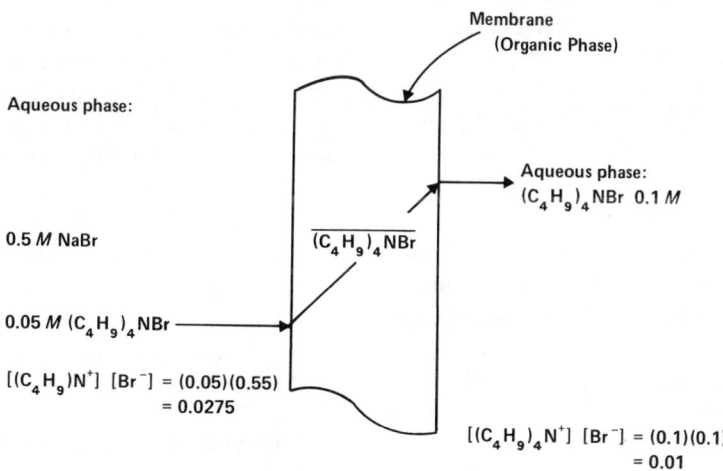

FIGURE 3. Movement of a solute (amine) against its concentration gradient by common ion effect (Br^-).

I' > 0 SYSTEMS

As mentioned earlier, there are many compounds which have very high partition coefficients in organic phase so that for practical purposes they can be considered trapped species; e.g., from Table 1, dibenzoyl methane, 5,7,-dibromo-8-hydroxyquinoline, and Arquad® 2HT-75.

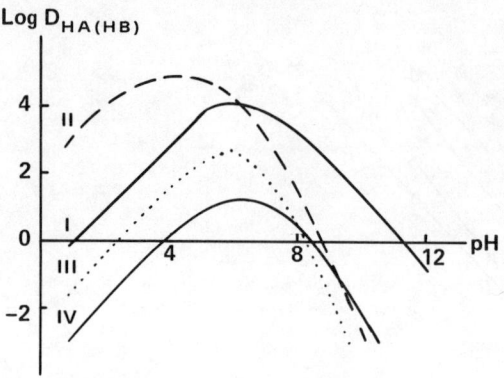

FIGURE 4. Effect of pH on extraction of ion pairs where both the anion and cation protanate. Organic phase: CH_2Cl_2, cation donor bromothymol blue (H_2B). Aqueous concentration of $H_2B = 10^{-3}$ M.

		Log E_{HAHB}
Curve I	Piribenzil	7.2
Curve II	Dextropropoxyphene	10.3
Curve III	Nicotine	5.8
Curve IV	Tryptamine	4.4

(From Schill, G., Isolation of drugs and related compounds by ion-pair extraction, *Ion Exchange and Solvent Extraction*, Vol. 6, Marinsky, J. A. and Marcus, Y., Eds., Marcel Dekker, New York, 1974. With permission.)

The compound selected can either form complexes with aqueous ions, or exchange ions to form new ion pairs. Typical of the ion exchange type are high molecular weight tertiary amines, and the reactions are of the form

$$H^+ + A^- + \overline{R'_3N} \rightleftharpoons \overline{R'_3NH^+A^-} \quad (40)$$

In the organic-membrane phase there is one compositional invariant ($S_p = 2$, $R_e = 1$, $F = 0^*$; $I' = [S_p - (F + R_e)] = 2 - (0 + 1) = 1$,), i.e., the amount of amine loaded into the membrane phase will have a direct effect on the transmembrane flux of anion (A).

The interfacial exchange reaction equilibrium can be represented by

$$K = \frac{[\overline{R H^+ A^-}]}{[\overline{R}][H^+][A^-]} \quad (41)$$

where R represents the amine. The total amine concentration in the membrane will be independent of position if the equal-diffusivity assumption is valid.

$$\overline{C}^°_{tot} \{R\} = [\overline{R H^+ A^-}] + [\overline{R}]$$

$$= [\overline{R H^+ A^-}] \left(1 + \frac{1}{K[H^+][A^-]}\right) \quad (42)$$

and the flux equation becomes

$$N_A = \frac{\mathcal{D}_R \overline{C}^°_{tot} \{R\}}{L} \Delta \left(\frac{1}{1 + \frac{1}{K[H^+][A^-]}}\right) \quad (43)$$

From this equation it can be seen that the behavior of these systems, $I' > 0$, differs from the previous examples in the important respect that they show saturation properties. A maximum limiting flux will be obtained when the amine is fully complexed at one side of the membrane and uncomplexed at the other side. Again, since it is the difference in the group

$$\left(\frac{1}{1 + \frac{1}{K[H^+][A^-]}}\right)$$

that determines the magnitude and direction of the anion flux, the anion can transport against its gradient and be concentrated by appropriately fixing the pH levels in the two aqueous solutions. Because of the saturating effect one may not obtain practical fluxes if the equilibrium constant is too large.[9] The detrimental effect of a strong binding constant was recently verified by Kobuke et al.[20] However, one might be able to choose an amine carrier with proper binding characteristics to promote selective transport once given a specification of the aqueous concentrations.

Some data for the ion-pair extraction of inorganic acids by amines are shown in Table 4. Metals can also be separated by amines if anionic metal complexes can form, e.g., MCl_4^-.[11]

COMPLEX FORMING SYSTEMS

Chelating agents, organophosphorous compounds, and long-chain fatty acids in organic solvents can form complexes with aqueous metal cations to result in the extraction of the cation into the organic phase.

Although the equlibria involved in complex formation are formally quite similar to those involved in ion-pair extraction, the chemistry is

---

*It is assumed that the organic phase is sufficiently nonpolar so that the free ions H^+ and A^- have negligible distribution coefficients.

TABLE 4

Equilibrium Constants for Amine-acid Ion-pairs in Different Solvents[11]

| Amine | Acid | \multicolumn{4}{c}{Equilibrium constants for different solvents} |
		CCl_4	Benzene	Cyclohexane	Nitrobenzene
		K*	K	K	K
Tri-n-octyl	Cl^-	0.20	0.35	0.0011	350
	Br^-	7.3	5.4	0.013	5,000
	I^-			0.9	220
	SO_4^-	0.5	25.0		
	NO_3^-	10.0	2.0		500
Tri-n-dodecyl	Cl^-		0.13		
	Br^-		0.6		
	I^-		28.0		

$$* K = \frac{[\overline{Amine - Salt}]}{[\overline{Amine}]^m [H^+][Salt^{-m}]}$$

different in that coordination rather than ionic bonds are involved. Some typical metal cation chelating agents are β-diketones, dithizones, and oxines; organophosphorous and carboxylic acid compounds also form polymeric structures in the organic and aqueous phases.

The reactions involved are of the following form:

$$M^{m+} + (m+n)\overline{HX} \rightleftharpoons \overline{MX_m (HX)_n} + mH^+ \quad (44)$$

The flux of a metal cation species across a carrier-mediated membrane of this type would be expected to follow the equation

$$N_M = \frac{\mathscr{D}_c}{L} \Delta[\overline{MX_m (HX)_n}] \quad (45)$$

where $\mathscr{D}_c$ is the diffusivity of the complex $\overline{MX_m(HX)_n}$ in the organic phase. With the boundaries at chemical equilibrium, the membrane concentrations are related to the aqueous phase by

$$[\overline{MX_m (HX)_n}] = K[\overline{HX}]^{m+n}[M^{m+}]/[H^+]^m \quad (46)$$

if the distribution of the extracting agent (HX) is primarily in the organic phase.

$$\overline{C}^\circ_{tot}(X) = [\overline{HX}] + (m+n)[\overline{HX_m(HX)_n}] \quad (47)$$

or

$$\overline{C}^\circ_{tot}(X) = [\overline{HX}] + \frac{(m+n) K[\overline{HX}]^{m+n}[M^{m+}]}{[H^+]^m} \quad (48)$$

Since (m + n) is usually greater than unity, these equations cannot be solved for $\Delta[\overline{MX_m(HX)_n}]$ explicitly and predictions for a given system are best obtained numerically. At low fractional saturation of the carrier species, i.e., $[\overline{HX}] \cong \overline{C}^\circ_{tot}(X)$, an approximate result is obtained.

$$N_M = \frac{\mathscr{D}_c}{L} K \, \overline{C}^\circ_{tot}(X)^{m+n} \Delta\left(\frac{[M^{m+}]}{[H^+]^m}\right) \quad (49)$$

In contrast to the first power dependency on carrier concentration exhibited by many of the carrier-mediated transport systems previously reviewed,[9] in these cases the flux could vary exponentially with carrier concentration.

Some typical values for distribution coefficients and solubilities of metal ion complexing agents in organic solvents are given in Tables 1 and 5. An important factor which may limit the capacity of a carrier-mediated membrane system using these compounds is the solubility of the complex in the organic phase; example values are given in Table 6. This phenomenon had been observed previoulsy in the study of oxygen transport through cobaltodihistidine solutions.[3] With the dithizonates, solubility of the complexes is much less than the solubility of the carrier, so that one is restricted to a relatively low fractional saturation of the carrier, a regime where equations of the type given in Equation 49 are valid.

The equilibrium constant for the extraction equilibria equation (Equation 46) is a function of the complex-forming agent, solvent, and metallic cation. Some values from the comprehensive tables

TABLE 5

Solubilities of Complexing and Ion-pairing Agents in Various Solvents[11]

Reagent	Solubility (g/l)			
	Water	CCl_4	$CHCl_3$	Benzene
Complexes at 25°C				
Dithizone	5×10^{-5}	0.64	17.8	1.64
p,p-Dimethyl-dithizone		1.79		
8-Hydroxy-quinoline	0.516		381	172
2-Methyl-8-hydroxyquindine			80	
5,7-Dibromo-8-hydroxyquindine			9.2	
Dimethylgloxime	0.63	0.007	0.052	6.5 (butyl alcohol)
n-Carboxylic acids at 20°C				
Octanoic	0.68	∞	∞	∞
Decanoic	0.15	2100	3260	3980
Myristic	0.02	176	325	292
Stearic	0.003	24	60	25
n-Aliphatic amines at 20°C				
$C_{10}H_{21}NH_2$		∞	∞	∞
$C_{12}H_{25}NH_2$		1480	3150	2770
$C_{18}H_{37}NH_2$		77	320	148
$(C_{12}H_{25})_2NH$		150	374	146
$(C_{12}H_{25})_3N$		∞	∞	∞
$(C_{18}H_{37})_3N$		130	135	42

TABLE 6

Solubilities of Complexes and Ion-pairs in Various Solvents, at Room Temperature[11]

Complex	Solubility (mol/l)			
	Water	CCl_4	$CHCl_3$	Benzene
Dithizonates				
$Cu(HDz)_2$	8×10^{-9}	1.2×10^{-3}	1.5×10^{-3}	
$Zn(HDz)_2$	1×10^{-4}	2.8×10^{-3}	1.6×10^{-3}	
$Co(HDz)_2$	1×10^{-4}	1.6×10^{-4}	1.4×10^{-3}	
$Pb(HDz)_2$		2.5×10^{-5}	4.3×10^{-4}	
Ag HDz		2×10^{-3}	4×10^{-2}	
Agz Dz		1×10^{-6}	1×10^{-6}	
Alkylamine HCl		(g/100 gm solvent)		
$C_{10}H_{21}NH_2 \cdot HCl$				0.13
$C_{12}H_{25}NH_2 \cdot HCl$			4.0	0.02
$(C_{10}H_{27})_2NH \cdot HCl$		0.76		0.09
$(C_{12}H_{25})_2NH \cdot HCl$		0.01		0.08

of Marcus[11] are given in Table 7. Again, the large range in binding affinity for different cations suggests that some very specific separations may be possible.

In addition to metallic cations, organic cations have been found to form complex compounds with these extracting agents. Modin and Johnson[17] have shown that a number of drugs classified as hydrophillic amines form complexes with an alkylphosphoric acid (bis(2-

TABLE 7

Distribution Equilibrium Constants, K, for Metal-complex Formation with Different Ligands[1]

Ligand (HX)	AgX	CaX$_2$	ZnX$_2$	CuX$_2$	CoX$_2$	NiX$_2$	MnX$_2$	PbX$_2$	AlX$_3$	FeX$_3$
Diketonates										
Benzoylacetone (Benzene)	−7.8	−18.3	−10.8	−4.2	−11.1	−12.1	−14.6	−1.2	−7.6	−0.5
Thenoyltrifluoroacetone (Benzene)		−12.0		−1.3	−6.7				−5.2	3.3
Dithozonates										
(CCl$_4$)	7.6		2.2	10.5	1.60	−1.2		−0.05		
(CHCl$_3$)	6.0		0.64	6.5	−1.5	−2.9		−1.2		
Oximates (HOX) (CHCl$_3$)	AgOX·HOX −4.51	CaOX$_2$·HOX −17.9	ZnOX$_2$(HOX)$_2$ −2.4	1.8	CaOX$_2$(HOX)$_2$ −2.2	−2.2	−9.3	−8.0	−5.2	4.1

Note: $K = \dfrac{[\overline{MX_m}][H^+]^m}{[\overline{HX}]^m[M^{m+}]}$ or $K = \dfrac{[\overline{MOX_m \cdot HOX_n}][H^+]^m}{[\overline{HOX}]^{m+n}[M^{m+}]}$

ethylhexyl)phosphoric acid). These reactions follow the pattern

$$HA^+ + n\overline{HX} \rightleftharpoons \overline{HAX(HX)}_{n-1} + H^+ \qquad (50)$$

where HA^+ represents the protonized amine and HX represents the alkylphosphoric acid. The composition of the organic phase complexes formed are shown in Table 8; the extraction constants are given in the original article.

Another type of interaction which can result in the transport of ionic species across an organic film is the solvation of ion pairs. Persson[18] has shown that 1-pentanol serves to transport amines into chloroform by the reaction

$$HA^+ + X^- + n\,\overline{ROH} \rightleftharpoons \overline{HAX(ROH)}_n \qquad (51)$$

in addition to simple ion-pair formation.

$$HA^+ + X^- \rightleftharpoons \overline{HAX} \qquad (52)$$

The formula of the compounds found in the organic phase are shown in Table 9. It is interesting that in cyclohexane simple ion-pair formation does not occur and that phosphates as the anion do not form simple ion pairs in either solvent.

Acidic species can also form neutral ion-pair adducts, as Modin and Schroder-Neilsen[15] have shown for the system penicillin-n-dodecylamine. In a neutral pH range, penicillins are ionized and n-dodecylamine is present mainly in the chloroform phase, and ion-pair adducts and polymers are formed according to the following stoichiometric equations:

$$X^- + H^+ + \overline{A} \rightleftharpoons \overline{HAX} \qquad (53)$$

$$2\overline{HAX} \rightleftharpoons \overline{HAX_2} \qquad (54)$$

$$\overline{A} + \overline{HAX} \rightleftharpoons \overline{HAX \cdot A} \qquad (55)$$

$$A + \overline{HAX_2} \rightleftharpoons \overline{(HAX)_2 \cdot A} \qquad (56)$$

where X^- is the penicillinate and A is the amine. At high pH values, the first reaction is driven towards the left so that the extent of penicillin distribution into the organic phase is diminished. The distribution coefficient varies over 6 orders of magnitude and goes through a maximum over the pH scale as shown in Figure 5, where

TABLE 8

Composition of Dialkylphosphate Ion-pairs in Organic Phases: Formation of Monomers and Dimers[17]

	Forms found in chloroform*	
Amine	Monomer HAX·HX	Dimer HAX·(HX)$_2$
Phenethylamine	+	+
Amphetamine	+	+
Norephedrine	+	+
Ephedrine	+	+
Norphenephrine		+
p-Hydroxynorephedrine		+
Synephrine	+	+
Epinephrine		+

*Plus signs (+) indicate which forms of the complex have been detected.

TABLE 9

Solvation of Ion-pairs by Alcohols, Composition of the Complex Species in the Organic Phase[18]

		Chloroform + 1-pentanol*			Cyclohexane + 1-pentanol		
Amine	Anion	HAX	HAX·(ROH)$_n$	n†	HAX	HAX(ROH)$_n$	n
Protriptyline							
	$H_2PO_4^-$		+	2.1	+		1.9
	Cl^-	+	+	1.5	+		2.0
	ClO_4^-	+	+	1.4	+		2.0
Iprindole							
	H_2PO_4		+	2.2			
	Cl^-	+	+	1.0			
	ClO_4	+	+	1.0			

*Plus signs (+) indicate which forms have been detected in the organic phase.
†n values shown are the average solvation numbers for the ion-pair alcohol complexes.

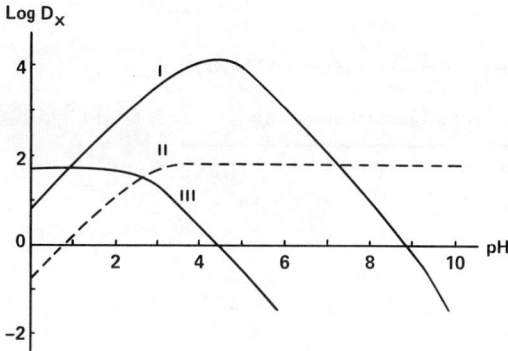

FIGURE 5. Effect of pH and adduct formation on the distribution of phenoxymethylpenicillin. Organic phase: $CHCL_3$. Curve I, penicillin ion-pair adduct with 0.1 M dodecylamine; Curve II, penicillin ion pair with 0.1 M tetrabutylammonium; Curve III, penicillin distribution as free acid. (From Schill, G., Isolation of drugs and related compounds by ion-pair extraction, in *Ion Exchange and Solvent Extraction*, Vol. 6, Marinsky, J. A. and Marcus, Y., Eds., Marcel Dekker, New York, 1974. With permission.)

$$D_x = \frac{\bar{C}_{tot}\{X\}}{C_{tot}\{X\}}$$

With the usual simplifying assumptions, the flux of penicillin across a chloroform-dodecylamine membrane is given by

$$N_x = \frac{\mathscr{D}_x}{L} \Delta(D_x C°_{tot}\{X\}) \quad (57)$$

Since D_x is not a fundamental quantity, it will vary with $\bar{C}°_{tot}\{X\}$, pH, and $\bar{C}°_{tot}\{A\}$. Figure 5 is representative of a condition where everything except pH is kept constant. These values of the distribution coefficient, D_x, can be used to predict the magnitude and direction of the penicillin flux across a membrane when there is only a difference in pH between the aqueous solutions on both sides.

Combined complex formation and adduct formation with different agents has been termed "synergistic extraction" by Marcus.[11] There are many examples where the distribution of a metal chelate into an organic phase is increased many orders of magnitude by an adduct formation with a neutral ligand dissolved in the organic phase. The results may be represented by an ion exchange reaction

$$M^{m+} + m\overline{HX} \rightleftharpoons \overline{MX_m} + mH^+ \quad (58)$$

$$K_{12} = \frac{[\overline{MX_m}][H^+]^m}{[M^{m+}][\overline{HX}]^m} \quad (59)$$

and an adduct reaction

$$\overline{MX_m} + x\bar{S} \rightleftharpoons \overline{MX_m \cdot S_x} \quad (60)$$

$$K_{11} = \frac{[\overline{MX_m \cdot S_x}]}{[\overline{MX_m}][\bar{S}]^x} \quad (61)$$

Some typical values for the equilibrium constants are given in Table 10.

In these systems, transport of the metal ion across the membrane is given by

$$N_M = \frac{\mathscr{D}_{MX}}{L} \Delta([\overline{MX_m}] + [\overline{MX_m \cdot S_x}]) \quad (62)$$

There are two compositional invariants ($I' = 2$), the preloading of the membrane with chelating agent, $\bar{C}°_{tot}\{X\}$, and adduct-forming agent, $\bar{C}°_{tot}\{S\}$, resulting in two material balance equations

$$\bar{C}°_{tot}\{X\} = [\overline{HX}] + m[\overline{MX_m}] + m[\overline{MX_m \cdot S_x}] \quad (63)$$

$$\bar{C}°_{tot}\{S\} = [\bar{S}] + x[\overline{MX_m \cdot S_x}] \quad (64)$$

These equations can be solved simultaneously to obtain the metal ion flux as a function of aqueous concentrations. An analytical result can be obtained in the limit of low complex formation in the membrane, i.e., $[HX] \approx \bar{C}°_{tot}\{X\}$, $[S] \approx \bar{C}°_{tot}\{S\}$ and when $[\overline{MX_m \cdot S_x}] \gg [\overline{MX_m}]$

$$N_M = \frac{\mathscr{D}_{MX}}{L} K_{12} K_{111} (\bar{C}°_{tot}\{X\})^m (\bar{C}°_{tot}\{S\})^x \Delta\left(\frac{[M^{m+}]}{[H^+]^m}\right) \quad (65)$$

In this dilute regime, different metal ions with the same complex composition would diffuse across the membrane independently, and relative separation rates could be predicted based on the relative values of the equilibrium constants. But if two metal ions formed complexes with different compositions, e.g., $\overline{MX_m \cdot S_x}$ and $\overline{MX_n \cdot S_y}$, then the relative transport rates of these two species across the membrane is given by the ratio

$$\frac{N_{M^{m+}}}{N_{M^{n+}}} = \frac{K_{m12} K_{m111}}{K_{n12} K_{n111}} (\bar{C}°_{tot}\{X\})^{m-n} (\bar{C}°_{tot}\{S\})^{x-y}$$

$$\frac{\Delta([M^{m+}]/[H^+]^m)}{\Delta([M^{n+}]/[H^+]^n)} \quad (66)$$

TABLE 10
Synergistic Effect of Adducts on Extraction of Complexes into Organic Phases[11]

		Added neutral ligand (S)		
		TBP	TOPO	MIBK
Adduct structure	log K_{12}	log K_{111}	log K_{111}	log K_{111}
In CCl_4				
$Ca(TTA)_2 \cdot S$	−13.4	4.14	5.6	1.8
$Ca(TTA)_2 \cdot S_2$	−13.4	8.22	10.68	2.66
$Cu(TTA)_2 \cdot S$	−1.08	2.27		0.47
$Zn(TTA)_2 \cdot S$	−8.64	4.4		1.22
In benzene				
$Cu(HFA)_2 \cdot S$		3.68	4.08	
$Cu(TTA)_2 \cdot S$		1.66	3.23	
$Cu(TFA)_2 \cdot S$		1.55	2.97	
$Cu(AA)_2 \cdot S$		1.65		

Note: TBP − Tributylphosphate; TAA − Thenoyltrifluoracetone; TOPO − Trioctylphosphine oxide; MIBK − Methyisobutyl ketone; HFA − Hexafluoroacetylacetone; TFA − Trifluoroacetylacetone; AA − Acetylacetone.

This equation shows that comparisons of binding constants alone are not sufficient to decide on the relative separation rates of metal ions.

At high concentrations of the synergistic agent in the membrane, the simple equilibria given above are no longer valid, due presumably to the formation of other complexes in the membrane, e.g.,

$$\overline{HX} + \overline{S} \rightleftharpoons \overline{HXS} \quad (67)$$

which reduces the pool of available carrier. Figure 6[11] shows the destruction of tributyl phosphate (TBP) synergism on the direct extraction of uranium and thorium from an aqueous solution into cyclohexane. In this example, the reduction in synergism at higher TBP levels is more sensitive to TBP concentration than the positive synergistic effect, which is directly proportional to TBP concentration at low levels.

SUMMARY

Specific membrane separations, concentration, and exchange processes can be developed based on the concept of carrier-mediated membranes.

A number of liquid-liquid (aqueous-organic) extraction systems have been reviewed that might form the basis of carrier-mediated transport membranes. Only systems which were globally nonreactive were selected for review. Estimates of transport behavior with changes in conditions are

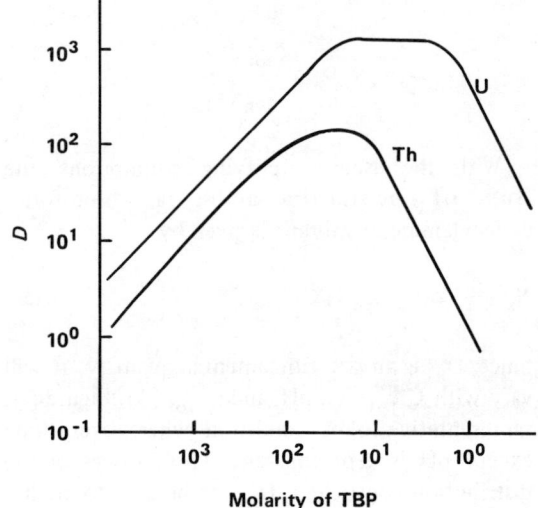

FIGURE 6. Synergistic and antagonistic adduct effects. At low concentrations TBP increases extraction of complex into cyclohexane. At higher concentrations of TBP, complexes between TBP and HTTA form, inhibiting extraction of metallic cation. Curve U, uranium (VI) +0.015 M HTTA in the aqueous phase with 0.01 N HCl; Curve Th, thorium + 0.01 M HTTA in the aqueous phase with 0.1 N HCl. (From Marcus, Y. and Kertes, A. S., *Ion Exchange and Solvent Extraction of Metal Complexes*, Interscience, New York, 1969. With permission.)

directly derived from equilibrium extraction information if one assumes equal diffusion coefficients for all forms of the transported radical, and that reaction rates are much faster than diffusion rates in the organic phase. Even in the absence of detailed knowledge of the extraction reactions, experimental phase distribution data on a particular species under conditions of interest can be used to predict transport rates across the membrane.

In this discussion, the problem of transport in the aqueous liquid film boundary layers was not addressed. If one can assume reaction equilibrium in the aqueous boundary layer, then estimating interfacial concentrations from bulk concentrations is rather straightforward, as shown by Olander.[21] Otherwise, a nonequilibrium approach will be required.[9]

NOMENCLATURE

$\overline{C}_{tot}\{i\}$ Total concentration of radical i contained in all species, mol/cm^3

  Diffusivity in organic phase, cm^2/s

D_i Distribution coefficient, equilibrium ratio of total organic phase concentration to total aqueous phase concentration of radical i

F Number of free or permeant species

I′ Number of fixed stoichiometric composition invariants

K Equilibrium constant

k Reaction rate constant

L Thickness of membrane, cm

N_i Flux across membrane of species i, mol/cm^2 − s

P Partition coefficient

R_e Number of independent chemical reactions

r_i Rate of reaction of species i, mol/cm^3 − s

S_p Number of chemical species in the membrane phase

x Distance coordinate, cm

Δ( ) Difference in membrane values evaluated at both interfaces

[ ] Concentration, mol/cm^3

— Organic phase

0,L At boundaries 0,L

REFERENCES

1. **Ciani, S., Eisenman, G., and Szabo, G.**, A theory for the effects of neutral carriers such as the marotetralide actin antibiotics on the electric properties of bilayer membranes, *J. Membr. Biol.*, 1, 1, 1969.
2. **Ward, W. J. and Robb, W. L.**, Carbon dioxide-oxygen separation: Facilitated transport of carbon dioxide across a liquid film, *Science*, 156, 1481, 1967.
3. **Bassett, R. J. and Schultz, J. S.**, Non-equilibrium facilitated diffusion of oxygen through membranes of aqueous cobaltohistidine, *Biochim. Biophys. Acta,* 211, 194, 1970.
4. **Otto, N. C. and Quinn, J. A.**, The facilitated transport of carbon dioxied through bicarbonate solutions, *Chem. Eng. Sci.,* 26, 949, 1971.
5. **Cussler, E., Evans, D. F., and Matesich, M.**, Theoretical and experimental basis for a specific countertransport system in membranes, *Science,* 172, 377, 1971.
6. **Ward, W. J.**, Immobilized liquid membranes, in *Recent Developments in Separation Science,* Vol. 1, Li, N. N., Ed., 1972, 153.
7. **Smith, D. R., Lander, R. J., and Quinn, J. A.**, Carrier-mediated transport in synthetic membranes, in *Recent Developments in Separation Science,* Vol. 3, Li, N. N., Ed., 1977.
8. **Moore, J. H. and Schechter, R. S.**, Liquid ion-exchange membranes, *AIChE J.,* 19, 741, 1973.
9. **Schultz, J. S., Goddard, J. D., and Suchdeo, S. R.**, Facilitated transport via carrier-mediated diffusion in membranes. Part 1. Mechanistic aspects, experimental systems and characteristic regimes, *Am. Inst. Chem. Eng. J.,* 20, 417, 1974.
10. **Goddard, J. D., Schultz, J. S., and Bassett, R. J.**, On membrane diffusion with near-equilibrium reaction, *Am. Inst. Chem. Eng. J.,* 25, 665, 1970.
11. **Marcus, Y. and Kertes, A. S.**, *Ion Exchange and Solvent Extraction of Metal Complexes,* Interscience, New York, 1969.
12. **Farrant, L.**, Computer-aided Design of Liquid Ion-exchange Processes, Ph.D. thesis, University of Michigan, 1974.
13. **Schill, G.**, Isolation of drugs and related compounds by ion-pair extraction, in *Ion Exchange and Solvent Extraction,* Vol. 6, Marinsky, J. A. and Marcus, Y., Eds., Marcel Dekker, New York, 1974.

14. Schill, G., Modin, R., and Perrson, B., Extraction of amines as complexes with inorganic anions, *Acta Pharm. Suec.*, 2, 119, 1965.
15. Modin, R. and Schroder-Nielsen, M., Quantitative determinations by ion-pair extraction. Part 9. Extraction of penicillins with ion-pairing and adduct-forming agents, *Acta Pharm. Suec.*, 8, 573, 1971.
16. Li, N., Permeation through liquid surfactant membranes, *AIChE J.*, 17, 459, 1971.
17. Modin, R. and Johansson, M., Quantitative determinations by ion-pair extraction. Part 8. Extraction of aminophenols and amino alcohols by bis(2-ethylhexyl) phosphoric acid, *Acta Pharm. Suec.*, 8, 561, 1971.
18. Persson, B., Extraction of amines as complexes with inorganic anions. Part 5. Studies of partition systems with organic phases containing two agents solvating ion pairs, *Acta Pharm. Suec.*, 7, 343, 1970.
19. Behr, J. P. and Lehn, J. M., Transport of amino acids through organic liquid membranes, *J. Am. Chem Soc.*, 95, 6108, 1973.
20. Kobuke, Y., Hanji, K., Horiguchi, K., Asada, M., Nakayama, Y., and Furukawa, J., Macrocyclic ligands composed of tetrahydrofuran for selective transport of monovalent cations through liquid membranes, *J. Am. Chem. Soc.*, 98, 7414, 1976.
21. Olander, D. R., Simultaneous mass transfer and equilibrium chemical reaction, *Am. Inst. Chem. Eng. J.*, 6, 233, 1960.

CHARGED MEMBRANE ULTRAFILTRATION

D. Bhattacharyya and R. B. Grieves

TABLE OF CONTENTS

Introduction 261

Membrane Rejection and Water Flux Models 263
    Solute Rejection 263
        Donnan Equilibrium Effects 263
        Rejection and Solute Flux 264
    Water Flux and Concentration Polarization 266

Ultrafiltration Results with Various Inorganic Salts 266
    Rejection Behavior of Oxyanions 267
    Rejection Behavior of Chloride and Sulfate Salts 268
    Water Flux Behavior with Sulfate and Chloride Salts 272

Scale-up Procedure: Process Design 273

Examples of Possible Applications 274
    Ultrafiltration of Plating Rinse Waters 274
    Ultrafiltration of Toxic Metal Constituents from Metal
        Manufacturing Wastewater 276
    Application to Dissolved Solids Reduction 278
    Ultrafiltration of Photographic Processing Water Constituents 278

Summary 280

Nomenclature 282

References 283

INTRODUCTION

Low-pressure ultrafiltration with charged, anisotropic (thin skin), non-cellulosic membranes is a promising technique for the separation and concentration of various inorganic salts present in aqueous solution. In many closed-loop processes involving water reuse, adequate rejections (up to 97%) with charged ultrafiltration membranes prevent the build-up of low molecular weight ionic solutes in the recycled water.[1-3] The process is termed ultrafiltration because of relatively low-pressure operation (2.5×10^5 N/m² to 7.0×10^5 N/m²) with moderately concentrated solutions of *effective* osmotic pressures (due to *rejected* solutes) less than 3.0×10^5 N/m². High water fluxes of 8×10^{-4} to 25×10^{-4} cm/sec are achieved, and specific ions can be selectively separated from mixtures of several ionic species. The Donnan exclusion mechanism is primarily responsible for the rejection of ionic solutes, and rejection is particularly high at low concentrations. Bhattacharyya et al.[1,3] have studied the rejection characteristics of such membranes for several alkaline-earth and heavy metal ions and oxyanions. Charged ultrafiltration membranes containing fixed sulfonic acid groups have been extensively studied by Gregor.[4] The simultaneous achievement of the adequate rejection of inorganic ions and the high rejection of organics has also been achieved with charged ultrafiltration membranes for complex wastes containing mixtures of inorganic electrolytes, modest-molecular-weight organic solutes (nonionic and anionic surfactants), organic macromolecules, and particulates.[5,6] Porter and Nelsen[7] have reviewed the applications of conventional (uncharged membranes) ultrafiltration to remove organic macromolecules and colloids present in various process streams.

Low-flux, tight, uncharged membranes (such as cellulose acetate) operated at pressures of 4.0×10^6 N/m^2 to 10^7 N/m^2 are commonly used to remove inorganic electrolytes from water by the reverse osmosis (hyperfiltration) process. The salt rejections of these tight membranes are generally higher than 98%; however, the high effective osmotic pressures (due to the high rejections of virtually all ionic solutes), together with the membrane tightness, necessitate high pressure operation which is often associated with membrane compaction problems and resulting water flux reduction. In addition, the selective separation of specific ionic solutes from mixtures is generally not possible. Rejections are described by various mechanisms such as the solution-diffusion model,[8,9] electrostatic repulsion of ions in the vicinity of membranes of low dielectric constant,[10] and preferential sorption-capillary flow.[11,12] The solution-diffusion model is most often used to describe solute and solvent transport through tight membranes, where each component at the high-pressure side is assumed to dissolve in the membrane according to an equilibrium distribution, and the concentration and pressure gradients produce diffusion through the membrane.

Charged (ion exchange) membranes used in electrically driven membrane processes are quite thick (>100 μm) and isotropic and thus are not suitable for pressure-driven processes. McKelvey et al.[13] reported that ion exchange membranes could be used in high-pressure desalination, but obtained very low water flux compared to conventional reverse osmosis membranes. High-flux, charged, reverse osmosis (hyperfiltration) membranes have since been developed; Yasuda, Lamaze, and Schindler[14] have experimentally proved that charged (negative- or positive-fixed-charge) membranes definitely provide higher salt rejections than uncharged membranes of similar resistance. The qualitative behavior of their salt rejection results is shown in Figure 1 in order to stress that loose-structure (low resistance and thus high water flux at a given pressure) charged membranes may provide adequate rejections in various applications involving solutions of moderate to low concentrations. High-pressure performance of sulfonated, 2,6-dimethyl polyphenylene oxide,[15,16] grafted polyethylene,[14,17] vulcanized copolymer of styrene and butadiene,[14,17] polyacrylic acid composite,[18,19] and polystyrene-polyisoprene-polystyrene block copolymer[20] charged membranes have been reported in the literature. The desired water permeability and solute rejection are obtained by controlling the ion exchange capacity of the polymer. LaConti et al.[16] have used cation-exchange, high-pressure membranes to treat brackish water, caustic feeds, secondary sewage, and synthetic wash water and obtained water fluxes of 2×10^{-4} cm/sec to 20×10^{-4} cm/sec at 5.6×10^6 N/m^2, depending on the ion exchange capacity.

Initial water fluxes as high as 10^{-2} cm/sec at 7×10^6 N/m^2 have been reported with dynamic, hyperfiltration, inorganic oxide membranes (with and without organic polymers) prepared by researchers at the Oak Ridge National Laboratory.[21,22] These membranes possess ion-exchange properties, and electrolyte rejections are due to Donnan exclusion mechanisms. Hydrous oxide membranes are generally formed (in situ) on microporous filters or on porous carbon tubes. Among the various oxide membranes reported in the literature,[21] hydrous zirconium (IV) oxide has received most attention. In basic solutions, hydrous oxide membranes behave as a cation exchanger with the opposite situation occurring in acidic solution; thus the rejections of most salts were reported[22] to be functions of pH. Further improvements in rejection and pH-dependent properties were reported with dual-layer zirconium oxide-polyacrylic acid (Zr-PAA) membranes. More reproducible properties and less sensitivity to the type of porous support have been obtained with the dual-layer membranes. Results[21] with paper mill wastes, brackish water, and dyeing wastes have been reported to be promising. Although the dynamic membranes have quite high initial water fluxes, severe concentration polarization (water flux drop) was observed[23] unless extreme (1500

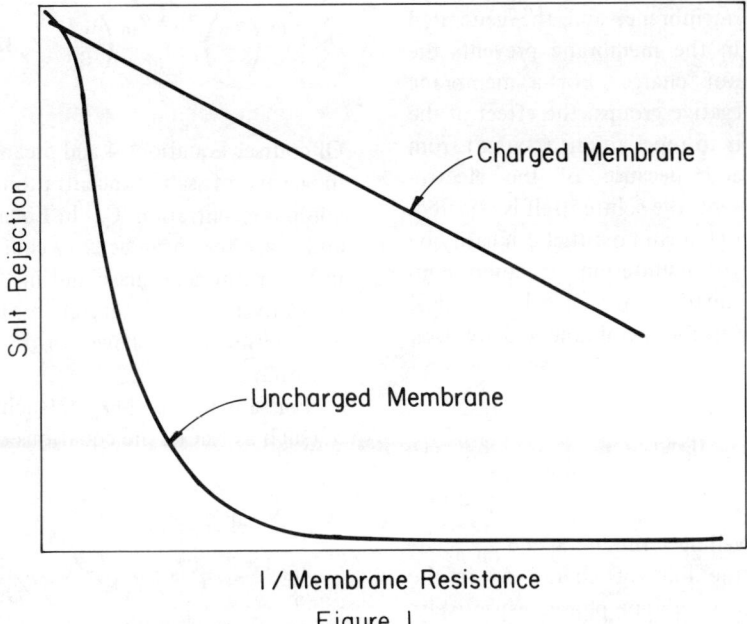

FIGURE 1. Salt rejection characteristics of charged and uncharged membranes.

cm/sec) circulation velocities on the high-pressure side were utilized. High-pressure membrane compaction problems (flux loss) are similar to those with conventional reverse osmosis membranes.

This chapter will primarily deal with the results of extensive experimental investigations of a commercially available, negatively charged, noncellulosic ultrafiltration membrane to separate various salts (including organic ions) from aqueous systems. The extent of separation (at transmembrane pressures less than 6×10^5 N/m^2) of several oxyanions, alkaline earth metal salts, and heavy metal salts is discussed. Ultrafiltration results of possible industrial significance involving dissolved solids reduction and metal manufacturing water and rinse water treatment, including metals recovery and water reuse, are also presented. A unique computer simulation model to predict the maximum extent of separation at high water recovery is utilized to scale up laboratory data obtained at very low water recoveries.

MEMBRANE REJECTION AND WATER FLUX MODELS

Solute Rejection

It is a well-known phenomenon that for membranes carrying fixed charged groups on the pore walls, even porous membranes can exclude salt to a considerable degree by the establishment of a Donnan equilibrium between the membrane phase and the aqueous solution. This is due to the long range of electrostatic forces: thus the repulsion of ions (coions) begins to be active at pore sizes several times larger than the diameter of the permeating species. The surface porosity of the ultrafiltration membranes used in these experiments was approximately 10 to 15 Å.

Donnan Equilibrium Effects

Although a membrane ultrafiltration process is not in static equilibrium, local equilibrium is substantially maintained, and in particular, the Donnan equilibrium salt distribution coefficient between the aqueous solution phase and the membrane phase can be used to predict the relative effects of certain membrane and solution variables and to evaluate rejections under limited operating conditions.[21,22,24] When a charged (ion exchange) membrane is immersed in a solution, the counterion concentration inside the membrane is higher than in the bulk solution, and the interstitial coion concentration is considerably lower (or zero) than that of the bulk. An electric field (Donnan potential) between the interior of the membrane and the bulk solution is created as the first few counterions begin to diffuse into the bulk solution. A similar situation also occurs under an applied pressure gradient which causes water flow through the membrane, and the generated electric potential in the membrane prevents the steady separation of charge. For a membrane containing fixed negative groups, the effect of the Donnan potential is to repel anions (coions) from the membrane, and because of the electroneutrality requirement, the solute itself is rejected.

The equilibrium Donnan potential can easily be written in terms of counterion or coion concentrations by equating the total chemical potential of an ion in the membrane and solution phases:

$$E_d = \frac{1}{Z_j F}(RT \ln \frac{a_j}{a_{j(m)}} - \Pi_s v_j) \tag{1}$$

where $\Pi_s v_j$ is a swelling term and a_j and $a_{j(m)}$ are the activities of the ion (of charge Z_j) in the solution and in the membrane phase, respectively. For a negatively charged membrane, the Donnan potential is negative in the membrane and since Z_j is negative for the coion, the activity (or concentration) $a_{j(m)}$ must be less than a_j in Equation 1.

For a salt of type $M_{Z_y} Y_{Z_m}$ which ionizes to M^{Z_m+} and Y^{Z_y-} (coion), the distribution of a particular ion (coion) between a solution and a negatively charged (cation exchanger) membrane of charge density (capacity) C_m^* can be obtained by combining Equation 1 (neglecting the swelling term) and the following electroneutrality equations:

(in membrane):

$$\Sigma Z_j C_{j(m)} - C_m^* = 0 \tag{2}$$

(in bulk solution):

$$\Sigma Z_j C_j = 0 \tag{3}$$

Thus the salt distribution between an aqueous solution and membrane is expressed as:

$$\left(\frac{C_{y(m)}}{C_y}\right)^{Z_m} \left(\frac{\gamma_m}{\gamma}\right)^{Z_y+Z_m} = \left(\frac{Z_y C_y}{C_m^* + Z_y C_{y(m)}}\right)^{Z_y} \tag{4}$$

Defining the salt distribution coefficient $C_{y(m)}/C_y$ as K^*, Equation 4 becomes:

$$(K^*)^{Z_m} \left(\frac{\gamma_m}{\gamma}\right)^{Z_y+Z_m} \left(\frac{C_m^*}{C_y} + Z_y K^*\right)^{Z_y} = (Z_y)^{Z_y} \tag{5}$$

Of course, Equations 4 and 5 can also be expressed in terms of salt concentration rather than the coion concentration, C_y. In Equations 4 and 5, γ_m and γ are the mean activity coefficients of the salt in the membrane phase and in the solution phase, respectively. Shor et al.[22] and Hwang and Kammermeyer[25] have also reported similar equations.

For salts of type $M_{Z_y} Y$ (such as Na_2SO_4) or M Y (such as NaCl), the equations can be reduced to:

($M_{Z_y} Y$ salt):

$$K^*\left(\frac{\gamma_m}{\gamma}\right)^{1+Z_y} \left(\frac{C_m^*}{C_y} + Z_y K^*\right)^{Z_y} = (Z_y)^{Z_y} \tag{6}$$

(M Y salt):

$$K^*\left(\frac{\gamma_m}{\gamma}\right)^2 \left(\frac{C_m^*}{C_y} + K^*\right) = 1 \tag{7}$$

With good coion exclusion and because $Z_y K^*$ and $K^* \ll C_m^*/C_y$, further simplifications[26] can be made; for example, Equation 7 for a univalent salt becomes:

$$K^* = \frac{C_y}{C_m^*} \frac{1}{(\gamma_m/\gamma)^2} \tag{8}$$

The relative effects of membrane charge density, bulk salt concentration, coion valency, and activity ratio on the equilibrium distribution can be estimated from the above equations.

Rejection and Solute Flux

The extent of separation of solutes by ultrafiltration can be defined in terms of a rejection parameter R^*:

$$R^* = 1 - \frac{J_s}{J_w C_i} = 1 - \frac{C_f}{C_i} \tag{9}$$

in which C_f is the concentration of a particular solute (or ion) in the ultrafiltrate, C_i is the concentration of the same solute (or ion) in the inlet (feed) stream, J_s is the solute flux, and J_w is the water flux. Various models developed for high-pressure, charged hyperfiltration membranes, which include the fixed charge model,[27,28] the ion association model,[24,29] and the extended Nernst-Planck model,[30] are equally applicable to the ultrafiltration case; the diffusion of solute through an ultrafiltration membrane can be considered negligible compared to high solute flux due to convection. With all charged membranes and for the condition of no concentration polarization, rejection increases with pressure (water flux). An asymptotic rejection is reached at high water flux where the entire process is governed by salt exclusion at the feed boundary, and the rejection can be related to the equilibrium distribution coefficient.[22,28]

At high water flux, K^* in Equations 4 to 8 is equal to C_f/C_i and

$$R^* = 1 - K^* \qquad (10)$$

Shor et al.[22] reported that the coupling coefficient between water flow and solute flow can be assumed to be unity. Hence from Equations 5 or 7 it can be shown that good rejection is possible even with $\gamma_m/\gamma = 1$, if the ratio of membrane charge density to the feed concentration is large. Rejection decreases with an increase in bulk salt concentration and with a decrease in membrane charge density. LaConti et al.[16] in their sulfonated polyphenylene oxide membranes showed that with 25 mM NaCl at 5.6×10^6 N/m^2 pressure, the rejection dropped from 0.96 to 0.89 when the membrane charge density decreased from 4.2 eq/liter to 2.5 eq/liter. Equations 4 and 5 also show that rejections of salts containing multivalent coions will be higher than those of univalent salts. The poor rejection of monovalent coions (such as Cl$^-$), particularly at high concentration, could be advantageous in some situations in which the selective separation or concentration of multivalent coions is desired. Hoffer and Kedem[28] have related rejection to the mobility (transport number) of ions, to the ratio of fixed charge density to feed solute concentration, and to ionic charge. Solute rejection is reported to be lower with coions of high transport number. Both Hoffer and Kedem[28] and Shor et al.[22] also noted that the actual salt rejection is lower in dilute solution than that predicted by Donnan equilibrium due to membrane nonhomogeneity and the tendency of counterions present in the membrane to be swept along with the permeate water, thus accelerating the coions to maintain electroneutrality. The quantitative effect of specific cations on the rejection of a common anion is difficult to predict when membrane swelling (or shrinkage) and/or membrane-ion association occurs. Also, in real systems, the effective charge, which includes both ion valency and hydrated radius, is probably a determining factor for rejection.

Finally, it should be recognized that the overall flux of an ion through the membrane during ultrafiltration or hyperfiltration is the sum of fluxes due to convection, diffusion, and electrical potential. The extended Nernst-Planck equation[30] for overall flux of an ion is (neglecting the change in activity coefficients),

$$(J_s)_j = B_j J_w \; C_{j(m)} - D_j \frac{dC_{j(m)}}{dy} - Z_j C_{j(m)} D_j \frac{F}{RT} \frac{d\psi}{dy}$$

(convection) (diffusion) (electric potential)

$$(11)$$

in which D_j is the diffusivity of the ion in the membrane, y is the distance coordinate in the membrane, F is the Faraday constant, and $\frac{d\psi}{dy}$ is the electrostatic potential gradient. B_j is a correction for the possibility that in the pure convection mode the ions may not be swept along with the velocity of the permeating water due to transitional binding of counterions with the fixed charges. The term $(J_s)_j$ is always equal to the total solution flux J_s for a univalent solute because of the electroneutrality requirement. Dresner[30] has integrated Equation 11 involving hyperfiltration of multicomponent solutions through charged membranes for the case of good coion exclusion. With ultrafiltration, particularly at operating pressures where the water flux is sufficiently high (thus neglecting the diffusion term) and for a single salt system, Equation 11 reduces[30] from a differential to an algebraic expression; the condition of no electric current flow is used to solve the potential term.

The dependence of solute rejection on concentration is indicated by all models. In the absence of concentration polarization and at constant transmembrane pressure, the ultrafiltrate quality C_f (or R^*), for continuous-flow, steady-state operation involving insignificant water recovery, can be related to C_i by simple functions of the form:[1,3]

$$C_f = K C_i^n \tag{12}$$

or

$$R^* = 1 - K C_i^{n-1}$$

in which the constants "K" and "n" include the effects of solute type, membrane charge density, and possible membrane-solute interactions. A solute containing di- or multivalent coions should yield smaller K and n compared to those with monovalent coions. In Equation 12, a value of n = 1 indicates constant rejection for any feed (inlet) stream concentration range.

Water Flux and Concentration Polarization

Water transport through charged ultrafiltration membranes follows a viscous flow mechanism. The water flux, J_w, can be computed from:

$$J_w = \frac{1}{R_m} (\Delta p - \Delta \Pi) \tag{13}$$

The membranes used in the study reported here had an average resistance $R_m = 3.5 \times 10^8$ N/m²/cm/sec at 25°C, in the absence of solutes. The viscous flow model was verified by water flux-temperature dependence with solute-free water which was accounted for by the variation of the viscosity of water with temperature. Yasuda and Schindler[17] and LaConti et al.[16] have related R_m to the volume fraction of water in the membrane, which in turn depends on the charge capacity of the membrane. The osmotic pressure difference $\Delta \Pi$ is equal to:

$$\Delta \Pi = \Pi_{at\ C = C_w} - \Pi_{at\ C = C_f} \tag{14}$$

The osmotic pressure for dilute solutions can be computed from the Van't Hoff equation; in the absence of concentration polarization, Equation 14 becomes,

$$\Delta \Pi = (i\ C_i\ R\ T)\ (R^*) \tag{15}$$

To prevent water flux limitations and a concomitant decrease in solute rejection, concentration polarization must be minimized. The accumulation of solute(s) on the membrane surface (concentration polarization) is caused by the rapid convection to the surface of rejected solutes. For systems containing ionic solutes in the presence of organic macromolecules and/or particulates, the actual water flux is considerably lower[6] than that given by Equation 13 due to an added resistance, and the conventional ultrafiltration model presented by Michaels et al.[31] and Porter and Nelsen,[7] can be used. For simple systems (involving osmotic pressure effects only), concentration polarization can be predicted[8] instead by the model used in reverse osmosis. During steady-state operation:

$$J_w = K_s \ln \frac{C_w}{C} \tag{16}$$

The value of the mass transfer coefficient, K_s, depends on the flow regime (laminar or turbulent) and various equations[7,31] have been used to predict the effects of channel geometry and channel velocity. To keep C_w/C close to unity, a high K_s must be obtained. For an ultrafiltration cell of fixed geometry, K_s is increased by increasing the channel velocity (or Reynolds number).

ULTRAFILTRATION RESULTS WITH VARIOUS INORGANIC SALTS

Continuous flow, steady-state, ultrafiltration experiments at very low water recovery were conducted with several salts (with pH adjustment when necessary) over a broad concentration range (0.1 mM to 16.0 mM), utilizing a thin-channel cell containing 50.3 cm² of commercially available (Millipore PSAL®) charged ultrafiltration membrane. The membranes used had the following characteristics:

Membrane: non-cellulosic skin with cellulosic backing
Pore width: approximately 15 Å
Membrane thickness: 150 μm (skin and backing)
Fixed charge: negative
Functional groups: sulfonate
Average membrane resistance: 3.5×10^8 N/m²/cm/sec
Typical flux at 2.8×10^5 N/m²: 8×10^{-4} cm/sec (16.1 gfd)
Operating pH: 2.5 to 11.0
Temperature limit: 35°C
Maximum pressure limit: 9.0×10^5 N/m²

Most experiments were performed at $\Delta p = 2.8 \times 10^5$ N/m² and at an average channel (1.27 cm width and 0.08 cm height) velocity of 167 cm/sec (Re = 2400). An average channel velocity above 80 cm/sec essentially eliminated concentration polarization, as indicated by no change in water flux or rejection with a decrease in velocity from 167 cm/sec to 80 cm/sec. The experimental unit used for the ultrafiltration study has been detailed by Bhattacharyya et al.[1-3]

Rejection Behavior of Oxyanions

Rejections by charged membranes are expected to decrease with an increasing concentration of the feed solution. Results[3] (single salt) with divalent (SO_4^{2-}, HPO_4^{2-}, or CrO_4^{2-}) or monovalent (HCO_3^-, $H_2PO_4^-$, or $HCrO_4^-$) oxyanions present as sodium salts are shown in Figure 2 in terms of ultrafiltrate concentration. For the divalent oxyanions, a single correlation gave a good fit and the rejection was a rather weak function of concentration:

$$C_f = 0.042 \, C_i^{1.10} \tag{17}$$

$$R^* = 1 - 0.042 \, C_i^{0.10}$$

With the monovalent oxyanions, the ultrafiltrate concentration and thus the rejection were not only stronger functions of concentration but also

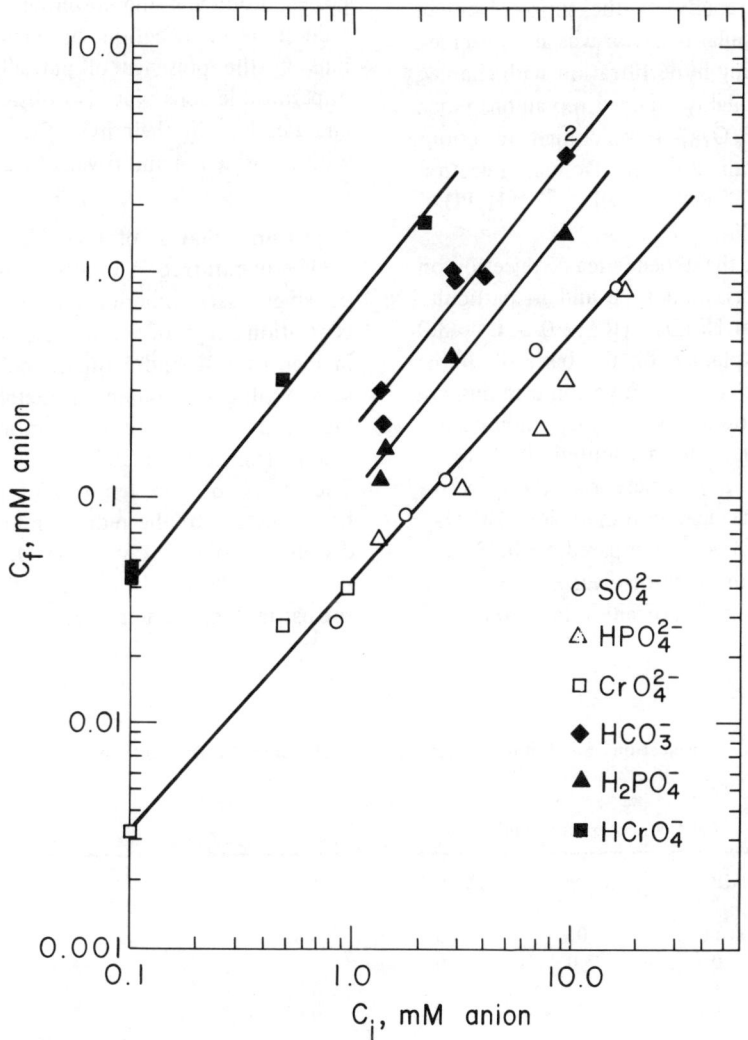

FIGURE 2. Relation between ultrafiltrate concentration and inlet stream concentration for sodium salts of mono- and divalent oxyanions.

depended on the type of ion. The relationship which gave a good fit for the monovalent species was

$$C_f = K_1 \, C_i^{1.25} \qquad (18)$$

$$R^* = 1 - K_1 \, C_i^{0.25}$$

The values of K_1 (ultrafiltrate concentration at unit inlet concentration) were 0.10 for $H_2PO_4^-$, 0.19 for HCO_3^-, and 0.73 for $HCrO_4^-$.

The poorer rejection of monovalent oxyanions compared to divalent oxyanions and the decrease in rejection with an increase in concentration are in agreement with Donnan equilibrium (Equations 6, 7, and 10). The rejection did not decline with concentration as rapidly as the simple Donnan theory predicts. Similar behavior was also observed by Johnson[21] during hyperfiltration with charged membranes. With highly charged oxyanions (such as PO_4^{3-} and $HP_3O_{10}^{4-}$), very high rejections (>0.98) were obtained.[5] The Donnan rejection trend of $HP_3O_{10}^{4-} > PO_4^{3-} > HPO_4^{2-} > H_2PO_4^-$ was found to be valid.[3,5]

In Equation 18, the dependence of rejection on the monovalent oxyanion type and in particular the low rejection of $HCrO_4^-$ ($R^* \to 0$ at $C_i \approx 3.0$ mM) cannot be explained on the basis of charge alone. Both charge and effective ionic radius are probably important with the monovalent oxyanions and may be accounted for[3] by a parameter such as Z^2/r where r is the effective ionic radius.[32] The acid chromate ion, $HCrO_4^-$ has a Z^2/r of 0.58 Å^{-1} compared to 0.73 Å^{-1} with $H_2PO_4^-$. In the case of divalent oxyanions, the charge plays the important role in terms of strong repulsion by fixed membrane sites, and thus differences in ionic radius would not be expected to have a significant effect on the relative rejections.

Ultrafiltration experiments with solutions containing mixtures of monovalent anions and divalent anions, with Na^+ as the common cation, showed a significant decrease in monovalent anion rejection, compared to single salt solutions. The results are shown in Table 1. A decrease in the rejection of HCO_3^- was observed in the presence of either SO_4^{2-} or HPO_4^{2-}. The divalent SO_4^{2-} or HPO_4^{2-} can be considered a partially membrane-impermeable ion because of its high rejection, and because HCO_3^- is more permeable, a special type[33] of Donnan equilibrium distribution takes place, producing the lowering of rejection. The reduction in rejection of membrane-permeable ions in the presence of partially or completely impermeable ions was also observed by Lonsdale and Pusch[33] in their hyperfiltration experiments with chloride ion and trivalent citrate ion.

Rejection Behavior of Chloride and Sulfate Salts

The ultrafiltration results of an extensive series of single salt experiments over the inlet concentration range of 0.3 to 16.0 mM are presented in Figures 3, 4, and 5 for the chloride and sulfate salts of alkali-alkaline earth metals (Na^+, Ca^{2+} and Mg^{2+}), of a trivalent counterion (La^{3+}), and of heavy (transition) metals (Cu^{2+}, Ni^{2+}, and Zn^{2+}). The study of the removal of potentially toxic heavy metals, of the reduction in hardness-causing dissolved solids (Ca^{2+}, Mg^{2+}), and/or of the recovery of valuable metals (Ni^{2+}) from various industrial processing waters is of obvious

TABLE 1

Rejection of a Monovalent Anion in the Presence of a Divalent Anion

Feed concentration (mM anion)			Rejection		
NaHCO$_3$	Na$_2$SO$_4$	Na$_2$HPO$_4$	HCO$_3^-$	SO$_4^{2-}$	HPO$_4^{2-}$
10.6	0	0	0.62	—	—
0	3.0	0	—	0.95	—
10.6	3.0	0	0.40	0.94	—
8.2	0	0	0.65	—	—
0	0	2.6	—	—	0.96
8.2	0	2.6	0.49	—	0.94

Note: $\Delta p = 2.76 \times 10^5$ N/m^2, U = 167 cm/sec.

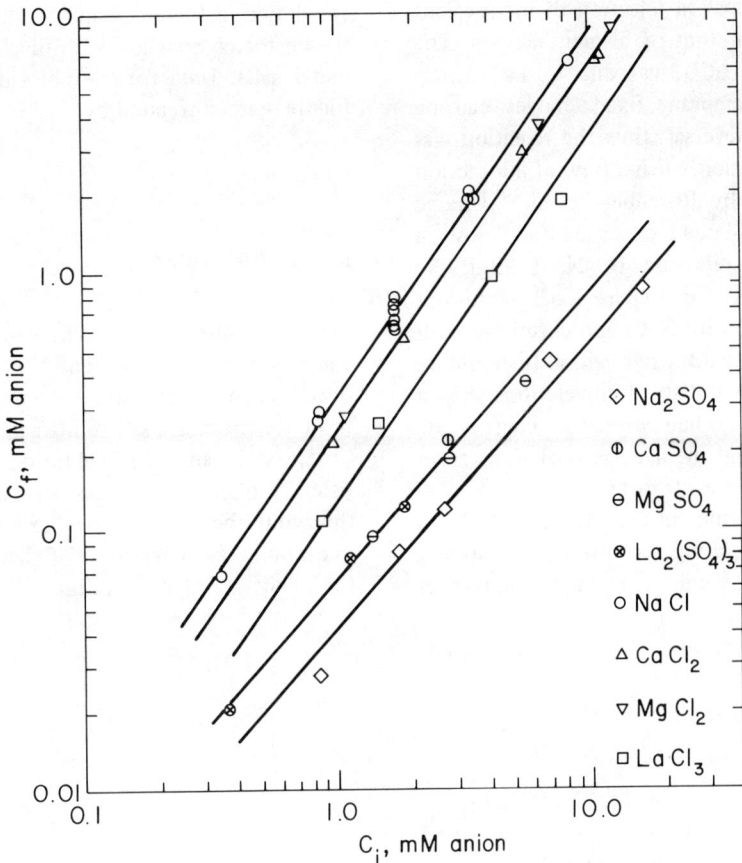

FIGURE 3. Relation between ultrafiltrate concentration and inlet stream concentration for chloride and sulfate with mono-, di-, and trivalent cations (counterions).

importance. The use of salts involving a highly charged cation (La^{3+}) was included to investigate possible membrane-counterion interactions. Similar to the oxyanions, it should be noted from Figures 3, 4, and 5 that power functions relating C_f to C_i (or R* to C_i) gave excellent fits for both chloride and sulfate with all cations.

From Figure 3, first noting the sulfate experiments involving Na$^+$, Ca^{2+}, Mg^{2+}, and La^{3+}, the rejection of SO$_4^{2-}$ with multivalent counterions was poorer (higher C_f) than in the presence of Na$^+$, consistent with the Donnan mechanism (Equation 5). For sulfate salts of Ca^{2+}, Mg^{2+}, and La^{3+}:

$$C_f = 0.065\, C_i^{1.10} \qquad (19)$$

$$R^* = 1 - 0.065\, C_i^{0.10}$$

Equation 18 applies for Na$_2$SO$_4$. For the chloride salts of Na$^+$, Ca^{2+}, Mg^{2+}, and La^{3+} (Figure 3), the rejection was different for each valency counterion. The rejection trend of LaCl$_3$ > CaCl$_2$ or MgCl$_2$ > NaCl is inconsistent with Donnan behavior due to membrane-cation interaction, as discussed below. The equations for chloride are

$$C_f = K_2\, C_i^{1.40} \qquad (20)$$

$$R^* = 1 - K_2\, C_i^{0.40}$$

in which the values of K_2 are 0.33 for Na$^+$, 0.24 for Ca^{2+} or Mg^{2+}, and 0.13 for La^{3+}.

The poor rejection of NaCl was due to membrane swelling which produced a decrease in membrane charge density, C_m^*. According to Equation 8, low C_m^* will increase K* (for a constant C_i and activity ratio) and thus will decrease the rejection. Membrane swelling is of course not desirable in most applications. Swelling should produce a loss in rejection of monovalent anions greater than that of divalent anions. The high rejection of LaCl$_3$ was due to La^{3+} interaction with the membrane fixed charges, causing membrane charge reversal; thus the rejection was due to cation exclusion. Such a type of interaction has been reported by Rosenberg et al.[34] Charge reversal was validated experimentally by a reduction in the rejection of NaCl after the membrane was exposed to the LaCl$_3$ solution. Membrane swelling with NaCl and shrinkage with LaCl$_3$ will be discussed further below. It should be mentioned that the statistical correlations shown in Equations 17 to 20 had correlation coefficients of 0.98 to 0.99$^+$, and the average deviations from experimental values were 9 to 18%.[3]

For the chloride and sulfate salts of Cu^{2+}, Ni^{2+}, and Zn^{2+}, results relating the ultrafiltrate stream concentration to the inlet stream concentration are shown in Figures 4 and 5. The dashed lines are shown for comparison with divalent alkaline-earth metal salts. Data for the chloride salts shown in Figure 4 are correlated by:

$$C_f = 0.093 \, C_i^{1.36} \tag{21}$$

$$R^* = 1 - 0.093 \, C_i^{0.36}$$

Figure 4 also includes multi-salt solutions of equimolar Cu^{2+}, Ni^{2+}, and Zn^{2+} chlorides, the rejections of which were essentially identical to those for the single salt solutions. The rejections of Cu^{2+}, Ni^{2+}, and Zn^{2+} chloride salts were consistently higher than those of Ca^{2+} or Mg^{2+} salts; this would be expected on the basis of the strong hydration characteristics of heavy metals. For Cu^{2+}, Ni^{2+}, and Zn^{2+} sulfate salts, again a single

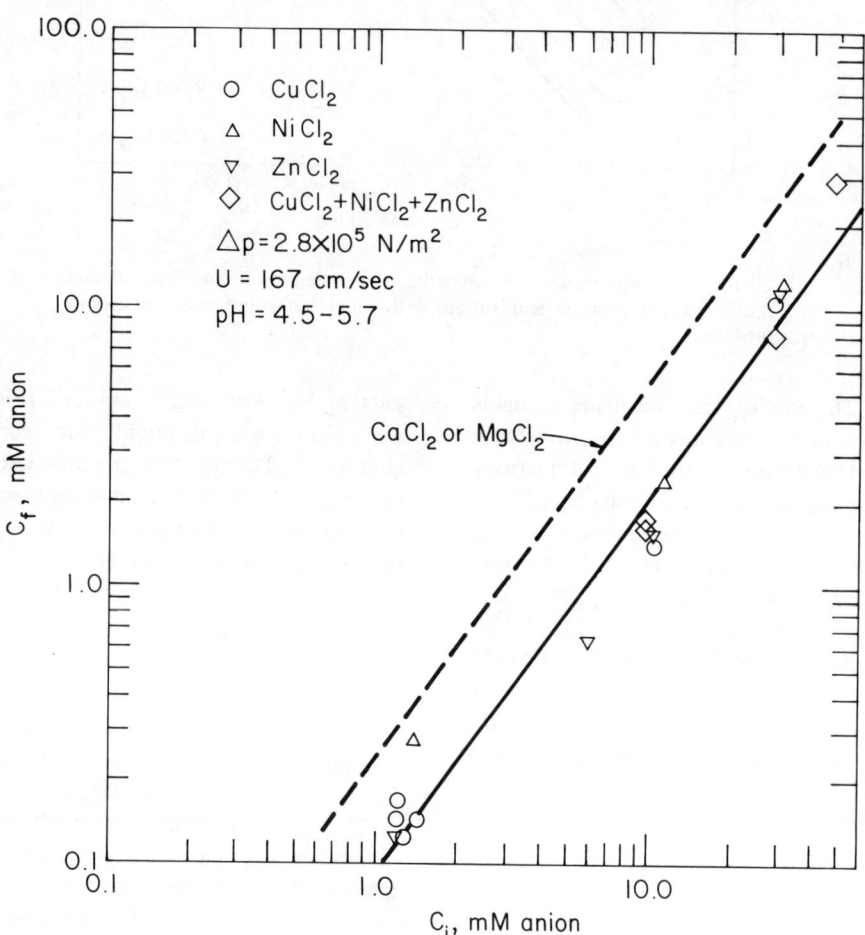

FIGURE 4. Relation between ultrafiltrate concentration and inlet stream concentration for chloride salts of heavy metals.

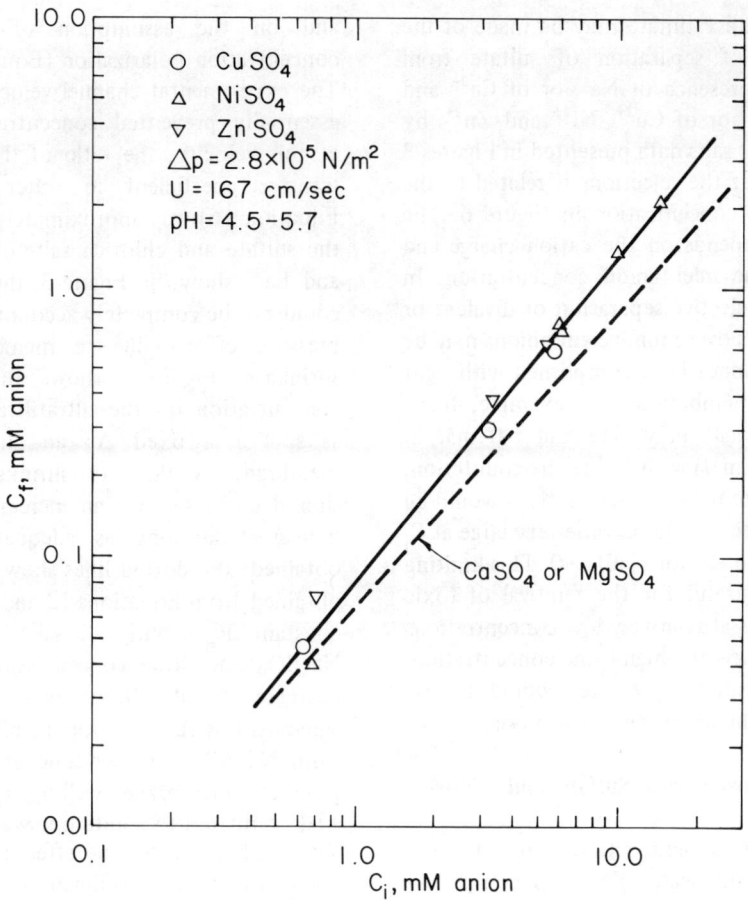

FIGURE 5. Relation between ultrafiltrate concentration and inlet stream concentration for sulfate salts of heavy metals.

power function (Figure 5) accurately describes the rejection dependence on concentration:

$$C_f = 0.082 \, C_i^{1.21} \quad (22)$$

$$R^* = 1 - 0.082 \, C_i^{0.21}$$

On the basis of hydration characteristics alone, the rejections of heavy metal sulfate salts should have been better than Ca^{2+} or Mg^{2+} sulfate salts; the reverse trend is due to incomplete ionization of the transition metal sulfate salts, particularly at concentrations above 1.0 mM. For example, with nickel sulfate at C_i = 10.0 mM (at pH 3.0 to 7.0), 50% of Ni will be present as unionized $NiSO_4$ at equilibrium.

From these studies with sulfate, chloride, and oxyanion salts, several conclusions can be drawn. First, the rejections of sulfate salts were consistently better than those of the corresponding chloride salts, and the rejections of all salts decreased with increasing inlet concentration, in accordance with the Donnan model (Equation 5). Second, the ultrafiltrate concentration of all chloride salts was a much stronger function of the inlet anion concentration than in the case of any oxyanions. Third, chloride rejections were lower than those of monovalent oxyanions (except $HCrO_4^-$). Fourth, the rejection behavior of chloride salts containing mono-, di-, and trivalent counterions was reversed compared to the Donnan model due to membrane swelling with the monovalent counterion and membrane charge reversal with the trivalent counterion. Fifth, comparing divalent metal salts, the rejections of those containing strongly hydrated cations (thus less effective charge for membrane-ion interaction) were greater.

An approximate estimate may be made of the possible extent of separation of sulfate from chloride in the presence of Na^+, or of Ca^{2+} and Mg^{2+}, or of La^{3+}, or of Cu^{2+}, Ni^{2+} and Zn^{2+}, by utilizing the single salts data presented in Figures 3 to 5. The ratio of the rejections is related to the total inlet anion concentration in Figure 6. The rejection ratio depends on the cation charge and increases with the inlet anion concentration. In some cases, the selective separation of divalent or multivalent anions over monovalent anions may be feasible, which cannot be accomplished with tight reverse osmosis membranes. For example, for a solution of 2.0 mM in sulfate and 6.0 mM in chloride (C_i = 8.0 mM) with Na^+ as the counterion, the ratio of sulfate to chloride rejections would be about 4.0; the ratio would become very large at $C_i \approx 18$ mM, because R^* for NaCl $\to 0$. This limiting effect might be useful for the removal of toxic multivalent anions at relatively low concentrations even in the presence of a high brine concentration, for which the osmotic pressures would be extremely high for the use of reverse osmosis.

Water Flux Behavior with Sulfate and Chloride Salts

With the chloride and sulfate salts of Cu^{2+}, Ni^{2+}, and Zn^{2+}, the water flux was completely predictable on the basis of osmotic pressure effects and on the assumption of the absence of concentration polarization (Equations 13 and 15). The experimental channel velocity of 167 cm/sec essentially prevented concentration polarization; at this velocity, the ratio of the calculated mass transfer coefficient to water flux (K_s/J_w in Equation 16) was approximately 8.0 to 10.0. For the sulfate and chloride salts of Na^+, Ca^{2+}, Mg^{2+}, and La^{3+} shown in Figure 3, the ultrafiltrate flux could not be completely accounted for by osmotic pressure effects due to membrane swelling or shrinkage; Figure 7 shows the effect of inlet concentration on the ultrafiltrate flux for these salts. For a fixed Δp and in the absence of membrane swelling or shrinkage behavior, J_w should decrease with an increase in C_i (since $\Delta\Pi$ increases) as long as adequate rejections are obtained: the dotted lines show the calculated J_w obtained from Equations 13 and 15 for the case of constant R_m. With all salts except NaCl and Na_2SO_4 at low concentrations, the percent decrease of water flux with C_i was in qualitative agreement with the drop in effective $\Delta p - \Delta\Pi$. Both Na_2SO_4 (at low concentrations) and NaCl produced membrane swelling; J_w increased about 17% compared to solute-free water. With chloride, swelling had a profound effect on rejection due to decreases in the Donnan potential and C_m^* (Equation 8). The rejection of NaCl $<$ $CaCl_2$ is in

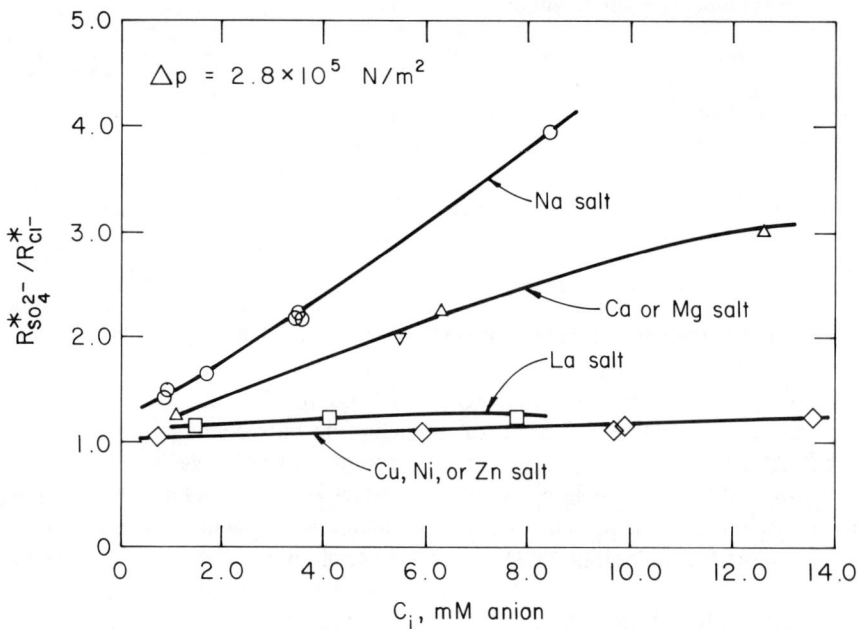

FIGURE 6. Effect of total anion (solute) concentration in inlet stream on relative rejections of sulfate ion and chloride ion.

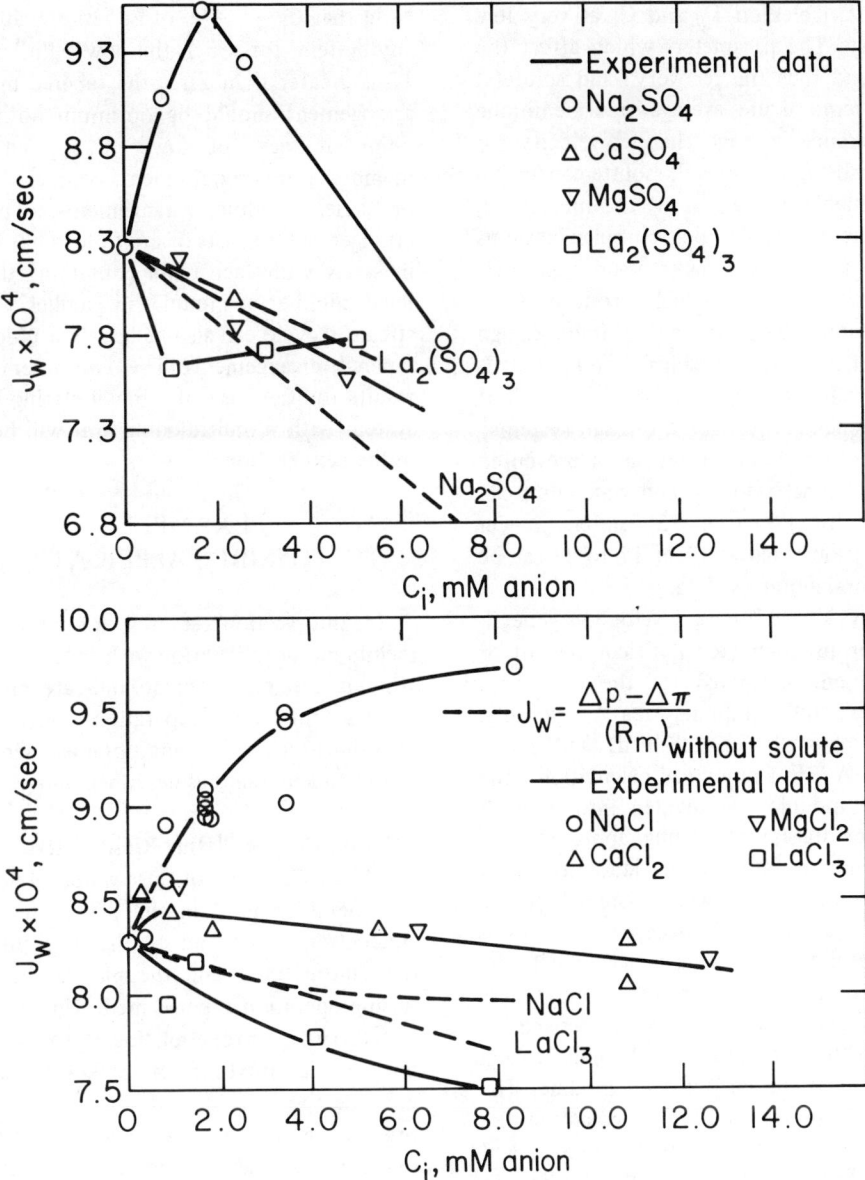

FIGURE 7. Dependence of water flux on inlet stream salt concentration.

accordance with the swelling behavior. The highly charged La^{3+} caused membrane pore shrinkage due to electrostatic cross-linking, which produced a permanent flux decrease. Bhattacharyya et al.[3] showed that cations of $Z^2/r > 3.0$ caused membrane shrinkage whereas cations of $Z^2/r < 1.0$ caused membrane swelling, as evidenced by the change in the values of R_m with solute.

SCALE-UP PROCEDURE: PROCESS DESIGN

For charged membrane ultrafiltration to be a feasible separation method to handle large inlet flow rates (F_i), and where water reuse and/or the recovery of valuable metals are desired, fractional water recoveries of the order of 0.8 to 0.95 must be achieved. The fractional water recovery, r', defined as the total ultrafiltrate stream flow rate divided by the inlet stream flow rate, can be increased by means of multiple membrane module arrangements in series, in parallel, or in series-parallel (tapered) combinations. The optimum arrangement can be predicted with parametric computer simulation procedures,[1,6,35] as long as laboratory data are available on the effects of the independent variables on J_w and C_f at very low water recovery. The parameters which affect the water flux (and thus the recovery) and solute(s) rejection or removal are average transmembrane pressure difference, average channel velocity (or Reynolds number), average bulk solute concentration on the high-pressure side of the membrane, and membrane resistance. For complex systems containing ionic solutes, organic macromolecules, and particulates, for which concentration polarization cannot be prevented, optimum design requires quantitative relationships[6] for J_w and C_f (at very low water recovery) as functions of all of the above independent variables. For salt systems[1] where concentration polarization can be prevented at moderate channel velocity, and water flux can be described by Equation 13, simple power function Equations such as 12 or 17 to 22 can be used to scale up laboratory data.

Bhattacharyya et al.[6] have developed a generalized computer program (for multicomponent or single component solutions) for the case of a module of arbitrary membrane area A, hydraulic diameter of channel (or tube) d, and cross-sectional area A_x. For numerical calculations, the length of the module was divided into N small sections (inner stages) containing approximately the same membrane area as the laboratory unit (A/N). The inner stages were assumed to be connected in series, and because the water recovery in each of these stages was small, the desired precision by an iterative method was obtained. Recursive equations were derived to simulate the actual conditions at the end of each inner stage and at the end of each module. The computer program was written to maintain the Reynolds number in a narrow specified range, and thus the recycle of concentrate or the decrease in the number of modules in any parallel bank was selected. Typical values of membrane module parameters are A = 5000 cm^2, N = 100, and d = 0.3 cm.

The overall removal of a particular solute in a multimodule unit is defined as:

$$\text{Overall removal} = 1 - \frac{C_{f\,avg}}{C_i} \quad (23)$$

$$C_{f\,avg} = \frac{\sum_{1}^{n} J_{wn} A C_{fn}}{F_i r'} \quad (24)$$

Note that $C_{f\,avg} = C_f$ of Equation 9 at $r' \to 0$. For any system, for which the power "n" in Equation 12 is greater than zero, the tapered multimodule arrangement should be optimum both from the point of view of lower $C_{f\,avg}$ and a lower membrane area requirement, compared to parallel or series module arrangements. The tapered arrangement consists of multiple banks of modules in series with each bank containing the same or fewer number of modules in parallel. Concentrate stream recycle can also be largely avoided with the tapered arrangement. The computer simulation results for the case of a metal-plating wastewater system with a multimodule unit will be discussed in the next section.

EXAMPLES OF POSSIBLE APPLICATIONS

In this section, several examples of charged membrane ultrafiltration with actual and synthetic systems are presented to indicate areas of usefulness for this separation technique. The experimental results were obtained with the same charged membranes as described above.

Ultrafiltration of Plating Rinse Waters

The constituents of rinse waters are dictated by the chemicals used in the plating bath, and the concentrations depend on the drag-out rate from the plating tank and the manner in which the rinsing operation is performed. Rinse water from a Watts-type nickel-plating process generally contains a mixture of nickel sulfate, nickel chloride, boric acid, and trace concentrations of organic brighteners. A series of ultrafiltration experiments was performed with actual rinse waters produced by diluting a Watts-type bright-nickel-plating solution. The inlet nickel concentration was varied from 0.6 mM (35 mg/l) to 16.0 mM (940 mg/l) at the unadjusted pH of 5.2 to 6.8. An excellent (correlation coefficient = 0.99[+]) power function correlation[1] for C_f vs. C_i at $\Delta p = 2.8 \times 10^5$ N/m^2 was obtained:

$$C_f = 0.164 \, C_i^{1.04} \quad (25)$$

$$R^* = 1 - 0.164 \, C_i^{0.04}$$

In this equation C_f and C_i are expressed as mM Ni. The approximate unity power on C_i indicates a constant nickel rejection of about 0.84 over the entire concentration range. Significant improvement in nickel rejection was observed at higher pressures: for example, at $\Delta p = 4.1 \times 10^5$ N/m^2 and $C_i = 6.0$ mM, the rejection increased to 0.89 indicating that the limiting maximum rejection was not achieved at the lower pressure. The increased rejection at the high Δp (thus high water flux) is in agreement with Equation 11, because the convective term predominates over the diffusion term.

Other parameters measured were rejections of total organic carbon, conductivity, and optical density (at 400 nm). For example, a rinse water containing 6.0 mM Ni had a total organic carbon concentration of 30 mg/l, conductivity of 1300 μmho/cm, and optical density of 0.04; at $\Delta p = 2.8 \times 10^5$ N/m^2, the rejections were 0.67, 0.73, and 0.90[+], respectively.

The water fluxes with the rinse waters were completely predictable on the basis of osmotic pressure effects and insignificant concentration polarization. Over the entire concentration range, the water flux (and rejection) were independent of channel velocity at and above 84 cm/sec (Re = 1200). A least squares correlation (correlation coefficient 0.99[+]) yielded:

$$J_w = 2.8 \times 10^{-9} (\Delta p - \Delta \Pi) \qquad (26)$$

From Equations 25 and 26 and the experimental observation that J_w and C_f were independent of channel velocity for velocities greater than 84 cm/sec, the computer simulation procedure described above was used to predict nickel separation and membrane area requirements at high water recoveries for inlet conditions of flow rate = 500 cm^3/sec and $C_i = 1.0$ mM Ni (59 mg/l). With the tapered arrangement, no recycle of the concentrate stream to the inlet (to maintain a minimum velocity) would be required. The tapered arrangement was found to be optimum, and the membrane area requirement increased linearly with recovery. Solute (nickel) separation as a function of water recovery is shown in Figure 8, in terms of the ratio of the average (Equation 24) nickel concentration in the total ultrafiltrate from all the modules to the inlet nickel concentration. The parallel arrangement consistently

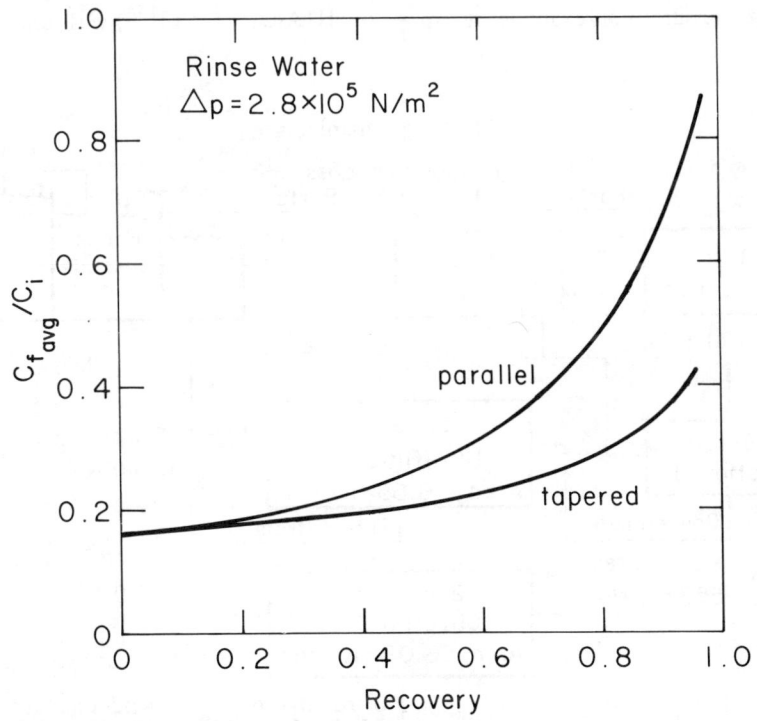

FIGURE 8. Dependence of nickel separation on water recovery for multi-module operation in parallel mode and in tapered mode.

yielded a poorer separation than the tapered arrangement.

Figure 8 (along with the membrane area requirement values, not shown here) can be used for the conceptual design of a closed-loop nickel-plating process, involving water recycle and simultaneous metal recovery. If 9 mg/l nickel (0.15 mM) is allowed[1] in the recycle to the final rinse, and for 59 mg/l (1.0 mM) nickel leaving the rinse tank, a maximum water recovery of 65% could be achieved and 35% fresh make-up water would be necessary; the concentrate stream would be too dilute to be recycled. If the simultaneous achievement of 9 mg/l nickel in the recycle to the final rinse tank, higher water recovery, and approximately 10:1 metal concentration is desired, a two-stage operation as shown in Figure 9 would be necessary. In the two-stage operation, the ultrafiltrate stream from the first stage would be ultrafiltered again, with each stage consisting of tapered module arrangements. In this mode of operation, an overall water recovery of 91% and minimum nickel separation of 83% would be achieved. The concentrate stream (9.4 mM Ni) would be used directly in the still rinse tank. A savings in membrane area (and further improvement in nickel separation) could be achieved at a higher operating pressure: for example at Δp = 4.1 × 10^5 N/m^2 ($\approx$ 4.1 atm), the total membrane area requirement would be 8.6 × 10^5 cm^2.

Ultrafiltration of Toxic Metal Constituents From Metal Manufacturing Wastewater

Nonferrous metal (such as Cu and Zn) production processes use large quantities of process water, and various wastewaters containing highly toxic metals (As, Se, Cd, etc.,) are produced. The metals are present as cations and/or oxyanions in acidic sulfate solutions. With arsenic and selenium (both present as oxyanions), experimental rejections at C_i = 6 mM (a concentration 10 to 30 times higher than that in actual waste streams), ranged between 80% and 97% depending on the pH and operating pressure. The cadmium (as sulfate) rejection was somewhat better (88%) than that observed with the sulfate salts of Cu^{2+}, Ni^{2+}, or Zn^{2+}.

A typical effect of pH involving a selenium solution (using H_2SeO_3 and pH adjustment by NaOH) is shown in Figure 10. The increase in rejection with an increase in pH is due to the conversion of monovalent $HSeO_3^-$ to divalent SeO_3^{2-}, as shown in the species distribution diagram (Figure 10 top). The pH behavior with As(V) oxyanions was similar. The rejection of $H_2AsO_4^-$ (at pH 5) was somewhat greater (R* =

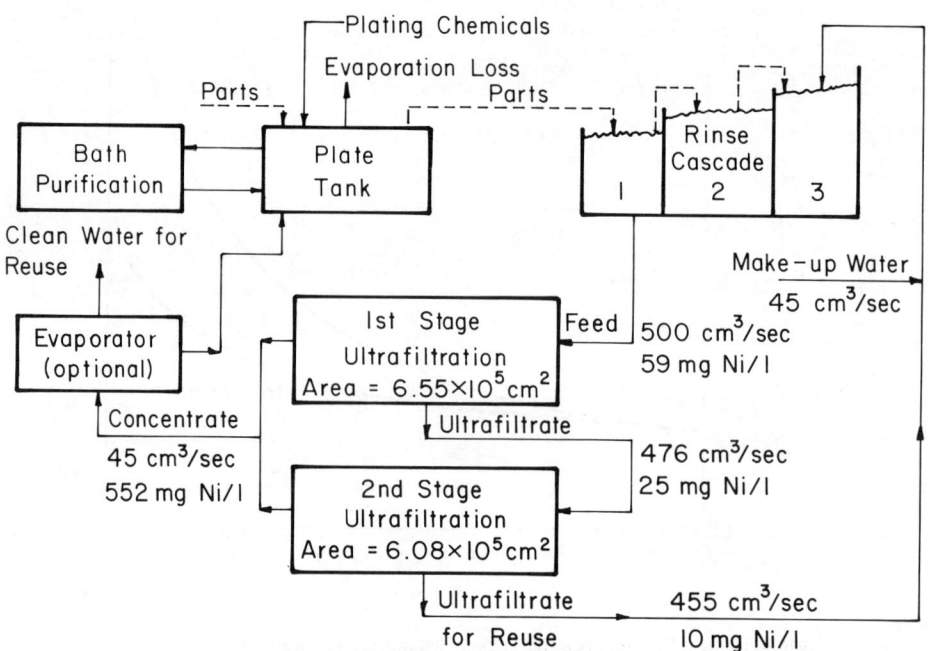

FIGURE 9. Two-stage ultrafiltration operation for a closed-loop, nickel plating process.

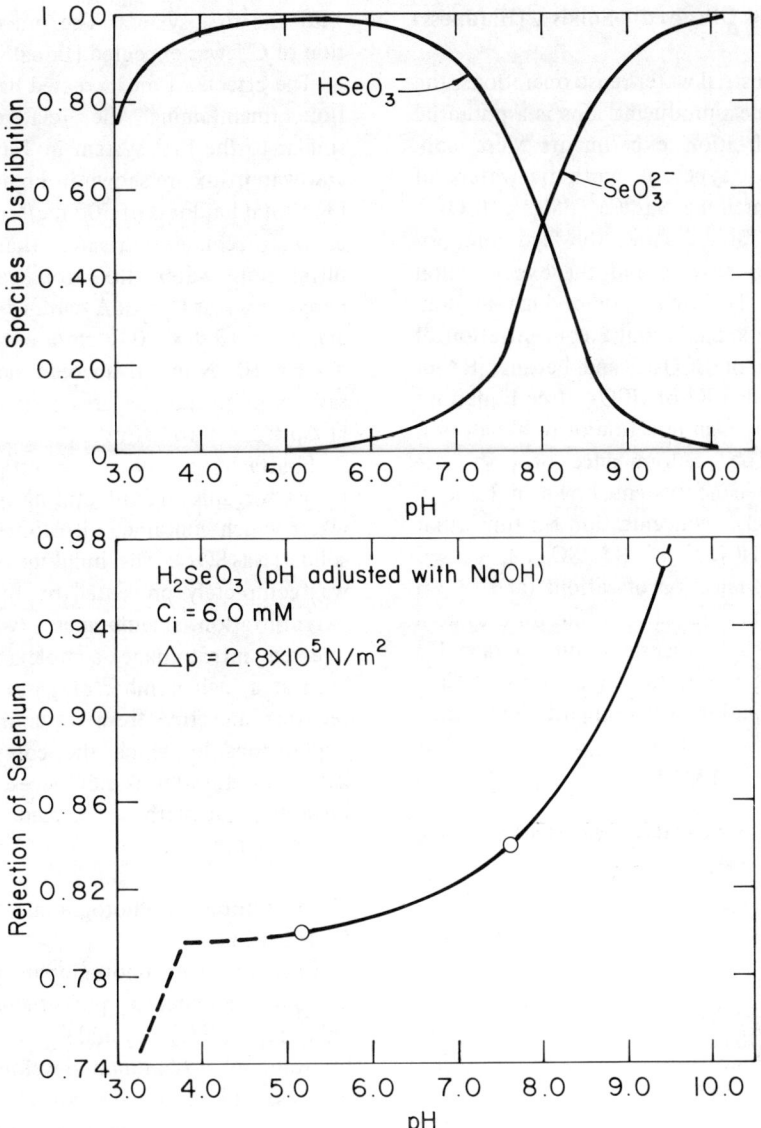

FIGURE 10. Effect of pH on selenium rejection, and dependence of selenium species distribution on pH.

0.88) than that of $HSeO_3^-$ ($R^* = 0.80$), whereas the rejections of SeO_3^{2-} and $HAsO_4^{2-}$ were similar to each other and to those of the other divalent oxyanions (SO_4^{2-}, CrO_4^{2-}, HPO_4^{2-}). No significant reduction in rejection of $H_2AsO_4^-$ ions was observed ($R^* = 0.86$) even in the presence of high sulfate concentration. The rejection of the unionized species H_3AsO_3 was zero as expected. The rejections of both As(V) and Se(IV) oxyanions increased with Δp, particularly at the acidic pH where the monovalent oxyanion predominates. Typical results with selenium are shown below:

Δp, N/m²	Species	Rejection
2.8×10^5	$HSeO_3^-$ (pH 5.0)	0.80
5.6×10^5	$HSeO_3^-$ (pH 5.0)	0.86
2.8×10^5	SeO_3^{2-} (pH 9.5)	0.97
5.6×10^5	SeO_3^{2-} (pH 9.5)	0.97

The separation of arsenic and selenium (which cannot be removed by hydroxide precipitation) and other heavy metals (Cd, Zn) can be achieved with ultrafiltration even at pH 4.5 to 5.0 and thus without the necessity of heavy metal precipitation.

Application to Dissolved Solids (Hardness) Reduction

In various industrial water reuse operations, the build-up of hardness-producing ions is a potential problem. Ultrafiltration experiments were conducted with three types of synthetic waters of high hardness containing Na^+, Ca^{2+}, Mg^{2+}, HCO_3^-, Cl^-, and SO_4^{2-}. Table 2 shows the feed composition of the three systems and the experimental rejection results. The first system, high in noncarbonate hardness, had a molar concentration of SO_4^{2-} twice that of HCO_3^-, and because R^* of SO_4^{2-} is greater than R^* of HCO_3^- (see Equations 17 and 18) a fairly high rejection of total hardness was obtained. The predominance of carbonate hardness in the second system shown in Table 2, which had a HCO_3^- concentration six times that of SO_4^{2-} (R^* HCO_3^- < R^* SO_4^{2-}), caused somewhat poorer rejection of cations (and thus a poorer total hardness rejection). The third system was similar to the first system but contained a higher chloride concentration; rejections of Ca^{2+} and SO_4^{2-} (thus hardness) were slightly lower than with the first system. The practically zero rejection of Cl^- was expected (Equation 20).

The effects of an increased hardness concentration (maintaining the relative concentrations similar to the first system in Table 2) on rejection and water flux are shown in Figure 11. Even at the high total hardness of 500 mg/l, rejection (divalent cations) remained greater than 0.80 and the ultrafiltrate water flux drop was only 7% compared to J_w at $C_i = 0$. A hardness rejection of 0.82 and $J_w = 13.0 \times 10^{-4}$ cm/sec was obtained at $\Delta p = 4.1 \times 10^5$ N/m^2; this would mean a considerable savings of membrane area compared to operation at $\Delta p = 2.8 \times 10^5$ N/m^2.

Bhattacharyya et al.[2] investigated the recycle to an organic manufacturing process of waste water which contained nitrotoluenes and dissolved solids (Na_2SO_4). The build-up of dissolved solids was completely prevented by the use of charged ultrafiltration membranes (which effectively rejected nitrotoluene complexes and Na_2SO_4), even at a high number of passes with 90% water recovery and 10% fresh water make-up. Thus, in applications in which the complete demineralization of a water is not necessary, low-pressure ultrafiltration with a charged membrane is an attractive process.

Ultrafiltration of Photographic Processing Water Constituents

Effluents from photographic processing plants contain a number of potentially toxic constituents, such as ferrocyanide (reduced bleach) and hydroquinone (common developer). Thiosulfate, borates, and phosphates are also present in processing rinse water. Bhattacharyya et al.,[5] in their studies with laundry waste constituents, obtained high rejections of borates and phosphates with charged membranes.

The rejection of hydroquinone as a function of pH is shown in Figure 12. Hydroquinone is a weak diprotic acid which ionizes in basic solution, and the increase in rejection with pH is approximately proportional to the extent of ionization. In order to test the hypothesis that rejection of phenolic compounds increases with ionization (phenol is completely ionized above pH 10), the rejection of phenol (although not present in photographic processing water) is shown in Figure 12 (top) as a function of pH. Cellulose acetate reverse osmosis membranes are incapable of adequately rejecting either hydroquinone or phenol.

TABLE 2

Composition of Three Synthetic Waters and the Rejections of Individual Ions

Feed composition mg/l as ion		Rejection
Total hardness 250 (as $CaCO_3$)		0.83
Ca^{2+}	72 (1.8 mM)	0.83
Mg^{2+}	17 (0.70 mM)	0.82
Na^+	30 (1.3 mM)	—
SO_4^{2-}	230 (2.4 mM)	0.93
HCO_3^-	82 (1.3 mM)	0.26
Cl^-	10 (0.28 mM)	—
Total hardness 171 (as $CaCO_3$)		0.76
Ca^{2+}	38 (0.95 mM)	0.75
Mg^{2+}	18 (0.74 mM)	0.77
Na^+	30 (1.3 mM)	—
HCO_3^-	152 (2.5 mM)	0.64
SO_4^{2-}	39 (0.41 mM)	0.90+
Cl^-	30 (0.84 mM)	—
Total hardness 200 (as $CaCO_3$)		0.79
Ca^{2+}	80 (2.0 mM)	0.79
Na^+	500 (21.7 mM)	—
SO_4^{2-}	150 (1.6 mM)	0.89
HCO_3^-	18 (0.30 mM)	—
Cl^-	780 (22.0 mM)	0.10

Note: $\Delta p = 2.76 \times 10^5$ N/m^2, $U = 167$ cm/sec, pH = 7.0 to 8.2.

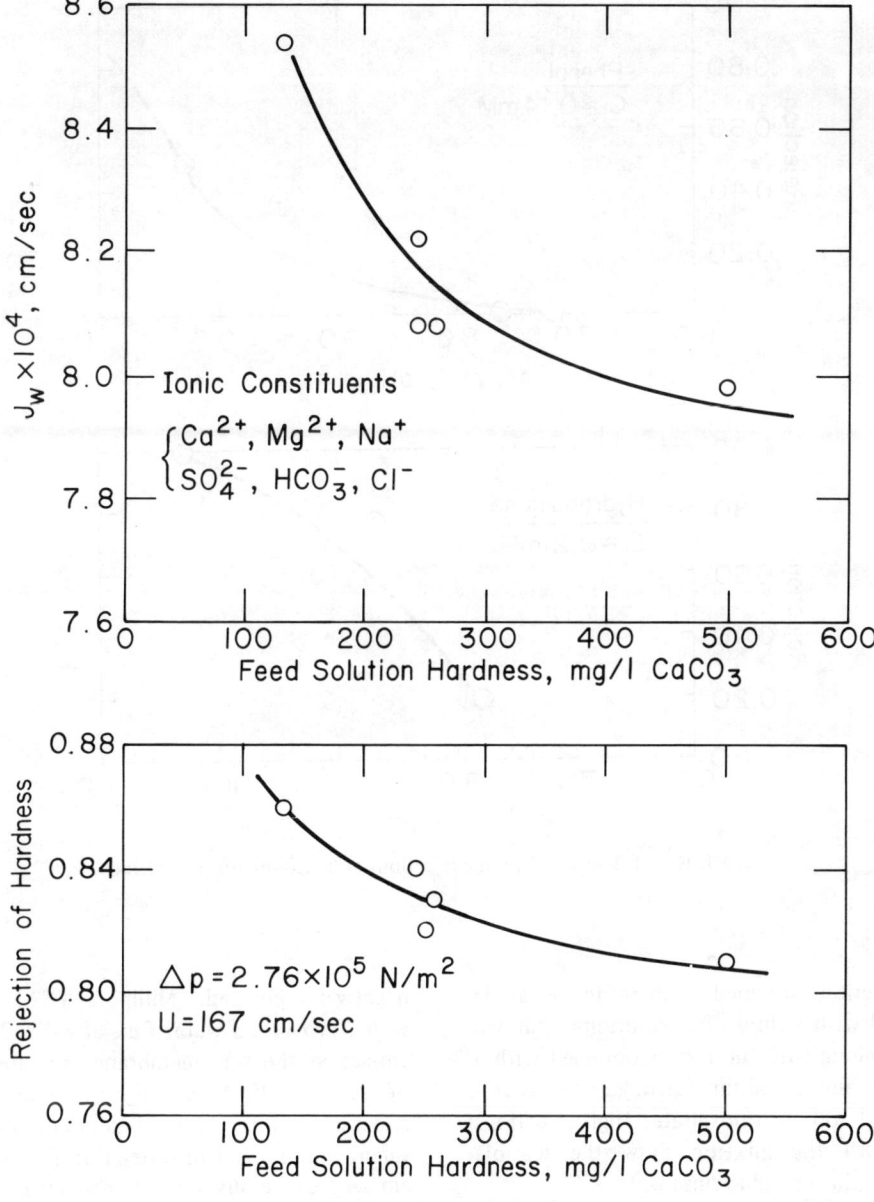

FIGURE 11. Dependence of hardness rejection and water flux on inlet stream hardness concentration.

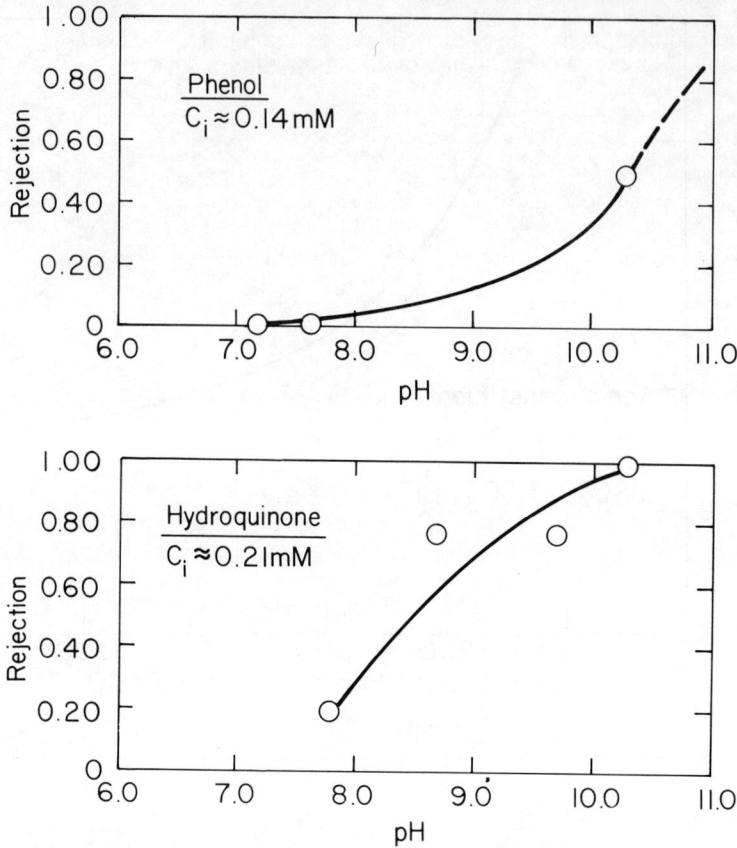

FIGURE 12. Effect of pH on the rejections of hydroquinone and phenol.

The rejection obtained with a ferrocyanide solution and with a thiosulfate solution are shown in Table 3 along with the results obtained with a mixture containing sodium ferrocyanide, hydroquinone, and sodium thiosulfate. High rejections obtained with the mixture show the feasible treatment of photographic rinse water.

SUMMARY

The use of charged, non-cellulosic membranes provides a novel and broadly applicable technique for the separation and concentration of various ionic solutes present in aqueous solution. These membranes provide good water flux at low pressures and adequate to excellent rejections of various inorganic salts, depending on the process application. Because di- or multivalent anions are rejected considerably better than monovalent anions, selective separations can be achieved. Table 4 is a summary of the rejection behavior of the negatively charged, Millipore PSAL membrane which provides a water flux of 8×10^{-4} to 10^{-5} cm/sec at the transmembrane pressure difference of 2.8×10^5 N/m^2. Further improvement in rejections, particularly of salts containing monovalent anions, and in water flux (up to 2×10^{-5} cm/sec) can easily be obtained at a higher operating pressure of 6.3×10^5 N/m^2.

Promising results have been demonstrated in the areas of valuable metal separation and concentration, water reuse involving closed-loop unit processes, the reduction of dissolved solids build-up in water reuse operations, the removal of toxic metals such as arsenic and selenium which cannot be removed by conventional hydroxide precipitation, the separation of low-molecular-weight organic ions, and the simultaneous achievement of the adequate rejection of inorganic ions and the high rejection of modest-molecular-weight organics (such as nonionic and anionic surfactants) from complex wastes. Other feasible applications may

TABLE 3

Photographic Waste Constituents

Constituent	Feed concentration, mM	Rejection
Hydroquinone (pH 10.3)	0.21	≈1.0
Sodium ferrocyanide (pH 7.2)	0.47	0.96
Sodium thiosulfate (pH 7.1)	0.32	0.93
Mixture*: (pH 9.5)		
Organic carbon	4.33	0.95
Conductivity (μmho/cm)	978	0.96
Optical density (at 400 nm)	0.22	0.91

*Contains $Na_4Fe(CN)_6$, hydroquinone, and $Na_2S_2O_3$.

TABLE 4

Rejection of Various Inorganic Ions by a Negatively Charged Ultrafiltration Membrane

Anion	Primary cation	Salt*	Rejection
PO_4^{3-}	Na^+	Na_3PO_4	0.98
HPO_4^{2-}	Na^+	Na_2HPO_4	0.95
CrO_4^{2-}	Na^+	$Na_2Cr_2O_7$	0.95
SO_4^{2-}	Na^+	Na_2SO_4	0.95
SO_4^{2-}	Ca^{2+}	$CaSO_4$	0.92
SO_4^{2-}	Mg^{2+}	$MgSO_4$	0.92
SO_4^{2-}	Cu^{2+}	$CuSO_4$	0.88
SO_4^{2-}	Ni^{2+}	$NiSO_4$	0.88
SO_4^{2-}	Zn^{2+}	$ZnSO_4$	0.88
SO_4^{2-}	Cd^{2+}	$CdSO_4$	0.90
SO_4^{2-}	La^{3+}	$La_2(SO_4)_3$	0.92
SeO_3^{2-}	Na^+	H_2SeO_3	0.97
$HAsO_4^{2-}$	Na^+	Na_2HAsO_4	0.94
$S_2O_3^{2-}$	Na^+	$Na_2S_2O_3$	0.93
$B_4O_7^{2-}$	Na^+	$Na_2B_4O_7$	0.92
$H_2PO_4^-$	Na^+	NaH_2PO_4	0.84
HCO_3^-	Na^+	$NaHCO_3$	0.70
$HCrO_4^-$	Na^+	$Na_2Cr_2O_7$	0.0
$HSeO_3^-$	Na^+	H_2SeO_3	0.80
$H_2AsO_4^-$	Na^+	Na_2HAsO_4	0.88
$H_2AsO_3^-$	Na^+	$NaAsO_2$	0.84
Cl^-	Na^+	$NaCl$	0.32

Note: Feed concentration = 6 mM (anion); transmembrane pressure = 2.8×10^5 N/m^2.
*pH adjusted to obtain proper species distribution.

include the selective rejections of undesirable solutes (toxic metals at low concentration) while passing others (such as sodium chloride), in which case only the rejected solutes contribute to the effective osmotic pressure and low pressure operation is satisfactory. Charged membrane ultrafiltration is not suitable for the desalination of sea water, and because rejection decreases (particularly for monovalent anions) with increasing concentration, the removal of salts present at very high concentrations is not feasible.

ACKNOWLEDGMENTS

The authors gratefully acknowledge the valuable contributions of Michaelee Moffitt, David P. Schaaf, and James M. McCarthy in the various phases of the experimental work.

NOMENCLATURE

a_j	Activity of jth ion in bulk solution, mM
$a_{j(m)}$	Activity of jth ion in membrane, mM
A	Membrane area of a module, cm^2
A_x	Cross-sectional area of a module, cm^2
C	Average concentration of a solute (or ion) in channel on high pressure side, mM
C_f	Concentration of a solute (or ion) in ultrafiltrate stream, mM
$C_{f\ avg}$	Average concentration of a solute (or ion) in ultrafiltrate stream (Equation 24) from multi-module unit, mM
C_{fn}	Concentration of a solute (or ion) in ultrafiltrate stream from the nth module, mM
C_i	Concentration of a solute (or ion) in inlet stream, mM
C_j	Concentration of jth ion in bulk solution, mM
$C_{j(m)}$	Concentration of jth ion in membrane, mM
C_m^*	Charge density (capacity) of membrane, meq/l
C_w	Solute concentration at the membrane-solution interface on high pressure side, mM
d	Hydraulic diameter of channel, cm
D_j	Diffusion coefficient of jth ion in membrane, cm^2/sec
E_d	Donnan potential, volts
F	Faraday constant
F_i	Flow rate of inlet stream, cm^3/sec
i	Ionization factor, Equation 15
J_s	Solute flux, mmol/(sec cm^2 membrane area)
$(J_s)_j$	Flux of jth ion, mmol/(sec cm^2 membrane area)
J_w	Water (ultrafiltrate) flux, cm^3/(sec cm^2 membrane area)
J_{wn}	Water (ultrafiltrate) flux from nth module, cm^3/(sec cm^2 membrane area)
K_s	Mass transfer coefficient, cm/sec
K^*	Solute (or ion) distribution coefficient, $C_{j(m)}/C_j$ or $C_{y(m)}/C_y$ for salt $M_{Z_y} Y_{Z_m}$
Δp	Average transmembrane pressure difference, N/m^2
r	Effective ion radius, Å
r'	Fractional water recovery (total ultrafiltrate flow rate per unit inlet flow rate)
R	Gas constant
R^*	Solute (or ion) rejection, Equation 9
R_m	Resistance of ultrafiltration membrane to water flux, N/m^2/cm/sec
T	Solution temperature, °K
U	Average channel velocity, cm/sec
v_j	Partial molar volume of jth ion, cm^3/mol
Z_j	Valence of jth ion
γ	Activity coefficient of solute in bulk solution
γ_m	Activity coefficient of solute in membrane
$\Delta\Pi$	Osmotic pressure difference, N/m^2
Π_s	Swelling pressure, N/m^2

REFERENCES

1. Bhattacharyya, D., Schaaf, D. P., and Grieves, R. B., Charged membrane ultrafiltration of heavy metal salts: application to metal recovery and water reuse, *Can. J. Chem. Eng.*, 54, 185, 1976.
2. Bhattacharyya, D., Garrison, K. A., and Grieves, R. B., Membrane ultrafiltration of nitrobodies from industrial wastes, *J. Water Pollut. Control Fed.*, in press, 1977.
3. Bhattacharyya, D., McCarthy, J. M., and Grieves, R. B., Charged membrane ultrafiltration of inorganic ions in single and multi-salt systems, *AIChE J.*, 20, 1206, 1974.
4. Gregor, H. P., Ultrafiltration membranes, in Membranes in Separation Processes — A Workshop Symposium, sponsored by the National Science Foundation, Case Western Reserve University, Cleveland, Ohio, 1973.
5. Bhattacharyya, D., Bewley, J. L., and Grieves, R. B., Ultrafiltration of laundry waste constituents, *J. Water Pollut. Control Fed.*, 46, 2372, 1974.
6. Bhattacharyya, D., Garrison, K. A., Jumawan, A. B., and Grieves, R. B., Membrane ultrafiltration of a nonionic surfactant and inorganic salts from complex aqueous suspensions: design for water reuse, *AIChE J.*, 21, 1057, 1975.
7. Porter, M. C. and Nelsen, L., Ultrafiltration in the chemical, food processing, pharmaceutical, and medical industries, in *Recent Developments in Separation Science,* Vol. 2, Li, N. N., Ed., CRC Press, Cleveland, Ohio, 1972, 227.
8. Lonsdale, H. K., Theory and practice of reverse osmosis and ultrafiltration, in *Industrial Processing with Membranes,* Lacey, R. E. and Loeb, S., Eds., Interscience, New York, 1971, 263.
9. Merten, U., Transport properties of osmotic membranes, in *Desalination by Reverse Osmosis,* Merten, U., Ed., M.I.T. Press, Cambridge, Mass., 1966, 15.
10. Glueckauf, E., On the mechanism of osmotic desalination with porous membranes, in Proceedings of the First International Symposium on Water Desalination, Vol. 1, Office of Saline Water, U.S. Dept. of the Interior, Washington, D.C., 1967, 143.
11. Sourirajan, S., *Reverse Osmosis,* Logos Press, London, 1970.
12. Matsuura, T., Pageau, L., and Sourirajan, S., Reverse osmosis separation of inorganic solutes in aqueous solutions using porous cellulose acetate membranes, *J. Appl. Polym. Sci.*, 19, 179, 1975.
13. McKelvey, J. G., Spiegler, K. S., and Wyllie, M. R. J., Ultrafiltration of salt solutions through ion exchange membranes, *Chem. Eng. Prog. Symp. Ser. No. 24*, 55, 199, 1959.
14. Yasuda, H., Lamaze, C. E., and Schindler, A., Salt rejection by polymer membranes in reverse osmosis, *J. Polym. Sci., Part A-2*, 9, 1579, 1971.
15. Kimura, S. G., Reverse osmosis performance of sulfonated poly (2,6-dimethylphenylene ether) ion exchange membranes, *Ind. Eng. Chem. Prod. Res. Dev.*, 10, 335, 1971.
16. LaConti, A. B., Chludzinski, P. J., and Fickett, A. P., Morphology and reverse osmosis properties of sulfonated 2,6-dimethyl polyphenylene oxide membranes, in *Reverse Osmosis Membrane Research,* Lonsdale, H. K. and Podall, H. E., Eds., Plenum Press, New York, 1972, 263.
17. Yasuda, H. and Schindler, A., Reverse osmosis properties of ionic and nonionic polymer membranes, in *Reverse Osmosis Membrane Research,* Lonsdale, H. K. and Podall, H. E., Eds., Plenum Press, New York, 1972, 299.
18. Milstead, C. E. and Tagami, M., Polyacrylic acid composite membranes for brackish water desalination, in *Reverse Osmosis Membrane Research,* Lonsdale, H. K. and Podall, H. E., Eds., Plenum Press, New York, 1972, 405.
19. Lonsdale, H. K., Filey, R. L., Milstead, C. E., LaGrange, L. D., Douglas, A. S., and Sachs, S. B., Research on Improved Reverse Osmosis Membranes, Office of Saline Water, Research and Development Progress Report No. 577, U.S. Dept. of the Interior, Washington, D.C., 1970.
20. van der Velden, P. M. and Smolders, C. A., Initial flux decline and initial rejection increase for swollen ionic membranes, *J. Appl. Polym. Sci.*, 20, 1153, 1976.
21. Johnson, J. S., Polyelectrolytes in aqueous solutions — filtration, hyperfiltration, and dynamic membranes, in *Reverse Osmosis Membrane Research,* Lonsdale, H. K. and Podall, H. E., Eds., Plenum Press, New York, 1972, 379.
22. Shor, A. J., Kraus, K. A., Smith, W. T., and Johnson, J. S., Salt rejection properties of dynamically formed hydrous zirconium (IV) oxide membranes, *J. Phys. Chem.*, 72, 2200, 1968.
23. Shor, A. J., Kraus, K. A., Johnson, J. S., and Smith, W. T., Hyperfiltration, *Ind. Eng. Chem. Fundam.*, 7, 44, 1968.
24. Vofsi, D. and Kedem, O., Hyperfiltration in Polyelectrolyte Membranes, Office of Saline Water, Research and Development Progress Report No. 787, U.S. Dept. of the Interior, Washington, D.C., 1972.
25. Hwang, S. and Kammermeyer, K., *Membranes in Separations,* Interscience, New York, 1975, 124.
26. Johnson, J. S., Dresner, L., and Kraus, K. A., Hyperfiltration (reverse osmosis), in *Principles of Desalination,* Spiegler, K. S., Ed., Academic Press, New York, 1966, 346.
27. Simons, R. and Kedem, O., Hyperfiltration in porous fixed charge membranes, *Desalination*, 13, 1, 1973.
28. Hoffer, E. and Kedem, O., Hyperfiltration in charged membranes: the fixed charge model, *Desalination*, 2, 25, 1967.
29. Lakshminarayanaiah, N., *Transport Phenomena in Membranes,* Academic Press, New York, 1969.
30. Dresner, L., Some remarks on the integration of the extended Nernst-Planck equations in the hyperfiltration of multicomponent systems, *Desalination*, 10, 27, 1972.

31. **Michaels, A. S., Nelsen, L., and Porter, M. C.,** Ultrafiltration, in *Membrane Processes in Industry and Biomedicine,* Bier, M., Ed., Plenum Press, New York, 1971, 197.
32. **Cobble, J. W.,** The entropies of the oxy-anions and related species, *J. Chem. Phys.,* 21, 1443, 1953.
33. **Lonsdale, H. K. and Pusch, W.,** Donnan-membrane effects in hyperfiltration of ternary systems, *J. Chem. Soc., Faraday Trans.,* 71, 501, 1975.
34. **Rosenberg, N. W., George, J. H. B., and Potter, W. D.,** Electrochemical properties of a cation-transfer membrane, *J. Electrochem. Soc.,* 104, 111, 1957.
35. **Grieves, R. B., Bhattacharyya, D., Schomp, W. G., and Bewley, J. L.,** Membrane ultrafiltration of a nonionic surfactant, *AIChE J.,* 19, 766, 1973.

WASTE WATER TREATMENT BY LIQUID ION EXCHANGE IN LIQUID MEMBRANE SYSTEMS

J. W. Frankenfeld and N. N. Li

TABLE OF CONTENTS

Introduction 285

Characteristics of Liquid Membranes 285

Removal of Anions from Waste Water Streams 286

Cleanup of Waste Waters Containing Chromates 287

Separation of Heavy Metal Ions and Metal Ion Complexes 288

Use of Biological Systems as Trapping Agents 289

Process and Economic Considerations 291

References 292

INTRODUCTION

The recent discovery of liquid membranes[1,2] has greatly extended the application of liquid ion exchange to the extraction of metals and other substances from aqueous systems. Oil-soluble liquid ion exchange (LIX®) reagents, themselves a fairly recent development,[3] are ideally suited for incorporation into liquid membrane films. Two of the most important applications for such membranes are in the fields of extractive hydrometallurgy and waste water treatment. The characteristics of liquid ion exchange reagents and their use in hydrometallurgy had been reviewed by Lewis.[3] This chapter will deal with the application of LIX reagents in liquid membranes to the treatment of waste water.

CHARACTERISTICS OF LIQUID MEMBRANES

Liquid membranes were invented by Li in 1968[1,2] and have been proposed as solutions to a number of waste disposal problems.[4-6] There are two major types of liquid membrane systems. One is a water-immiscible emulsion dispersed in water and the other is an oil-immiscible emulsion dispersed in oil. For waste water treatment and minerals concentration from mine leaching solutions, the former is used. The water-immiscible emulsion consists of an oil phase, composed of surfactants and various additives in a hydrocarbon solvent, which encapsulates microscopic droplets of an aqueous solution of appropriate reagents for removing (stripping) and trapping waste water contaminants or minerals from mine leachates. A conceptualized drawing of a liquid membrane "capsule" is given in Figure 1. The example shown is that of removal of Cu^{+2} ion according to the equations[7,8]

Extraction:

$$2\ RH + Cu^{+2} \rightleftharpoons R_2Cu + 2H^+ \qquad (1)$$
Org   Aq   Org   Aq

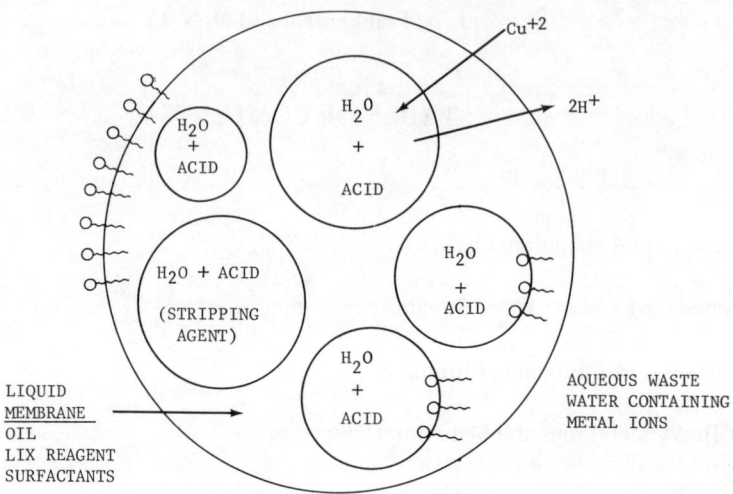

FIGURE 1. Schematic diagram of liquid membrane capsule for Cu^{+2} extraction.

Stripping:

$$2H^+ + R_2Cu \rightleftharpoons Cu^{+2} + 2RH \quad (2)$$
Aq   Org   Aq   Org

where RH represents a liquid ion exchange agent. Among the most useful for copper extraction are the oxime type produced by General Mills.[7,9]

Extraction (Equation 1) occurs at the membrane-external aqueous phase interface, while stripping (Equation 2) occurs at the membrane-internal aqueous phase interface. The overall reaction represents an exchange of a copper ion for two hydrogen ions. The copper is effectively trapped in the interior of the liquid membrane by the large excess of hydrogen ions. The effectiveness of this process for removal of copper from waste water is discussed later in this chapter.

The above is a good example of one of the two types of "facilitated transport" through liquid membranes.[10] That is, the transport of an oil-insoluble ion is facilitated by incorporating a carrier such as a liquid ion exchange agent in the membrane. In the cases of removal of oil-soluble contaminants such as ammonia,[8,13] organic acids,[5] and phenol,[11] no such additive is necessary. The mechanism of unassisted transport through liquid membranes has been discussed by Cahn and Li[11] and Li and Shrier.[5] This discussion is restricted to the removal of metals from waste water by facilitated transport using liquid ion exchange agents.

Liquid membrane processes are not limited to metal cations as shown above. Metal anions and complex anions are also removed as well by proper choice of liquid ion exchange and stripping agents.

REMOVAL OF ANIONS FROM WASTE WATER STREAMS

Many anions, difficult to remove by other methods, can be extracted from waste water by liquid membrane - liquid ion exchange processes. A typical example is phosphate ion, which can be scavenged from aqueous systems with oil-soluble amines or quaternary ammonium salts in liquid membrane films according to the equations[4,5]

Extraction:

$$3[R_3N-HCl] + PO_4^{-3} \rightleftharpoons [R_3N-H]_3PO_4 + 3Cl^-$$
Org   Aq   Org   Aq

(3)

Stripping:

$$2[R_3N-H]_3PO_4 + 3CaCl_2 \rightleftharpoons 6(R_3N-HCl) + Ca_3(PO_4)$$
Org   Aq   Org   Aq

(4)

Again, extraction occurs at the membrane-external phase interface, while stripping is accomplished in the central, aqueous phase of the

TABLE 1

Extraction of Phosphates from Waste Waters with the Liquid Membrane-Liquid Ion Exchange System[13]

Contact time (min)	Wt % PO_4^{-3} in waste stream	
	Run no. 1*	Run no. 2**
0	0.565	0.273
2	0.265	0.123
5	0.200	0.073
18	0.050	0.016
44	–	0.004

*$CaCl_2$ plus NH_4OH as internal stripping agent.
**$CaCl_2$ plus $Ca(OH)_2$ as internal stripping agent.

membrane capsules. The phosphate is effectively trapped by conversion to insoluble $Ca_3(PO_4)_3$ which can neither permeate back through the oil film nor be picked up by the amine salt. The overall reaction is exchange of phosphate and chloride ions. Counter-ions other than chloride can, of course, be used if desired. Removal rates in a typical batch experiment are shown in Table 1.[13] Nitrate ion, another contaminant which is difficult to remove from aqueous systems by purely chemical means, was similarly extracted.[13] These systems have the capability of removing several anionic contaminants at once. In addition, Li and Shrier have demonstrated the simultaneous extraction of organic and inorganic waste materials with a single liquid membrane formulation.[5]

CLEANUP OF WASTE WATERS CONTAINING CHROMATES

Chromate removal from waste water is a serious problem both in the United States and elsewhere. The Takuma Co., Ltd., of Japan in conjunction with Exxon Research and Engineering Company has recently reported the use of liquid membrane-liquid ion exchange processes for cleanup of metal plating baths containing chromates.[8,12,22] Chromates are extracted from aqueous systems by employing either tertiary amines or quaternary ammonium salts:[8,14]

Extraction by amine neutralization:

$$[2R_3N] + [2H^+ + Cr_2O_7^=] \rightleftharpoons [(R_3NH)_2 Cr_2O_7] \quad (5)$$
$$\text{Org} \quad\quad \text{Aq} \quad\quad\quad \text{Org}$$

Extraction by salt formation:

$$[(R_3NH)_2 X] + [2H^+ + Cr_2O_7^=] \rightleftharpoons$$
$$\text{Org} \quad\quad \text{Aq}$$

$$[(R_3NH)_2 Cr_2O_7] + [2H^+ + X^=] \quad (6)$$
$$\text{Org} \quad\quad\quad \text{Aq}$$

$X = SO_4$ or Cl_2

The exact extraction mechanism depends upon the stripping agent employed. Stripping of the chromate ion is accomplished in the aqueous, internal phase of the liquid membrane capsules by one of two methods as shown in Equations 7 and 8:

Basic stripping:

$$[(R_3NH)_2 Cr_2O_7] + [4(Na^+ + OH^-)] \rightleftharpoons$$
$$\text{Org} \quad\quad\quad \text{Aq}$$

$$[R_3N] + [2(2Na^+ + CrO_4^=)] + [3H_2O] \quad (7)$$
$$\text{Org} \quad\quad\quad \text{Aq}$$

Acid stripping:

$$[(R_3NH)_2 Cr_2O_7] + H_2SO_4 \rightleftharpoons$$
$$\text{Org}$$

$$[(R_3NH)_2 SO_4] + [2H^+ + Cr_2O_7^=] \quad (8)$$
$$\text{Org} \quad\quad\quad \text{Aq}$$

Stripping with a base regenerates the free amine in the membrane which then re-extracts additional $Cr_2O_7^{-2}$ by Equation 5. Acid stripping regenerates the amine salt for extraction as shown in Equation 6. Both methods have been used successfully in liquid membranes. One of the important advantages of the liquid membrane process lies in the extraction and stripping in a single stage, rather than two separate stages as required by solvent extraction. In addition, by simultaneously extracting and stripping, the liquid membrane process drives the equilibrium of extraction as shown in Equations 5 and 6 to the right by trapping of the complexed ions as formed. This removes the equilibrium limitation inherent in some solvent extraction methods.

Some of the typical results of laboratory experiments are given in Figure 2 and Table 2.[8,12] Figure 2 (Curve 2) shows that in a batch run, nearly complete chronium removal was achieved in 10 min with a feed containing an initial chromium concentration of 400 ppm. The results of experiments at two different external pH values are given in Table 2. These data indicate that in order to transfer chromium from the external feed to the internal reagent phase of the liquid membrane

TABLE 2

Effect of External pH on Chromium Removal from Waste Water by Liquid Membrane Emulsions*

pH of external phase	Results in ppm Cr^{+6} contact time (min)					
	0	5	10	15	20	30
3.5	390	250	80	12	3	2
7.0	390		(no removal)			390

*Conditions: reagent phase = 2% NaOH; feed = 0.1% $K_2Cr_4O_7$; membrane/reagent wt ratio = 1.2; feed/emulsion = 3.1; temperature = 25°C.[8]

emulsion, the pH of these two phases must be controlled to within certain critical limits. At a constant pH of the internal phase, given by the 1% NaOH aqueous solution, the transfer of chromium was very fast when the external phase was acidic. The Cr^{+6} content of the external phase was reduced from 390 ppm to 12 ppm (97% removal) in 15 min under these conditions. However, under basic conditions, no transfer of chromium was observed.[8]

SEPARATION OF HEAVY METAL IONS AND METAL ION COMPLEXES

Heavy metal ions may be extracted from waste water streams either as cations or as anionic complexes by employing liquid membrane systems

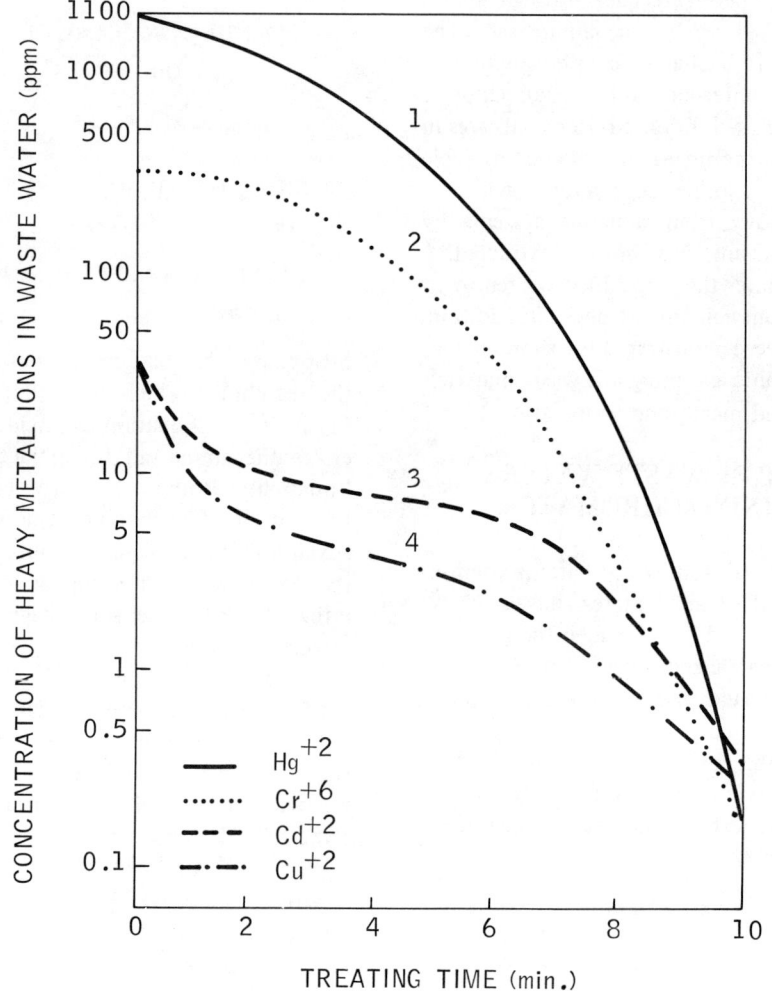

FIGURE 2. Removal of heavy metal ions from waste water streams by liquid membrane-liquid ion exchange technique.[1,2]

containing various liquid ion exchange agents. The removal of cations was illustrated, in the case of copper, by Equations 1 and 2 above. Frequently, heavy metal ions are present in the form of soluble anionic complexes with halogens or other anions. Solvent extractions of these complexes using liquid ion exchangers have been reported.[3,15-17] For example, Moore has extracted mercury-halogen complexes from brine solutions.[15,16] In such solutions mercury is present as the soluble complex:

$$Hg^{+2} + 4Cl^- \rightleftharpoons HgCl_4^{-2} \quad (9)$$

Extraction is accomplished by contacting the aqueous solution with a high molecular weight amine or quaternary ammonium salt in a suitable organic diluent:

$$2(R_4NX) + HgCl_4^{-2} \rightleftharpoons (R_4N)_2HgCl_4 + 2X^- \quad (10)$$
$$\text{Org} \quad \text{Aq} \quad \quad \text{Org} \quad \text{Aq}$$

Regenerative stripping was accomplished in a second step by contacting with strong acids, inorganic bases, or organic bases such as ethylenediamine or propylenediamine.[16] High molecular weight primary, secondary, and tertiary amines are effective extractants for acidic brines but quaternary ammonium salts must be used at high pH.

Extractants of this type have been successfully incorporated into liquid membranes for extraction of Cu^{+2} and Hg^{+2}.[8,12] The liquid membrane system, of course, combines extraction and stripping into a single step and numerous advantages accrue. Some typical results are shown in Figure 2. (Curves 1 and 4). The Hg^{+2} concentration was rapidly reduced from 1100 ppm to less than 0.2 ppm, while Cu^{+2} was reduced from 50 ppm to less than 0.3 ppm.

The removal of cadmium from metal plating baths by a liquid membrane-liquid ion exchange technique has also been reported.[8,12] Although the general principle is the same as for Hg^{+2}, the removal of Cd^{+2} presents a somewhat different problem.[8] In the effluent from metal plating baths, cadmium exists as the tightly bound cyanide complex:

$$Cd^{+2} + 4CN^- \rightleftharpoons Cd(CN)_4^{-2} \quad (11)$$

This complex is readily extracted from aqueous solutions by liquid ion exchange agents such as Aliquat 336® (methyltrioctylammonium chloride).[17] However, stripping is quite difficult because of the great stability of the $Cd(CN)_4^{-2}$ complex under basic conditions ($K_{ST} = 10^{19}$).[18] The usual stripping agents, salts and mineral acids, are ineffective even at high concentrations. Ethylenediaminetetraacetic acid (EDTA) is known to chelate strongly with cadmium complexes.[19] In the form of its disodium salt, EDTA successfully removed cadmium from Aliquat 336 with regeneration of the ion carrier so long as the pH of the internal phase of the emulsion was maintained in the range of 4 to 6.

This procedure was studied by Kitagawa and the Exxon group[8,12] who report the results shown in Figure 2 (Curve 3). The cadmium content of an actual plating bath effluent was rapidly reduced from 50 to less than 0.5 ppm.

USE OF BIOLOGICAL SYSTEMS AS TRAPPING AGENTS

Mohan and Li[6,20,21] reported a novel use of biological enzyme systems to trap nitrates and nitrites in liquid membrane formulations. This process has promise for large-scale continuous operations of secondary and tertiary waste water treatment. Various amines were used as liquid ion exchange agents to transfer the NO_2^- or NO_3^- across the membrane film. Particularly effective was Amberlite LA-2®, an oil-soluble secondary amine manufactured by Rhom and Haas, Philadelphia, Pa. This amine readily extracted the anions from weakly acidic solutions. In one modification of the process stripping in the internal phase of the liquid membrane emulsions was accomplished by buffered salt solutions of either purified enzymes or cell-free extracts from *Micrococcus denitrificans*.[6] The extracts rapidly reduced the nitrate or nitrite to products which were not reextracted by the liquid ion exchange agent. Thus, the reduction products were efficiently trapped within the membrane. Nitrite was shown to be an intermediate product in the reduction of nitrate. The presumed final product was elemental nitrogen.[6]

This concept was carried one step further by encapsulating intact cells of *M. denitrificans* in liquid membrane emulsions.[20,21] Again, amines were used as transferring agents. Both NO_2^- and NO_3^- were readily reduced, probably to N_2. The rapid reduction of NO_3^- achieved is illustrated in Figure 3.[20] As NO_3^- was reduced in the internal phase of the membrane, NO_2^- was liberated and was carried to the external phase, where its appearance was monitored (Figure 3). However, the rate of NO_2^- appearance was much slower than NO_3^- reduction, indicating that most of the nitrite was reduced further as formed. This sequential reduction is illustrated by the data in Table 3. The NO_3^- was completely reduced in 120 min contacting, while 90% of the NO_2^- formed was reduced in the same period.

The encapsulated cells remained intact and active for periods of at least 5 days. It was claimed that this period could be extended even further by encapsulating special nutrients along with the cells.[20] In addition, such known enzyme inhibitors as $HgCl_2$ in the external phase had no effect on the encapsulated cells, indicating complete protection by the surrounding membrane. Nonencapsulated cells were inactive in the same medium. This has considerable importance in developing waste water treatment processes, since many effluents to be treated contain such inhibitors.[20,21] Mohan and Li also showed that the process could be carried out continuously. A

TABLE 3

Simultaneous Reduction of NO_3^- and NO_2^- by Encapsulated Viable Whole Cells of *M. Denitrificans* ATCC 21909[20]

Time (min)	NO_3^- % reduced	NO_2^- % in aqueous phase	NO_3^- % reduced beyond NO_2^-
0	0	0	0
15	20	0	20
30	32	1.6	30
45	45	3.2	42
60	54	4.8	49
75	63	5.8	57
90	73	8.3	65
105	88	9.2	79
120	100	10.0	90

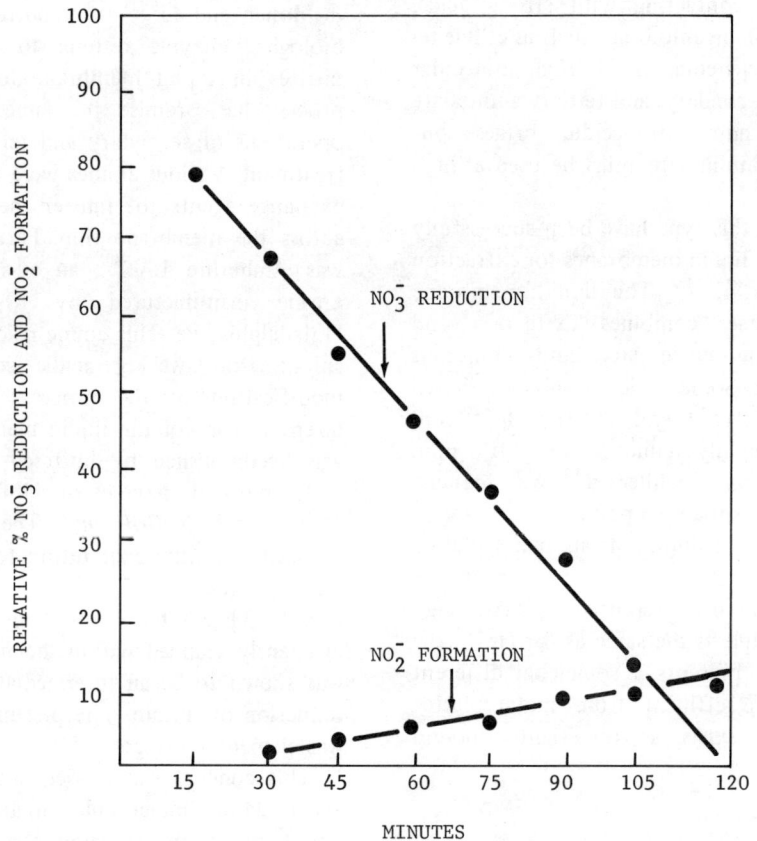

FIGURE 3. Reduction of NO_3^- and formation of NO_2^- by liquid membrane encapsulated whole viable cells of *Micrococcus denitrificans* ATCC 21909.[20]

TABLE 4

Removal of Cr^{+6} by Two-Stage Pilot Plant[1,2]

Test no.	1	2	3
L.M. emulsion			
Membrane (A)			
Reagent solution (B)	10% NaOH	20% NaOH	20% NaOH
B/A	1/2	1/2	1/2
Flow rates (cc/min)			
Waste water	300	300	300
Emulsion	280	105	064
Treat ratio	1/4.85	1/4.85	1/3.40
Mixing speed (rpm)	200	200	300
Concentration of Cr^{+6}			
After treatment			
In effluent	1 ppm	1 ppm	1 ppm
In emulsion	35,800 ppm	108,000 ppm	182,000 ppm

Note: Feed to test unit: waste water containing 100 ppm Cr^{+6} from plating process.

special apparatus was designed and has been described.[20]

PROCESS AND ECONOMIC CONSIDERATIONS

The Takuma Co. of Japan has evaluated liquid membrane processes for heavy metals removal in a continuous, two-stage countercurrent pilot plant, and they have reported the results in two recent publications.[8,12] These studies included the scavenging of NH_4^+ as well as Cr^{+6}, Hg^{+2}, Cd^{+2}, and Cu^{+2} by processes similar to those described above. The pilot plant equipment was also detailed.[12] Typical results, illustrated by the removal of Cr^{+6} from the effluent of a metal plating process, are shown in Table 4. In all cases, the concentration of Cr^{+6} in the waste water was reduced, under continuous flow conditions, from 100 ppm to less than 1 ppm.

Takuma estimate that the treatment cost of Cr^{+6} cleanup will be about 15¢/ 1 m^3 of waste water (57¢/1000 gal) for a plant with a 100 m^3/hr capacity with an emulsion recycling capability. Plant cost, installation, chemicals, personal expenses, and other necessary expenditures were included in this estimate.[12] This cost is quite attractive compared to most industrial waste water cleanup processes. Furthermore, there are no good alternates for achieving the low levels of metal required by Japanese law.

This new concept, the combination of liquid membrane and liquid ion exchange techniques, is just emerging as a potentially useful processing technique. It would appear that such processes may play an important role in solving future waste water treatment problems.

REFERENCES

1. **Li, N. N.,** U.S. Patent 3,410,794, 1968.
2. **Li, N. N.,** Permeation through liquid surfactant membranes, *AIChE J.,* 17, 459, 1971.
3. **Lewis, C. J.,** Liquid ion exchange in hydrometallurgy, in *Recent Developments in Separation Science,* Vol. 2, Li, N. N., Ed., CRC Press, Cleveland, 1972, 47.
4. **Li, N. N., Cahn, R. P., and Shrier, A. L.,** U.S. Patent 3,617,546, 1971.
5. **Li, N. N. and Shrier, A. L.,** Liquid membrane water treating, in *Recent Developments in Separation Science,* Vol. 1, Li, N. N., Ed., CRC Press, Cleveland, 1972, 163.
6. **Mohan, R. R. and Li, N. N.,** Reduction and separation of nitrate and nitrite by liquid membrane-encapsulated enzymes, *Biotechnol. Bioeng.,* 16, 513, 1974.
7. **Merigold, C. R. and House, J. E.,** The Application of Liquid Ion Exchange Technology to the Recovery of Copper, presented at Copper Technology Seminar, U.S. Dept. of Commerce, Bureau of East-West Trade, Washington, D.C., December 1975.
8. **Kitagawa, T., Nishikawa, Y., Frankenfeld, J. W., and Li, N. N.,** Waste water treatment by a liquid membrane process, accepted for publication in *Environ. Sci. Technol.*
9. **Kordosky, G. A., MacKay, K. D., and Virnig, M. J.,** A New Generation Copper Extractant, presented at AIME Annual Meeting, Las Vegas, February 22 to 26, 1976.
10. **Matulevicius, E. S. and Li, N. N.,** Facilitation transport through liquid membranes, *Sep. Purif. Methods,* 4, 73, 1975.
11. **Cahn, R. P. and Li, N. N.,** Separation of phenol from waste water by the liquid membrane technique, *Sep. Sci.,* 9, 505, 1974.
12. **Kitagawa, T. and Nishikawa, N.,** Waste Water Treatment by Liquid Membrane Process, presented at Am. Chem. Soc. Centennial Symposium on Separation and Encapsulation by Liquid Membranes, New York, April 6, 1976.
13. **Li, N. N., Cahn, R. P., and Shrier, A. L.,** U.S. Patent 3,779,907, 1973.
14. **General Mills Corp.,** Chromium, Technical Bulletin CDSI-61, 1961.
15. **Moore, F. L.,** Liquid-liquid extractions of mercury with high molecular weight amines from iodide and bromide solutions, *Sep. Sci.,* 7, 505, 1972.
16. **Moore, F. L.,** Solvent extraction of mercury from brine solutions with high molecular weight amines, *Environ. Sci. Technol.,* 6, 525, 1972.
17. **McDonald, C. W. and Moore, F. L.,** Liquid-liquid extraction of cadmium with high molecular weight amines from iodide solutions, *Anal. Chem.,* 45, 983, 1973.
18. **Cotton, F. A. and Wilkinson, G.,** *Advanced Inorganic Chemistry,* 3rd ed., Interscience, New York, 1966, 514.
19. **Irving, H. M. N. H. and Al-Jariah, R. H.,** The extraction of various metals as their ionic complexes with EDTA by solutions of Aliquat-336 chloride in 1,2-dichloroethane, *Anal. Chim. Acta,* 74, 321, 1975.
20. **Mohan, R. R. and Li, N. N.,** Nitrate and nitrite reduction by liquid membrane encapsulated whole cells, *Biotechnol. Bioeng.,* 17, 1137, 1975.
21. **Frankenfeld, J. W., Asher, W. J., and Li, N. N.,** Biochemical and Biomedical Separations Using Liquid Membranes, presented at the Am. Chem. Soc. Symposium on Recent Developments in Separation Science, San Francisco, California, August 31, 1976.
22. **Schiffer, D. K., Hochhauser, A., Evans, D. F., and Cussler, E. L.,** *Nature,* 250, 484, 1974.

REMOVAL OF SOLVENT AND MONOMER RESIDUALS FROM GLASSY POLYMERS

A. R. Berens and H. B. Hopfenberg

TABLE OF CONTENTS

Introduction 293

Experimental Characterization of VCM Transport and Equilibria in PVC 294
    Materials 294
    Equipment and Procedure 295

Theoretical Model for Data Analysis Describing VCM Transport in PVC Powders 295

Experimental Results and Data Analysis 296
    General Character of Sorption Curves 296
    Diffusion Coefficient Determined for Emulsion PVC 297
    Diffusion Coefficient Determined for Suspension PVC 299
    Concentration Dependence of the Diffusion Coefficient 300
    Temperature Dependence of the Diffusion Coefficient 300
    Results on Rigid PVC Film 300
    Later Stages of Sorption/Desorption 301

Migration of VCM Residuals from PVC Pipes and Bottles into Noninteracting Fluid Contents 303
    Case I — VCM Loss During Storage 304
    Case II — VCM Migration from Freshly Formed PVC Products into Finite Fluid Contents . . . 305
    Case III — VCM Migration in Closed-system Service from Previously Aged PVC Products . . . 308
    Case IV — VCM Migration from Thin-walled Bottles into Finite Closed Contents 311

Conclusions 311

References 312

INTRODUCTION

In the early days of polymer manufacture, the removal of solvents and monomers from rubbers and plastics was performed primarily because of the economic advantage in recovering and recycling these starting materials. Johnson and Otto in 1949 described in detail a process for the recovery of the monomers, styrene and butadiene, used in GR-S manufacture.[1] In many cases, however, the direct economic incentive associated with raw material recovery and recycling is not sufficient to justify the design and implementation of sophisticated recovery techniques for monomer and/or solvent. In some cases, monomer removal has been required to improve the properties of the resulting formed polymer. For example, rather elaborate techniques involving extraction of residual monomers with alcohols have been carried out on polystyrene beads to reduce the residual monomer concentration and, in turn, raise the softening temperature of the resulting polymer.[2] The historical and classical motivation of monomer and solvent removal, therefore, focused upon economic processing of high quality product.

An additional motivation is related to the

requirement to remove residual solvent and monomers which are suspected of providing significant health hazards to both production workers and, possibly, ultimate users of polymers.[3,4] The recent history of the PVC industry is a case in point.

In January 1974, the B.F. Goodrich Company voluntarily revealed to federal and state regulatory officials that, in the preceding 3 years, three workers at its poly(vinyl chloride) polymerization plant in Louisville, Kentucky had succumbed to an angiosarcoma of the liver. Maltoni[5] independently reported that test rats exposed to vinyl chloride monomer (VCM) developed an angiosarcoma of the liver. By May 1974, a cause-effect relationship between long-time exposure to high levels of VCM and the development of angiosarcoma in humans was generally accepted. Confirmed cases of cancer in workers at PVC plants had risen to 19.

These findings prompted stringent regulations by the Occupational Safety and Health Administration (OSHA) and intense activity by industrial researchers to develop means for minimizing exposure of production workers to solvent and monomer vapors. These efforts were supplemented by research related to removal of VCM from the polymer. It was feared that residual monomer might ultimately migrate from formed polymers, typically used as potable water piping or as packaging for foods and pharmaceutical products.

The removal of these low molecular weight materials from polymers is an unsteady state process ideally controlled by diffusion in the polymeric phase. The diffusion taking place in the polymer phase, characterized by relatively small diffusion coefficients compared with liquid or gas phase diffusion, is fixed by the temperature and penetrant composition. Since this process is relatively slow, it is imperative to insure that gas and liquid phase mass transfer processes are not rate-limiting. For the case of successful VCM removal from PVC, transport processes in the polymer phase completely control the overall observed removal kinetics.

The case study involving successful removal of VCM from PVC is useful not only because of the potential hazard but also because of the complex nature of the physicochemical system comprised by VCM in PVC. Specifically, poly(vinyl chloride) at room temperature is a glassy polymer containing low, albeit discernible, levels of somewhat ill-characterized crystalline material. The equilibria and kinetics describing sorption and desorption of vinyl chloride monomer in the parent polymer are confounded by all of the complex behavior associated with transport of low molecular weight penetrants in glassy polymers, as well as those complications peculiar to transport in semi-crystalline materials. In addition, the polymer is subject to both long-term and short-term relaxation processes which further complicate the characterization of a true equilibrium and, in turn, contribute to the observed kinetics of the transport in the polymer.[6]

Growing public concern and government regulations have recently required efforts to remove potentially hazardous components from food-packaging materials and potable-water pipe. The selection of the vinyl chloride/PVC case study as an appropriate model for penetrant removal in general is motivated not only by the reality of the problem but also by the effective and immediate success which was the result of cooperative industrial efforts to comply with the stringent regulations for product purity. Moreover, the physicochemical details of the VCM/PVC system provide the basis for process design and development in similarly complex systems as well as for systems which are not nearly as complicated. This review will, therefore, focus on the recent literature relating to the characterization of the VCM/PVC system and, most importantly, on the general techniques which can be brought to bear for the effective removal of low molecular weight molecules from glassy, semi-crystalline polymeric materials in particular and polymer resins and products in general.

EXPERIMENTAL CHARACTERIZATION OF VCM TRANSPORT AND EQUILIBRIA IN PVC

Materials

Polymer samples used in this study included a number of commmercial and experimental PVC homopolymers made by suspension, mass, and emulsion polymerization techniques. Virtually all of the samples were studied in the powder form, recovered directly from the polymerization process by normal drying procedures. Particle structures were characterized by optical and electron microscopy and by nitrogen-adsorption surface-area measurements. The resin powders used in this study are described in Table 1. A few

TABLE 1
PVC Samples Used

Sample	Type	Intrinsic viscosity	Surface area m^2/g	Glassy-particle content
A	Suspension	1.10	2.3	~nil
C	Suspension	0.70	<0.1	very high
D	Emulsion	1.58	9.8	nil
F	Suspension	1.10	0.32	high
G	Suspension	0.65	0.14	moderate
H	Suspension	0.65	0.35	~nil
J	Suspension	0.75	0.21	high
K	Suspension	0.95	0.69	low
L	Suspension	0.95	0.70	very low
M	Suspension	1.10	1.3	~nil

measurements were also made on a sample of ~2-mil rigid PVC film.

Equipment and Procedure

Most of the measurements reported here were made with a recording Electrobalance® (Model RG, Cahn Division, Ventron Instruments Corp., Paramount, California). The balance was mounted in a glass vacuum-chamber with connections to a source of VCM vapor and to vacuum pumps. PVC samples were suspended from the balance in a light aluminum pan near the bottom of a 40-cm Kovar® hangdown tube, which was surrounded by a circulating liquid jacket for temperature control. Sample weights were from 100 to 500 mg; use of the balance at 1 μg sensitivity thus could detect weight changes as small as 2 ppm. The sample on the balance was evacuated to constant weight, then VCM vapor was admitted to a preselected pressure and the weight change was recorded as function of time on a strip-chart recorder (Hewlett-Packard 7100B). Pressures were measured to ±0.1 mmHg with a strain-gauge transducer and digital voltmeter. Because of the large vapor volume and small sample size, sorption caused negligible change in pressure after an addition of VCM vapor.

Pressure changes (increases) were made instantaneously by bringing a small vessel to a measured VCM pressure, then quickly opening this vessel to the balance chamber. The pressure could be increased to a predetermined value in a few tenths of a second. Correspondingly sharp decreases in pressure were not so easily achieved. Using oil diffusion and mechanical pumping, liquid-nitrogen traps, and large vacuum lines, pressure could be reduced from, say, 200 mm to ~30 μm in 1 min. In most experiments, the half-evacuation time was considerably less than the half-desorption time, so no corrections were made for the noninstantaneous pressure change. The response of the balance and recorder provided useful data 3 to 5 sec after a pressure change.

THEORETICAL MODEL FOR DATA ANALYSIS DESCRIBING VCM TRANSPORT IN PVC POWDERS

The mathematical statement of Fick's first law of diffusion in rectilinear coordinates is

$$F = -D \frac{\partial C}{\partial X} \tag{1}$$

where F is the flux of diffusant (quantity passing through a unit plane area in unit time), $\partial C/\partial X$ is the concentration gradient, and D is the diffusion coefficient. The minus sign is conventional to indicate that transport occurs toward the region of lower concentration. The conventional units of D are cm^2/sec.

From Equation 1, it can be shown that

$$\frac{\partial C}{\partial t} = D \frac{\partial^2 C}{\partial X^2} \tag{2}$$

The application of Fick's laws to experimental situations requires solution of Equation 2 for the particular geometry and boundary values involved. Many of the useful solutions are collected in Crank's text.[7]

As a model to approximate the sorption of VCM by PVC resins, Crank's solution for the non-steady state diffusion in a sphere was used.

Specifically, the solution of Equation 2 conforming to the experimentally respected boundary conditions of uniform initial concentration through the sphere, constant concentration at the surface, and constant diffusion coefficient is

$$M_t/M_\infty = 1 - \frac{6}{\pi^2} \sum_{n=1}^{\infty} \frac{1}{n^2} e^{-4Dn^2\pi^2 t/d^2} \qquad (3)$$

where M_t is the weight of penetrant entering or leaving the sphere in time t after an instantaneous change in surface concentration, M_∞ is the total weight change after infinite time at the new surface concentration, n is an integral counter, D is the diffusion coefficient, and d is the diameter of the sphere.

The following assumptions are implicit in applying Equation 3 to VCM sorption or desorption experiments on PVC powders:

1. Diffusion obeys Fick's laws.
2. D is constant over the concentration range covered by the experiment.
3. The solution for a sphere is applicable to a sample consisting of many spherical particles of uniform size, i.e., diffusion occurs in each particle independently, with no interparticle diffusion.
4. The PVC powder does indeed consist of uniform-sized spherical particles.
5. At the start and end of the experiment, the VCM concentration is uniform through the particles.
6. The VCM concentration at the particle surface is always at equilibrium with the surrounding vapor phase.
7. At time zero, the VCM pressure is instantaneously changed to a new value, thus immediately changing the surface VCM concentration of the particles.
8. The VCM concentration in the particles is zero initially in the case of sorption starting at zero VCM pressure, and zero finally for desorptions to zero pressure.
9. There is no concentration gradient in the vapor phase, and no surface layer effect, i.e., transport is limited only by diffusion within the PVC particles.

If all these assumptions are met experimentally, then Equation 3 will describe the weight gain (or loss) as a function of time in a VCM sorption (or desorption) experiment. This equation (the Fickian, uniform sphere model) predicts that the time required for a given fraction of the ultimate weight change to occur is independent of the initial or final VCM concentration (or pressure), and is proportional to the ratio d^2/D. This ratio, then, can be determined by fitting Equation 3 to the experimental data, but evaluation of D requires independent determination of the particle diameter.

EXPERIMENTAL RESULTS AND DATA ANALYSIS

General Character of Sorption Curves

The nature of sorption vs. time curves depends upon the partial pressure of VCM, P_m. Two typical curves for Sample D at room temperature are presented in Figure 1. In the lower P_m range, there is an initial rapid sorption to a VCM content which then remains nearly constant; at a somewhat higher P_m range, a similar rapid initial VCM uptake is followed by a slower further sorption which continues to very long times. Somewhat similar results were reported by Bagley and Long[8] for the system acetone/cellulose acetate, although two-stage sorption was not observed at such low penetrant concentrations.

Presumably, the Bagley-Long explanation of two-stage sorption is also applicable to the VCM/PVC system. In the first stage, a uniform concentration of penetrant is reached throughout the polymer, i.e., diffusion equilibrium is attained quite rapidly. The second stage involves a slow relaxation of the developed swelling stress, permitting further VCM sorption; since diffusion here is much faster than stress relaxation, the second-stage sorption involves a very low concentration gradient from surface to center of the PVC particles.

These results suggest that one advantage to the use of finely powdered polymer samples is the attainment of difffusion equilibrium before appreciable stress relaxation has occurred. Thus, the two stages may be clearly separated. The extent of first-stage sorption then measures the polymer in a state essentially unperturbed by penetrant swelling, and may be used to assess effects of prior history.

The experiments relevant to the removal of trace residuals involve sufficiently low P_m values that only the first-stage sorption is evident. These experiments will, therefore, be emphasized in this

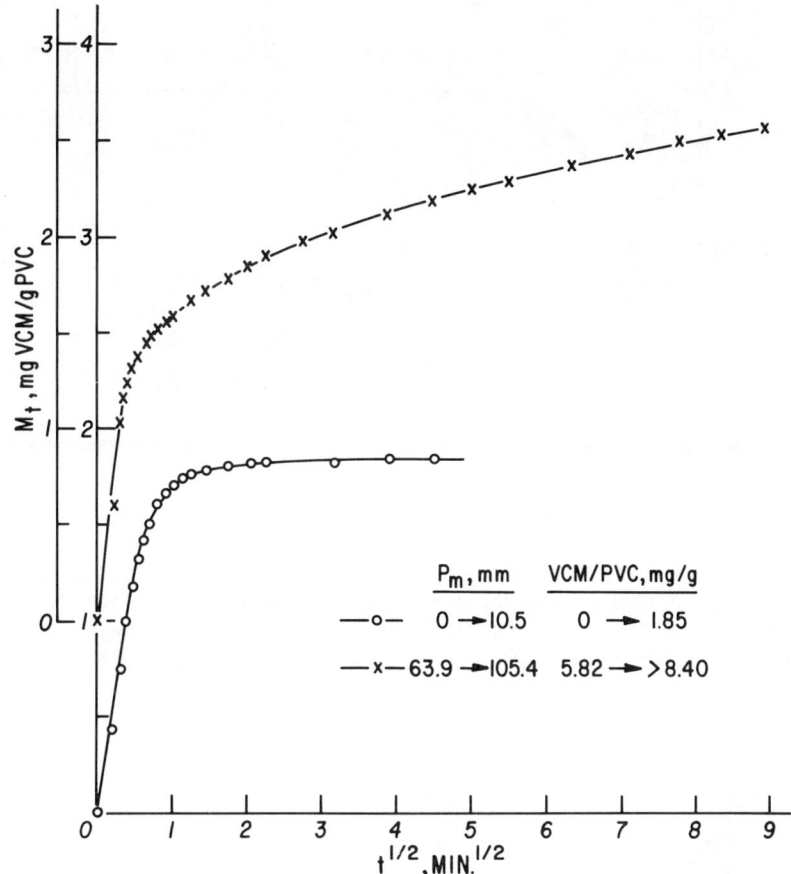

FIGURE 1. VCM sorption kinetics in emulsion PVC (Sample D, see Table 1) at moderate and low VCM pressure and 23°C.

review although characterization and analysis of the concomitant relaxations have also been carried out.[6]

To test the agreement between experimental data and the Fickian, uniform sphere model, two graphic representations of Equation 3 are useful. A plot of M_t/M_∞ (or of M_t) against the square root of time is initially nearly linear, then bends over to approach a well-defined limit (see Figures 1 and 2). Plots of $\log(1 - M_t/M_\infty)$ vs. t (Figure 3) show initial curvature, then approach straight lines. Each of these plots offers a means of evaluating D/d^2 from experimental sorption data. From Equation 3 it can be shown that $t_{0.5}$, the time at which $M_t/M_\infty = 0.5$, or the "half-sorption time," is

$$t_{0.5} = 7.66 \times 10^{-3} \, d^2/D \tag{4}$$

where d is expressed in cm and D has the units cm²/sec. Thus $t_{0.5}$ may be interpolated from a plot of M_t/M_∞, or M_t, vs. $t^{1/2}$ and D/d^2 evaluated by Equation 4. Alternately, D/d^2 may be obtained from the limiting slope of the $\log_{10}(1 - M_t/M_\infty)$ vs. t plot, e.g.,

$$\frac{d[\log_{10}(1 - M_t/M_\infty)]}{dt} = -\frac{4\pi^2 D}{2.303 \, d^2} \tag{5}$$

Note that the determination of D/d^2 from half-sorption time depends mainly on the data in the early stages of sorption, while the $\log(1 - M_t/M_\infty)$ method emphasizes the later-stage data.

Diffusion Coefficient Determined for Emulsion PVC

M_t/M_∞ vs. $t^{1/2}$ is plotted in Figure 2 for a typical VCM sorption experiment on a monodisperse emulsion PVC of 0.44 μm particle diameter. The agreement between the experimental points and Equation 1 (fitted to the observed $t_{0.5}$) shows that the sorption closely follows Fickian behavior. Equation 4, using the particle diameter

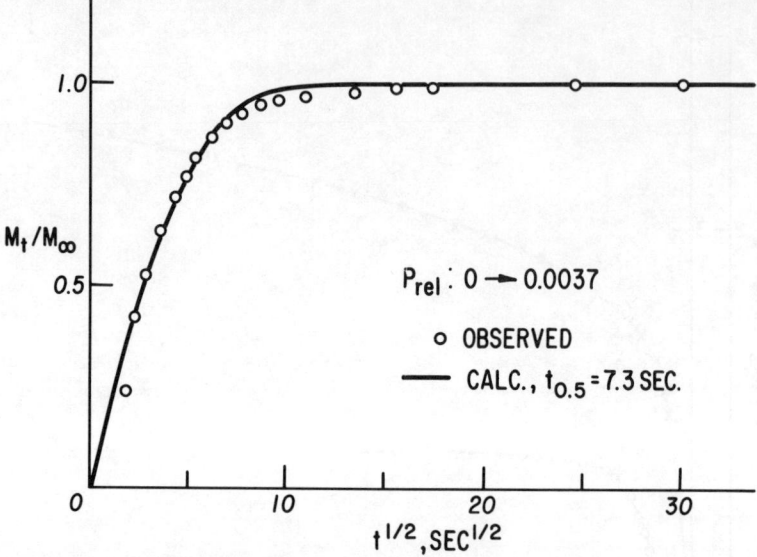

FIGURE 2. Comparison of the experimental and theoretical sorption behavior for VCM transport in PVC (Sample D, see Table 1) at low VCM pressure and 23°C.

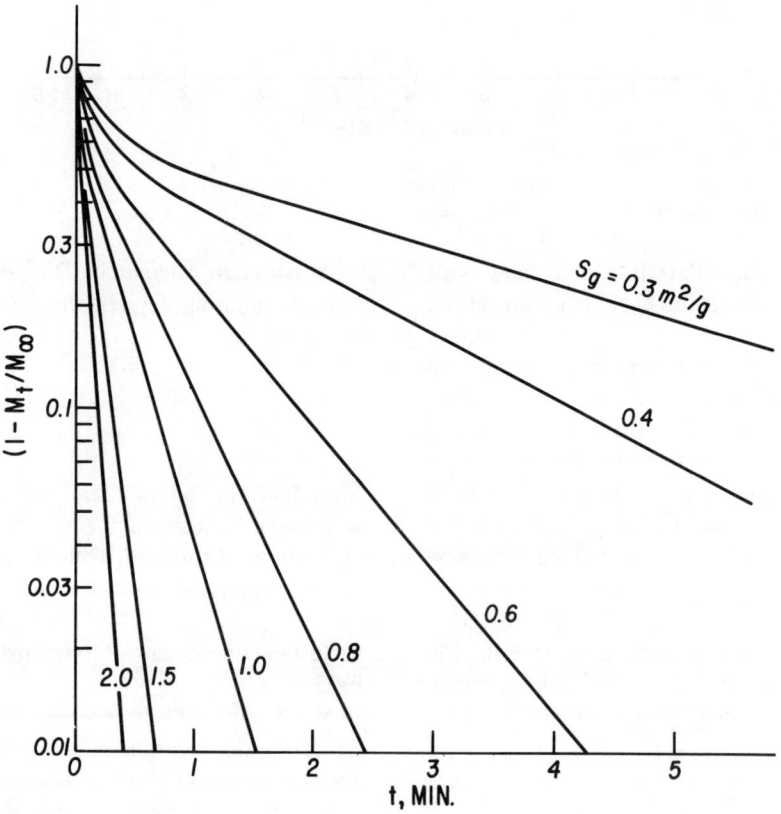

FIGURE 3. Theoretical late stage desorption curves for spherical PVC particles of various diameters using $D = 2 \times 10^{-10}$ cm^2/sec at 90°C.

measured by electron microscopy, gives $D = 1.9 \times 10^{-12}$ cm^2/sec.

Diffusion Coefficient Determined for Suspension PVC

Several studies by electron microscopy[9-11] have shown that the ~100 μm particles of suspension PVC are porous agglomerates of primary or subparticles of about 1 to 5 μm diameter. Scanning electron micrographs indicate that the primary particles in a given sample are fairly uniform in size. This observation suggests the possibility of applying the uniform-sphere model to diffusion in suspension PVCs.

The average size of the primary particles of the suspension PVC can be obtained from determination of specific surface areas measured by N_2-adsorption. For a sphere, of density ρ, the surface-to-weight ratio is $6/\rho d$. For PVC spheres, $\rho \cong 1.4$, and, therefore

$$\bar{d}_s = 4.29/S_g \tag{6}$$

where S_g is the specific surface area in m^2/g and $\bar{d}_s$ is the surface-average particle diameter in micrometers. Quite reasonable agreement is observed between $\bar{d}_s$ determined in this way and the primary particle diameter estimated by scanning microscopy on sectioned granules of suspension PVC.

Sorption curves on porous suspension PVCs conform quite well to the shape predicted by Equation 1, as illustrated in Figure 4. Application of Equation 2 to sorption data on Sample A ($\bar{d}_s$ = 1.85 μm) at 25° given $D = 2.0 \times 10^{-12}$ cm^2/sec, in good agreement with the results on monodisperse emulsion PVC. The agreement of diffusion coefficients and amounts sorbed among PVC samples of widely varied particle size and surface area provides assurance that surface adsorption is not a significant factor in the sorption process.

The uniform-sphere model seems applicable to porous suspension PVC, and the dimension controlling sorption rates appears to be the size of the primary particles, not the external diameter of the agglomerate. This means that diffusion of VCM through the pores, and through any pericellular membrane around the gross particle, must be very rapid. Use of the external particle size in calculations of D from sorption data would increase D by several orders of magnitude. The incorrect selection of the rate-determining dimension presumably accounts for the discrepancy between these results and those recently reported by Wolf and Kreter.[12]

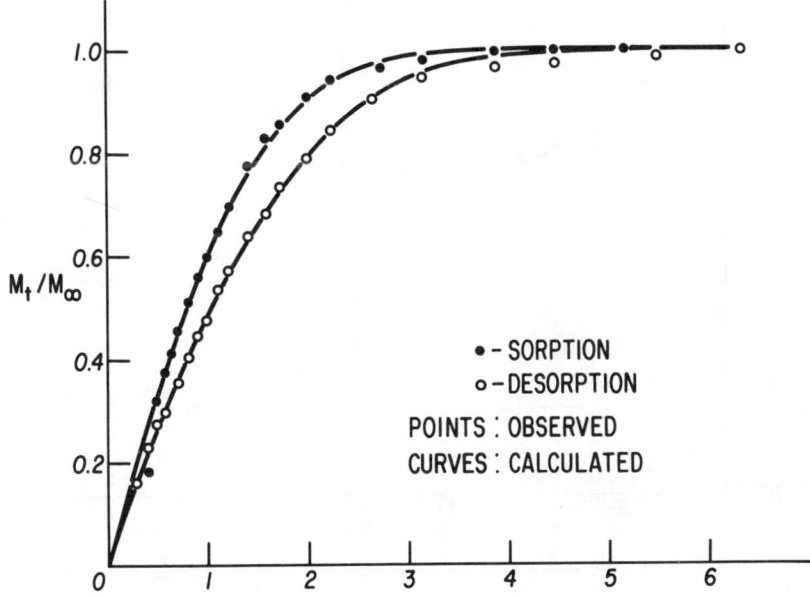

FIGURE 4. Comparison of sorption and desorption kinetics for VCM transport in suspension PVC (Sample A, see Table 1) at 40°C.

Concentration Dependence of the Diffusion Coefficient

In most of the VCM sorption/desorption experiments on PVC, the desorption is somewhat slower than sorption, as illustrated in Figure 4. This is the expected result when D increases with increasing concentration of the penetrant.[7] Such an effect seems likely in this system, due to the plasticization of PVC by VCM. Over the limited range of VCM concentrations considered here, the variation of D is small; an average of sorption and desorption values for D is probably a satisfactory figure for most practical applications.

Temperature Dependence of the Diffusion Coefficient

Sorption and desorption kinetics were characterized at temperatures from 25 to 90°C for several suspension PVC resins. Samples of varied $\bar{d}_s$ were required to bring $t_{0.5}$ into an experimentally accessible or convenient time range. An Arrhenius plot summarizing the data is presented in Figure 5. The agreement among samples is satisfactory; most of the scatter may be attributed to uncertainty in $\bar{d}_s$ or in particle geometry, so there is no clear evidence that D differs significantly from one sample to another. The samples covered the range of molecular weights (or intrinsic viscosities) of normal commercial PVCs, so it also appears that D is insensitive to molecular weight in this range. D varies with temperature from about 2×10^{-12} cm^2/sec at 30° to 2×10^{-10} at 90°C. Analysis of the data of Figure 5 provides a value of 17 kcal/mol for the activation energy for diffusion. Limited results at 110°C show no significant change in activation energy above T_g.

Results on Rigid PVC Film

A single sorption experiment at 25°C on a specimen of rigid PVC film 0.05-mm thick revealed a linear relationship between M_t and $t^{1/2}$, although the system was still far from diffusion equilibrium after 1 week. Assuming that the final solubility (M_∞) was similar to that in PVC powders, an estimated value of $D = 3 \times 10^{-12}$ cm^2/sec is obtained. It appears, therefore, that the values of D may also be applicable to fused (molded or extruded) rigid PVC products as well as to emulsion or suspension resins.

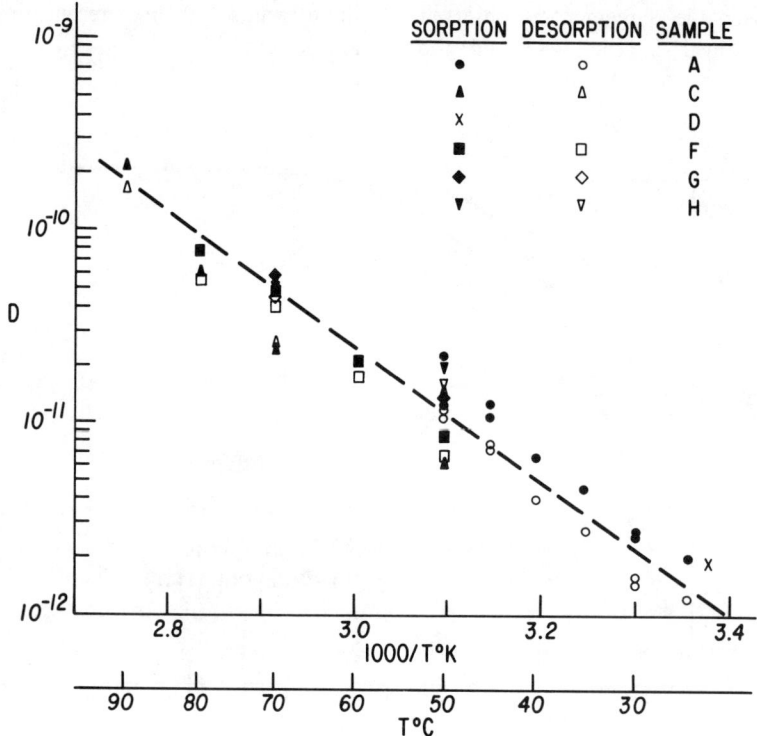

FIGURE 5. Arrhenius plot of the diffusion coefficient for VCM in several PVC samples (see Table 1) between 20°C and 95°C.

Later Stages of Sorption/Desorption

Plots of M_t/M_∞ vs. $t^{1/2}$ tend to focus attention on the first 90% of the sorption or desorption process, but the removal of unreacted VCM after polymerization concerns also the final few percent of sorbed monomer. For this region, it is instructive to examine plots of $\log(1 - M_t/M_\infty)$ vs. t. Figure 3 exhibits plots calculated from Equation 1 using $D = 2 \times 10^{-10}$ cm^2/sec and various values of $S_g = 4.29/\bar{d}_s$. These curves, therefore, represent the predictions of the uniform-sphere, Fickian model for sorption or desorption of VCM from PVC powders at 90°C; they are essentially linear beyond $M_t/M_\infty = 0.5$.

Experimental $\log(1 - M_t/M_\infty)$ vs. t plots for several representative suspension PVC samples are shown in Figure 6. Deviations from the uniform-sphere model are obvious and quite substantial in some cases. The initial parts of the curves, to $M_t/M_\infty \cong 0.8$, show slopes similar to those predicted from the S_g values. Some of the curves show a pronounced decrease in slope, and others a more gradual curvature. The slope in the latter stages does not correlate with S_g.

These results may be attributed to nonuniformity of particle structure. Under the optical microscope, Samples K and L, for example, appear to consist of opaque, porous grains but also contain a number of translucent to clear, "glassy" particles. In such granules the pores between the original primary particles have probably been largely filled with PVC during polymerization; the effective dimension controlling sorption rate then more nearly approximates the external particle size of perhaps 50 to 150 μm than the few-micrometer primary particle size. The effect of these particles on the sorption/desorption curves can be estimated through Equation 1 simply by adding the curves calculated for various weight fractions of different effective particle size. This procedure is illustrated in Figure 7, where the effect of varied weight percent of 40-μm ("glassy") particles in a sample otherwise comprising uniform 4-μm particles is demonstrated. Sorption measurements on deliberately prepared mixtures of porous and glassy particles give results closely resembling the calculated curves, as illustrated in Figure 8. For the samples of Figure 6,

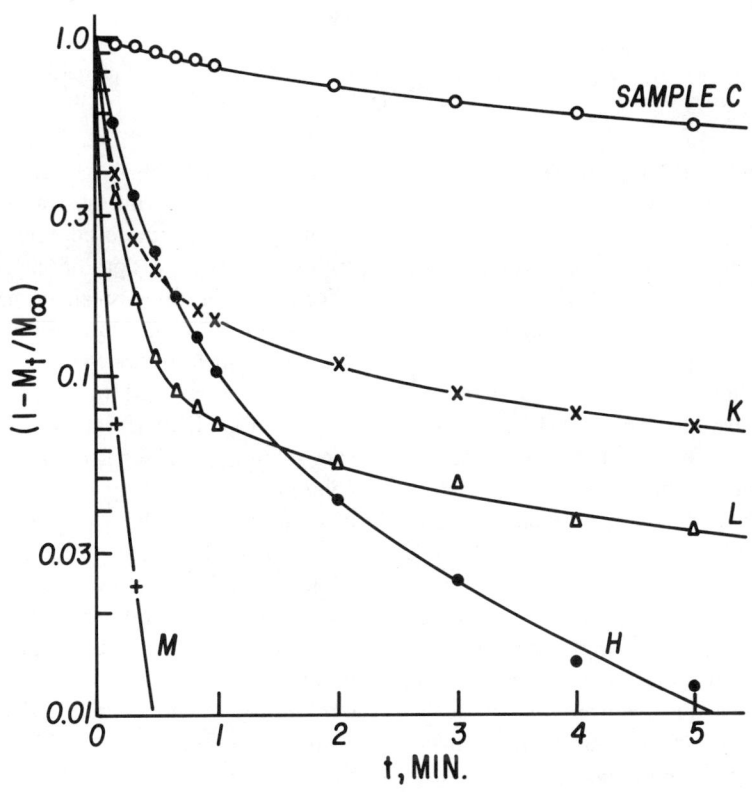

FIGURE 6. Experimental late stage desorption curves for VCM diffusion in various suspension polymerized PVC samples (see Table 1) at 90°C.

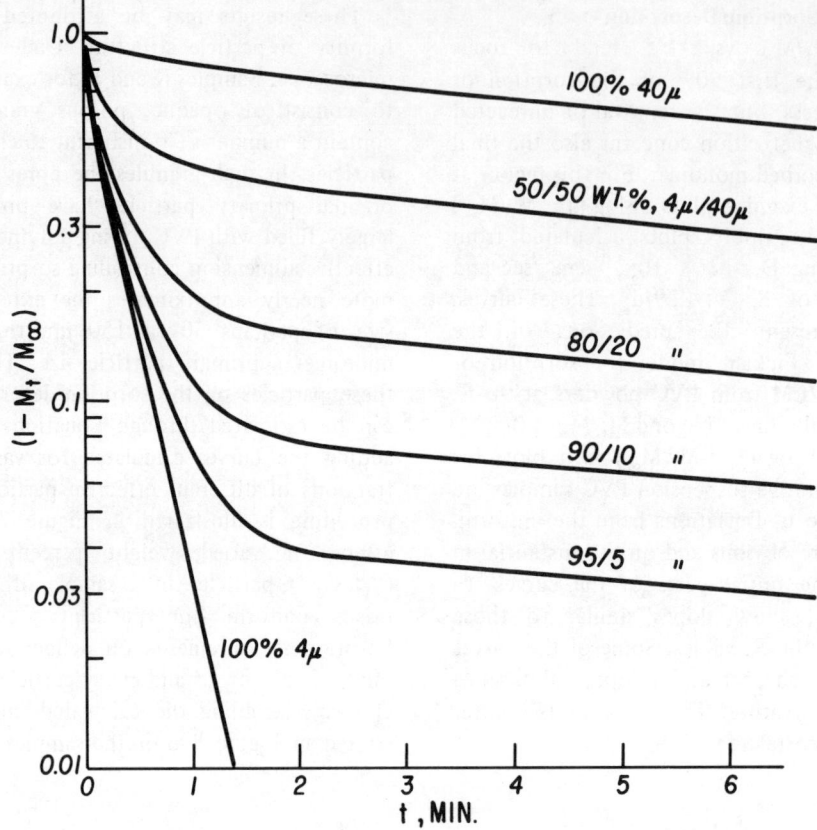

FIGURE 7. Theoretical late stage desorption for mixtures of PVC particles of different diameter using $D = 2 \times 10^{-10}$ cm^2/sec at 90°C.

the extents of sorption at longer times are in good qualitative accord with the relative contents of glassy particles.

The more gradually curved lines of Figure 6 may similarly be interpreted as the result of a narrower, more continuous distribution of effective particle sizes. Thus the shape of sorption curves can provide much useful information about the uniformity of particle structure. In principle, once D is known, it seems possible to obtain a complete measure of particle size and size distribution from sorption measurements on polymer powders.

These results suggest that removal of residual VCM from PVC resins could be accomplished in feasibly short times and at low temperatures, provided the polymer has a relatively small primary particle size and is free from nonporous or glassy particles. The presence of even a small percentage of such nonporous particles will seriously retard the final stages of VCM removal. Similarly, VCM removal will become very slow once PVC particles have become fused into a continuous mass during fabrication.

In practice, improved schemes for VCM removal from emulsion or suspension PVC are effected by ensuring small primary particle size. Moreover, countercurrent steam-stripping columns, which ensure that the PVC particles are continuously exposed to a maximum VCM concentration gradient (from particle to surrounding vapor) are used. The performance of the stripping columns relates directly to rates determined in the laboratory microbalance desorption experiments. The Fickian diffusion model is applicable to the VCM removal process without significant complication by relaxation-controlled diffusion. Low VCM concentration in the vapor and stripping at temperatures above T_g minimize any relaxation-controlled contribution. Consequently, the stripping process can be modeled very well by Equation 3 since virtually all the assumptions embodied in the simple Fickian uniform-sphere model are valid for the VCM-PVC system.

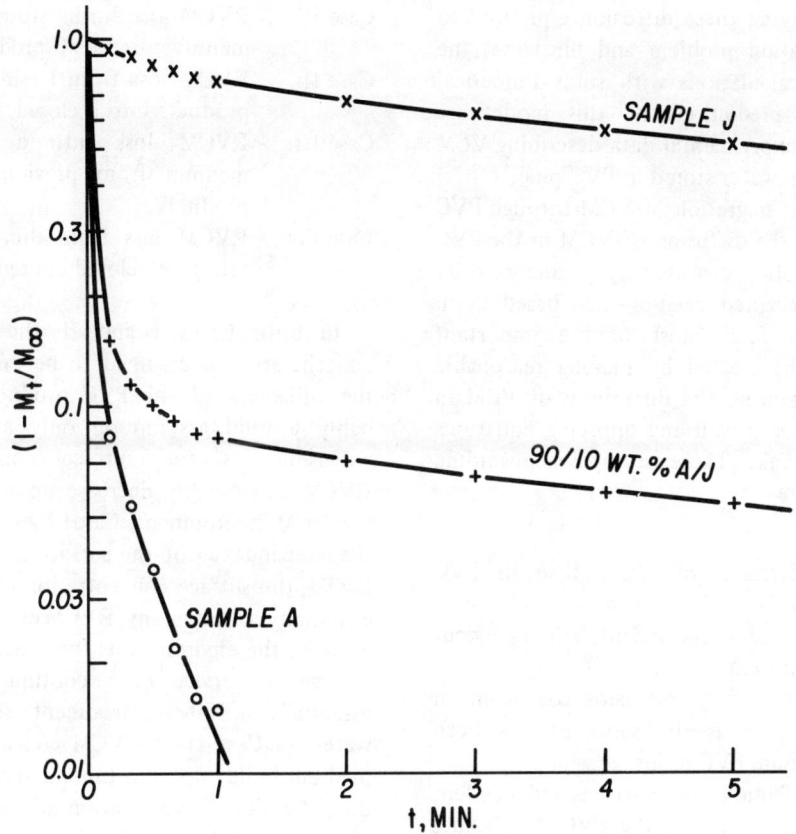

FIGURE 8. Experimental late stage desorption curves for VCM diffusion in two different diameter PVC samples and a mixture of the two samples (see Table 1) at 90°C.

Deviations accruing from nonuniform particle size distributions can be accommodated by straightforward geometric corrections.

Assuming that the diffusion coefficients obtained with emulsion or suspension polymers effectively characterize VCM transport in molded or extruded rigid PVC products, escape of residual VCM from such products will be an extremely slow process. Numerical calculations for products of varied geometry and dimensions can readily be made with available equations. These considerations are explored in depth in the following sections for the important applications of PVC bottles and water pipes, respectively.

MIGRATION OF VCM RESIDUALS FROM PVC PIPES AND BOTTLES INTO NONINTERACTING FLUID CONTENTS

Concern regarding the migration of residual VCM from finished PVC products into the environment or into liquids transported in PVC vessels has been prompted by the discovery of the potential toxic hazards attributed to VCM. Direct analysis for VCM in water, which has been stored in PVC pipe containing trace residuals of VCM, poses difficult analytical problems, since pipe containing even as much as 20 ppm of VCM will generate VCM concentrations in water only in the parts per billion range.

A reliable model for predicting the amount and rate of VCM migration from available basic transport data and theory would, therefore, be very useful to circumvent the difficult chemical analysis required by determining ppb quantities. A suitable predictive model has been developed by combining the solubility and diffusion data obtained for VCM in uncompounded PVC resin powders[13,14] with the solutions to the Fickian diffusion equations given by Crank.[7] The following treatment summarizes the assumptions and approximations involved in applying these diffusion equations to the VCM migration problem and illustrates the utility of the calculations with some numerical examples. The predictions of this model are compared with experimental data describing VCM concentrations in water stored in PVC pipe.[15]

Assuming that migration of VCM through PVC is controlled by the diffusion of VCM in the PVC phase, the extraction process may be analyzed by applying well-accepted relationships based upon Fick's laws. In fact, most of the important problems may be treated by making reasonable assumptions regarding the diffusivity of VCM in PVC pipe or bottle wall and applying equations developed by Crank.[7] These assumptions regarding the diffusivity are

1. The diffusion of VCM through PVC obeys Fick's law.
2. The diffusion coefficient is independent of VCM concentration.
3. The value of the diffusion coefficient in rigid PVC products is the same as has been determined for pure PVC resins.
4. The diffusion coefficient is independent of the medium surrounding the PVC, i.e., values determined by vapor sorption/desorption also apply to migration into a liquid phase.
5. Transport resistances are confined entirely to the polymer phase.

Assumptions 1 and 2 have been demonstrated to be very satisfactory approximations at quite low residual vinyl chloride monomer (RVCM) concentrations by the work on PVC powder.[14] More limited experiments on thin, rigid PVC films also support assumption 3.[14] Assumption 4 seems justified by experimental data describing VCM migration into water from PVC pipe. It is quite conceivable, however, that the presence of organic components in the fluid contents could result in swelling of the polymer and in altered transport kinetics. The model developed here is therefore limited to noninteracting fluid contents. Assumption 5 is virtually assured for all reasonable geometric ratios and fluid contents because of the vast difference in diffusivities in the liquid and polymer phases.

To describe the diffusion of RVCM from PVC products through Crank's equations, four situations have been considered:

Case I: RVCM loss during storage of a freshly manufactured PVC product.
Case II: RVCM loss from freshly formed PVC products into a closed, finite medium.
Case III: RVCM loss into a closed, finite medium from previously aged PVC products.
Case IV: RVCM loss from thin-walled bottles into finite closed contents.

In both Cases I and II, the initial RVCM concentration is assumed to be uniform through the thickness of the PVC product; this is probably a valid assumption only at the time of extrusion, as the surface concentration of RVCM will quickly decrease upon exposure to a low-VCM environment. Cases I and II differ in the time-dependence of the surface concentration: In Case I, the surface concentration of VCM remains essentially zero, as any RVCM escaping is carried away in the environment; this case may represent storage or service in a continuously renewed, essentially infinite environment, such as flowing water. In Case II, the VCM concentration in the medium builds up with time, and consequently so does the VCM concentration at the surface of the PVC. Case III is the general situation in a realistic application: VCM loss into a closed medium follows a variable storage period and thus proceeds from a product in which the surface VCM concentration is depleted. The aging period between manufacture and closed-system service is quite important in determining the rate of VCM migration into the contents of a PVC pipe. Case IV describes the long-time situation for thin-walled bottles wherein migration of the VCM from the bottle contents into the surrounding environment depletes the VCM concentration previously developed in the bottle contents.

It is worthwhile to consider the details and quantitative implications of each of these four cases.

Case I — VCM Loss During Storage

Consider a freshly extruded PVC product, quickly cooled to ambient temperature and stored in an atmosphere of essentially zero VCM content. The initial RVCM concentration in the product is C_0 and may be assumed to be uniform through the product. At the surface, equilibrium is quickly established with the environment and the RVCM concentration is zero. It is assumed that VCM leaving the PVC is carried away (e.g., good air circulation) so that the surface concentration remains zero. It is useful to calculate the amount of VCM which leaves the PVC and the concentration profile which is developed within the PVC, both as functions of time, temperature, and sample thickness.

Crank[7] gives solutions to this problem for several simple geometries — plane sheets, solid and hollow cylinders, and solid spheres. The predictions for hollow cylinders are virtually identical to those for plane sheets, provided the wall thickness is less than the inside diameter. Thus for all practical PVC products (pipes, bottles, sheets, films), consideration is restricted to the mathematical solutions for plane sheets. Equations for the amount of VCM escaping from the sheet may be written in terms of M, the fraction of the original VCM which escapes in time t. The general expression, valid at all times, is

$$M = 1 - \sum_{n=0}^{\infty} \frac{8}{(2n+1)^2 \pi^2} e^{-D(2n+1)^2 \pi^2 t/L^2} \quad (7)$$

where n is the series of integers (0,1,2 . . .), D the diffusion coefficient, and L the sheet thickness. For the late stages of the process (M > ~0.6), terms beyond n = 0 become insignificant, and Equation 7 becomes

$$M = 1 - \frac{8}{\pi^2} e^{-D\pi^2 t/L^2} \quad (8)$$

For M < ~0.6, a very good approximation is given by

$$M = 4 \left(\frac{D}{\pi L^2} \right)^{1/2} t^{1/2} \quad (9)$$

Thus the initial loss of VCM is proportional to the square root of the storage time after extrusion. For numerical calculations, only the sheet thickness and the diffusion coefficient values are required. From the measurements on PVC resins, the value of D at various temperatures between 20 and 90°C is given by:[14]

$$D = 3.7 \exp(-17{,}000/RT) \quad (10)$$

for D in cm^2/sec, T in °K, and R = 1.987 cal/mol °K.

Equations 8, 9, and 10 permit prediction of the fractional loss of RVCM from rigid PVC products for different sheet thicknesses, times, and storage temperatures. Examples of numerical results are given in Figure 9 as plots of M vs. $t^{1/2}$ for several sheet thicknesses at 30°C, where $D = 2 \times 10^{-12}$ cm^2/sec.

The concentration of RVCM remaining at time t at various distances from the sheet surface (i.e., the concentration profiles) can also be calculated from the same parameters. Crank gives the general solution as

$$\frac{C - C_0}{C_1 - C_0} = \sum_{n=0}^{\infty} (-1)^n \operatorname{erfc} \left[\frac{(2n+1)\ell - x}{2(Dt)^{1/2}} \right]$$
$$+ \sum_{n=0}^{\infty} (-1)^n \operatorname{erfc} \left[\frac{(2n+1)\ell + x}{2(Dt)^{1/2}} \right] \quad (11)$$

where C is the concentration at time t at distance x from the center of the sheet, C_0 is the initial (uniform) concentration, C_1 is the constant concentration at the surface (zero for this case), and ℓ is the half-thickness of the sheet; "erfc" stands for error function complement, defined as

$$\operatorname{erfc} z = 1 - \operatorname{erf} z \quad (12)$$

For times short enough that the concentration at the center of the sheet does not decrease significantly below C_0 only the first term of Equation 11 is necessary; then, using Equation 12, we have simply

$$\frac{C}{C_0} = \operatorname{erf} \left[\frac{\ell - x}{2(Dt)^{1/2}} \right] \quad (13)$$

Using Equation 13 or 11 as appropriate, concentration profiles at 30°C ($D = 2 \times 10^{-12}$ cm^2/sec) were calculated for a PVC sheet 1/8-in. thick (e.g., a pipe or bottle with a 1/8-in. wall). The results at various times are plotted in Figure 10. The VCM lost in the first month comes only from the 100 μm of PVC near the surface. It takes over 20 years in this case for the VCM concentration near the center of the sheet to decrease appreciably.

Case II — VCM Migration from Freshly Formed PVC Products into Finite Fluid Contents

The Case I calculations may be applied whenever VCM leaving the PVC product is carried away by the environment (storage in circulating air,

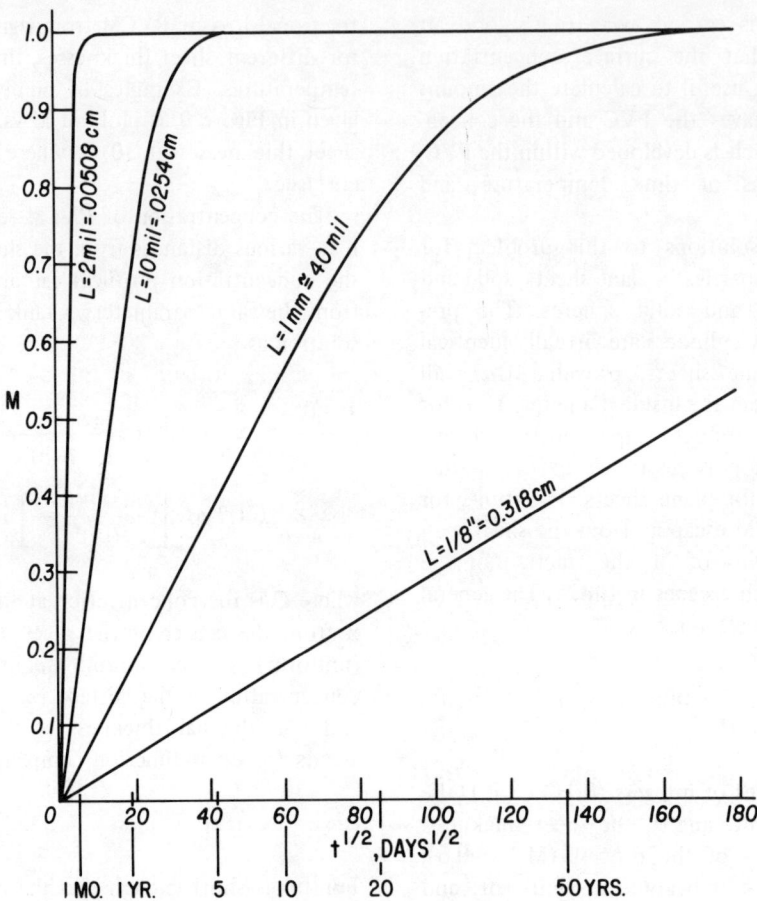

FIGURE 9. Theoretical curves of fractional VCM loss (M) versus the square root of time for PVC sheets of various thicknesses (L) for $D = 2 \times 10^{-12}$ cm^2/sec at 30°C (Case I).

water-pipe service with flowing water, etc.), so that the surface concentration of RVCM at both boundaries remains essentially zero. For PVC pipes or liquid food containers in ordinary service, the outer surface is generally exposed to a low-VCM environment, i.e., a Case I situation. Conversely, on the inside of pipes or bottles with finite and therefore, limiting contents, the VCM concentration in the contents increases appreciably. Consequently, the inside surface concentration of RVCM in the PVC, assumed to remain in equilibrium with the contents, also increases with time. At very long times, the RVCM in the inner half of the PVC wall and in the contents will diffuse outward to the environment. Diffusion of VCM from the bottle contents into the external environment will occur after the VCM concentration at the inner wall exceeds the concentration at all other positions in the PVC. Since this effect occurs at extremely long times for pipes of normal wall thickness, it has not been considered in the Case II model, but instead Case I and Case II calculations have been applied independently to the outer and inner halves of the PVC walls, respectively. This more complicated effect, involving long-time desorption of VCM from the PVC and contents into the environment, has been treated, however, by Daniels and Proctor for the specific case of VCM migration from relatively thin-walled bottles.[16] Their results will be summarized as Case IV.

The rate and amount of VCM leaving the pipe and entering the liquid contents may be calculated through equations given by Crank.[7] The maximum VCM concentration established in the contents is determined by the equilibrium between the VCM in the PVC and the contents. Specifically, the partition coefficient and the ratio of volumes of PVC and contents determine the maximum VCM concentration in the contents consistent with the

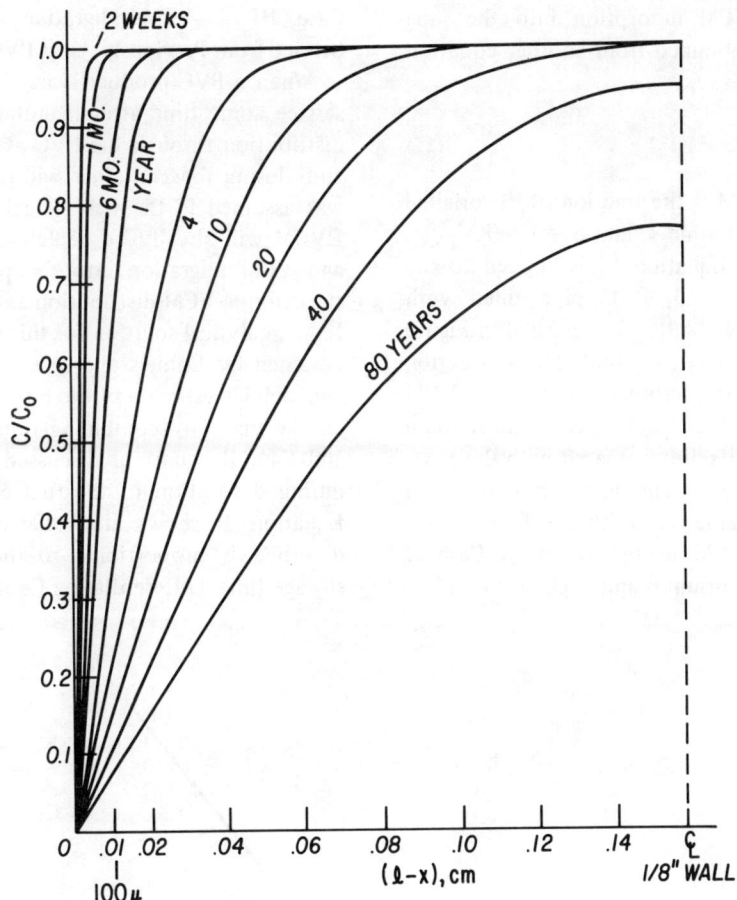

FIGURE 10. Relative VCM concentration (C/C_0) vs. depth below the sheet surface (ℓ-x) at various times, calculated for 1/8 in. thick PVC sheet for $D = 2 \times 10^{-12}$ cm²/sec at 30°C (Case I).

constraining assumptions assigned to this case. The partition coefficient, K, is defined as the ratio of RVCM concentration in the PVC to that in the contents at equilibrium (both expressed in the same units, e.g., g/liter). The volume of PVC supplying VCM to the container contents, V_{PVC}, is one half the total PVC volume, as RVCM in the outer half diffuses outward in the situation considered here. The maximum VCM concentration, $C_{m,max}$, (ppm by weight) in the liquid contents of a PVC pipe or thick-walled bottle originally containing C_0 ppm VCM, is

$$C_{m,max} = \frac{C_0 d_{PVC}}{\left(K + \frac{V_m}{V_{PVC}}\right) d_m} \qquad (14)$$

where d_{PVC} and d_m are densities of the PVC and contents, and V_m is the volume of the contents.

From the data[13] on VCM solubility as a function of VCM pressure over PVC and water, an estimated value of K = 49 for the distribution coefficient of VCM between PVC and water at 30°C appears reasonable. The estimate of K assumes that the solubility of VCM in PVC is not affected by contact with water. It also involves a somewhat arbitrary selection of a value for VCM solubility in PVC, since it has been shown that this system shows non-ideal and history-dependent solubility.[6] Using this value of K, Equation 14 predicts that the maximum VCM concentration in water in a 1-in. I.D., 1/8-in. wall, PVC pipe containing 1 ppm residual VCM will be 0.027 ppm.

The rate of VCM desorption into the pipe contents may be obtained from another equation presented in Crank's text:[7]

$$M = \alpha[1 - e^{T/\alpha^2} \text{erfc}(t/\alpha^2)^{1/2}] \quad (15)$$

where $T = Dt/l^2$, M is the fraction of the original RVCM desorbed at time t, and $\alpha = V_m/KV_{PVC}$. As an illustration, Equation 15 is applied to the 1-in. I.D., 1/8-in. wall, PVC pipe filled with standing water. The results are presented in Figure 11 with ordinates presenting both M, the fraction of original RVCM desorbed, and the VCM concentration in the water per original ppm RVCM. These results are compared with the M vs. $t^{1/2}$ plot for Case I. The initial rate of VCM desorption is the same for both Cases I and II, but the buildup of VCM in the water in Case II decelerates as equilibrium is approached.

Case III — VCM Migration in Closed-system Service from Previously Aged PVC Products

When a PVC product is put into closed-system service some time after manufacture, the RVCM distribution through the PVC at the time of filling and closing the container will not be uniform, as was assumed in the Case II calculations. Rather, RVCM will already be depleted near the surface, and VCM migration into the pipe contents will start from a VCM distribution as calculated in Case I. An analytical solution for this situation has been obtained by Daniels and Proctor,[16] but a useful and simpler estimate of the rate of VCM migration can be made by combining results of the Case I and Case II equations discussed here. For Case I, during desorption of the first 60% of the VCM, Equation 9 shows that the amount of VCM desorbed is proportional to the square root of storage time. Differentiating Equation 9 gives

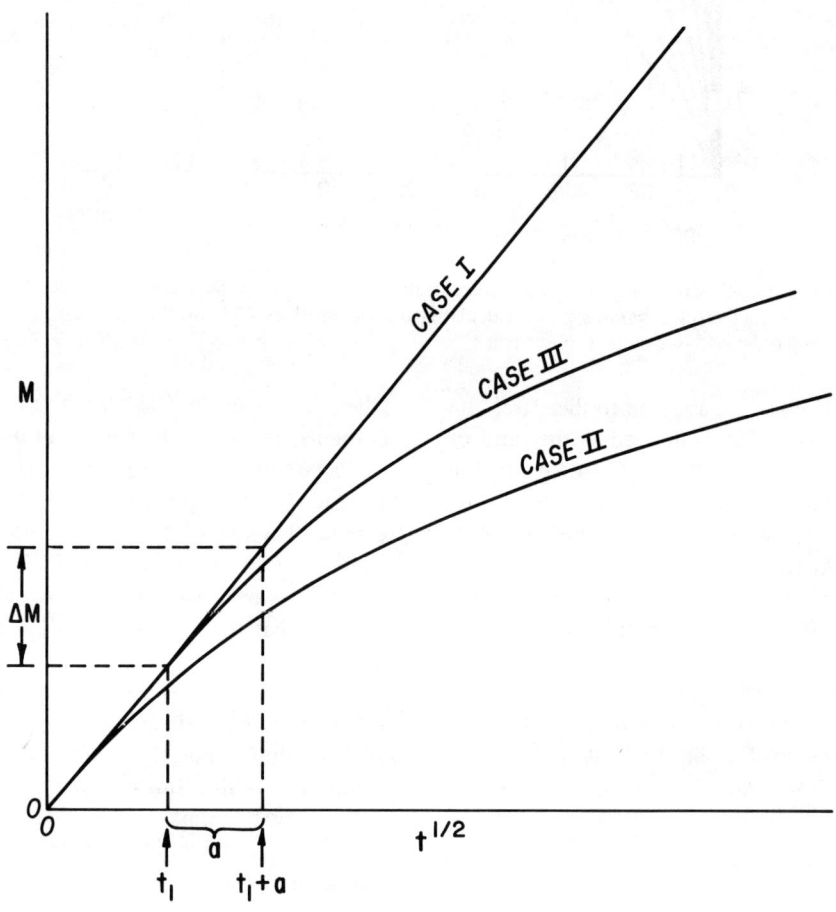

FIGURE 11. Comparison of fractional VCM loss (M) versus the square root of time curves for Cases I, II, and III.

$$\frac{dM}{dt} = 2\left(\frac{D}{\pi L^2}\right)^{1/2} / t^{1/2} \qquad (16)$$

Thus the *rate* of VCM loss is inversely proportional to the square root of storage time. For Case II, the *initial* rate of VCM migration into the medium in a closed system is the same as in Case I. Thus Equation 16 also gives the initial rate of VCM migration into the closed system when t is the storage age of the PVC product at the start of closed-system service. The rate of RVCM desorption drops very sharply in the first few weeks of storage after manufacture.

It is also possible to estimate the amount of VCM which will migrate from a PVC product during a given period of closed-system service following various "Case I" storage periods. This estimation may be explained with reference to Figure 11, which illustrates the VCM-loss (M) vs. $t^{1/2}$ curves for the three cases. Case III is approximated by shifting the origin of the Case II curve to point t_1 along the Case I line, where t_1 is the age of the PVC product at the start of closed-system service. Then, ΔM, the VCM lost from the PVC in the time interval a during continued open-system storage would be, from Equation 9

$$\Delta M = 4\left(\frac{D}{\pi L^2}\right)^{1/2} [(t_1 + a)^{1/2} - t_1^{1/2}] \qquad (17)$$

Equation 17 is useful for calculating the maximum fraction of the original RVCM which will migrate into the contents of a PVC pipe or bottle during closed-system service for any time period as a function of the age of the container at the time of filling and closing. The actual VCM concentrations at long times may, in fact, be somewhat lower due to effects described in detail in Case IV.

This analysis reinforces the intuitive notion that the age of a PVC product at the start of an extraction test is an important factor in determining the amount of VCM subsequently extracted. Since information concerning sample history is seldom available in reported extraction data, direct comparisons between these predictions and experimental data are possible for only a few cases.

For the data recently obtained by O'Mara and DeCapita[17] on the VCM content of water stored in 1-in. I.D., 1/8-in. wall PVC pipes at 23°C, the approximate age of the pipe samples, between extrusion and start of the extraction test, was known. A "headspace" GC analytical method was used to provide sensitivity to low VCM levels in the water (on the order of 1 to 2 ppb). Calculations of the expected VCM concentrations in the water were made using Equation 17 with diffusion coefficients from Equation 10. The calculated and experimental results are presented in Table 2. The agreement is quite satisfactory in view of the application of diffusivities obtained from resin powders to pipe compounds, the somewhat uncertain age and storage conditions of the pipe samples, and the difficulty of analysis of water for low ppb levels of VCM.

The simple approach discussed here appears quite adequate to describe and predict the migration of RVCM from PVC pipe into water. The reasonable agreement between predicted and observed migration data for this application of rigid PVC supports the premise that the diffusion coefficient determined for pure PVC resins is applicable to rigid PVC pipe compounds, and that contact of PVC with water produces little change in the diffusivity of VCM compared to the diffusivity measured by vapor sorption/desorption techniques. Moreover, for relatively thick-walled products, such as pipes, the diffusion into the environment from the outer half of the wall, and into the contents from the inner half, may be treated as independent processes over normal service lifetimes.

This simple predictive model may thus be used with some confidence for estimating the concentrations of VCM in water which might arise during actual service of PVC water pipe systems. To illustrate, Equation 17 has been used to calculate the VCM concentrations in water resulting from exposure to varying diameter PVC pipes containing 1 ppm of VCM consequent to storage periods between 30 and 90 days. Some results of these calculations are given in Table 3 for 2-, 6-, and 8-in. SDR21 pipes in service at room temperature (~23°C).

Even for the most extreme conditions (e.g., new installations of recently manufactured small diameter PVC pipe after long stagnation periods), the predicted VCM-in-water concentrations are well below the level of 2 ppb when the original residual VCM content of the pipe is 1 ppm or less. In actual installations, stagnation times of more than a few days are rarely encountered. A typical residence time for water in PVC pipes is believed

TABLE 2

Comparison of Predicted and Experimental Extraction Results, Water-filled 1-in. I.D., 1/8-in. Wall PVC Pipe Samples, 23°C.

VCM in pipe mg/kg	Pipe age t	Extraction time a, days	VCM in water, mg/kg	
			Expt.	Calc.
292	~6 months	3	0.021	0.0257
292	~6 months	7	0.0414	0.0598
292	~6 months	14	0.113	0.118
177	~6 months	3	0.0173	0.0156
177	~6 months	7	0.0335	0.0362
177	~6 months	14	0.056	0.0717
22	~6 months	3	0.0006	0.0019
22	~6 months	7	0.0022	0.0045
22	~6 months	14	0.0046	0.0089
29	~1 year	14	0.0105	0.0084

TABLE 3

Calculated Concentrations of VCM in H_2O for Storage of Water in PVC Pipes of Varied Size and Age

Initial Conditions: 1 mg/kg Residual VCM Uniformly Distributed through Pipe Wall

Pipe size (SDR21)	Warehouse age, days	Service age, years	mg/kg of VCM in H_2O after various storage times		
			2 days	2 weeks	1 month
2 in.	30	0 (new)	0.00007	0.00044	0.00087
	60	0	0.00005	0.00033	0.00067
	90	0	0.00004	0.00027	0.00056
	90	1	0.000018	0.000125	0.000265
	90	2	0.00001	0.00009	0.000199
	90	5	0.00001	0.00006	0.00013
6 in.	30	0	0.00002	0.00015	0.00029
	60	0	0.00002	0.00011	0.00022
	90	0	0.00001	0.00011	0.00022
	90	1	0.000006	0.000042	0.000088
	90	2	0.000004	0.00003	0.000066
	90	5	0.000003	0.00002	0.00004
8 in.	30	0	0.00002	0.00011	0.00022
	60	0	0.00001	0.00008	0.00016
	90	0	0.00001	0.00007	0.00014
	90	1	0.0000045	0.000031	0.000066
	90	2	0.000003	0.00002	0.00005
	90	5	0.000002	0.00001	0.00003

to be about 2 days[18,19] and in this situation, the predicted VCM-in-water concentration falls in the parts per trillion range. Therefore, PVC pipe containing ≤1 ppm residual VCM will result in nondetectable VCM concentrations in water under any expected service conditions.

Case IV — VCM Migration from Thin-walled Bottles into Finite Closed Contents

For the important case of relatively thin-walled vessels such as blow-molded bottles, the assumptions regarding independent diffusional processes at each of the bottle-wall boundaries introduces a significant error into the predictive calculations. Daniels and Proctor[16] have treated the case involving diffusion of VCM into the environment and contents over a time scale sufficient to invalidate this assumption and to require treatment of the net back diffusion of the VCM from the contents and bottle wall into the environment. The assumptions embodied in Cases I to III regarding rate-limiting transport processes occuring in the bottle wall, perfect mixing in the environment and bottle contents, and diffusion controlled transport in the PVC are all adopted in this more comprehensive case.

Typical time response curves describing extraction of VCM from PVC bottles resulting from this model are presented in Figure 12. Most importantly, the curves all reach maximum values of concentration over time periods which may encompass a reasonable shelf life of the stored contents. Clearly, at infinite time the concentration of VCM in the bottle wall and contents will fall to zero.

In this regard, Daniels and Proctor[16] have treated cases involving prior storage as well as bottles filled immediately following manufacture. In all cases, their models satisfactorily predict in-use extraction behavior of VCM from PVC bottles.

CONCLUSIONS

The detailed mathematical and experimental analyses regarding removal of VCM from PVC resins and products should be directly applicable to the characterization of removal schemes for virtually any penetrant, in low concentration, dissolved in a glassy or rubbery polymer. The complications associated with glassy state anomalies and reordering and rearrangement of crystalline material, although important to the comprehensive study of, say, VCM in PVC,[6] are not germane to the removal of trace residuals from polymeric resins or products. Clearly, at higher concentrations (e.g., > 1%) drifts in apparent sorption equilibria and relaxation phenomena, discernable by dimensional or mechanical relaxations, would complicate the simplifying assumptions embodied in the analyses described here which successfully describe removal of residuals from polymers.[6]

Most applications requiring accurate prediction of the kinetics of removal of monomers, solvents, additives, or low molecular weight degradation

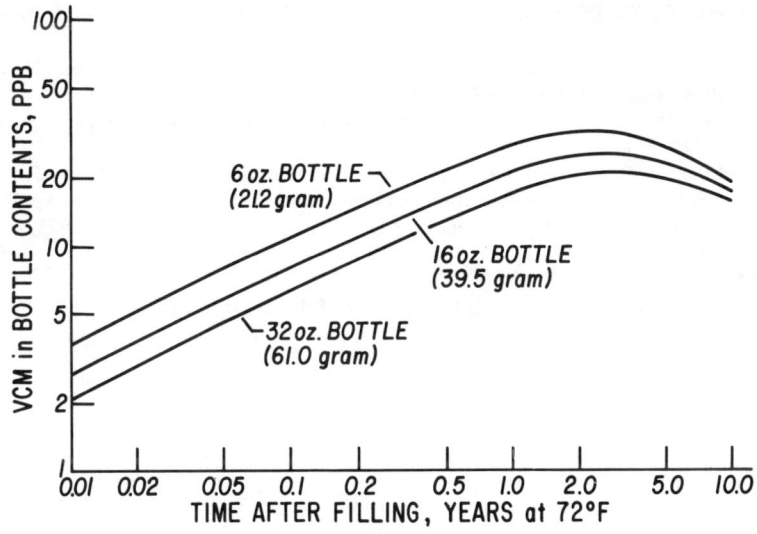

FIGURE 12. VCM concentration in water contents of various size PVC bottles initially containing 1 ppm of VCM, uniformly distributed through the bottle wall.

products from polymers involve initial concentrations of significantly less than 1%. Moreover, the manifold complications associated with transport at high concentrations in the complex system VCM-PVC did not, in fact, compromise the simplifying assumptions used so successfully in the case studies described here involving removal of low concentrations of VCM from PVC. Presumably, the mathematics, assumptions, criteria, and results described for this system are, therefore, directly translatable to a wide variety of applied systems involving removal schemes for low molecular weight residuals from polymer resins and products.

ACKNOWLEDGMENTS

The assistance of the National Science Foundation, Grant No. ENG75-22437, supporting Professor Hopfenberg's research related to removal of monomers and solvents from glassy polymers, is gratefully acknowledged. Appreciation is expressed to the B.F. Goodrich Company for permission to publish this work. The authors are also grateful to Mr. David J. Enscore, who helped proofread this manuscript, and to Mr. S. A. Oxenrider and Miss Ronna Gander who carried out much of the experimental work.

REFERENCES

1. Johnson, C. R. and Otto, W. M., Monomer recovery in GR-S manufacture, *Chem. Eng. Prog.*, 45, 407, 1949.
2. Grim, J. M., Extraction Process for Polymer Particles, U.S. Patent 2,691,008, 1954; British Patent 686,579.
3. Mantell, G. J., Barr, J. T., Chan, R. K. S., Stripping VCM from PVC resins, *Chem. Eng. Prog.*, 71, 54, 1975.
4. Mack, W. A., VCM reduction and control, *Chem. Eng. Prog.*, 71, 41, 1975.
5. Maltoni, C. and Lefemine, G., Carcenogenicity bio-assays of vinyl chloride, *Environ. Res.*, 7, 387, 1974.
6. Berens, A. R., Effects of Time, Sample History, and Temperature on the Sorption of Monomer Vapor by PVC, paper presented at 2nd Int. Symp. on Poly (vinylchloride), Lyon, France, July 1976.
7. Crank, J., *The Mathematics of Diffusion*, Oxford University Press, London, 1956.
8. Bagley, E. and Long, F. A., Two-stage sorption and desorption of organic vapors in cellulose acetate, *J. Am. Chem. Soc.*, 77, 2172, 1955.
9. Berens, A. R. and Folt, V. L., The significance of a particle-flow process in PVC melts, *Polym. Eng. Sci.*, 8, 5, 1968.
10. Glass, J. E. and Fields, J. W., Performance relationships in PVC-plasticizer by blending, *J. Appl. Polym. Sci.*, 16, 2269, 1972.
11. Tregan, R. and Bonnemayre, A., Microstructure of polyvinyl chloride by bulk polymerization, *Plast. Mod. Elastomeres*, 23(7), 220, 1971.
12. Wolf, F. and Kreter, E., Zur Diffusion von monomeren Vinylchlorid in Polyvinylchlorid, *Plaste Kautsch.*, 21, 27, 1974.
13. Berens, A. R., The solubility of vinyl chloride in poly(vinylchloride), *Polym. Prepr. Am. Chem. Soc. Div. Polym. Chem.*, 15, 197, 1974; *Angew. Makromol. Chem.*, 47, 97, 1975.
14. Berens, A. R., The diffusion of vinyl chloride in poly(vinylchloride), *Polym. Prepr. Am. Chem. Soc. Div. Polym. Chem.*, 15, 203, 1974.
15. Berens, A. R. and Daniels, C. A., Prediction of VCM migration from rigid PVC pipe, *Polym. Eng. Sci.*, 16, 552, 1976.
16. Daniels, G. A. and Proctor, D. E., VCM extraction from PVC bottles, *Mod. Packag.*, p. 45, April 1975.
17. O'Mara, M. M. and DeCapita, E. G., Internal report, B.F. Goodrich Chemical Company, 1975.
18. Ginn, H. W., Design Parameters for Rural Water Distribution Systems, *J. Am. Water Works Assoc.*, p. 1595, December 1966.
19. A Study of Residential Water Use, U.S.D.H.U.D., U.S. Govt. Printing Office, February 1967.

TREATMENT OF FERMENTATION WASTE WATERS, ITS PROBLEMS AND PRACTICES

H. Samejima, T. Koike, S. Noguchi, and M. Nagashima

TABLE OF CONTENTS

Introduction 314

Characteristics of Fermentation Waste Waters 314

Technology for Treatment of Fermentation Waste Waters 315
   General 315
   Preceding Measures for Treatment 315
      Changing Manufacturing Processes 316
      Classification of Waste Waters 316
      Survey of the Seasonal and Daily Changes of Waste Flows and Equalization of Wastes 316
      Consideration of the "Closed" System 316
   Biological Treatment 316
      Activated Sludge Treatment 316
      Anaerobic Digestion (Methane Fermentation) 318
      Miscellaneous Treatments 319
   Chemical and Physical Treatments 319
      Chemical Coagulation and Sedimentation 322
      Activated Carbon Adsorption 324
      Chemical Oxidation 324
      Ammonia Stripping 325
      Miscellaneous Treatments 325

Technology for Recovery and Utilization of Useful Waste Components 326
   General 326
   Utilization as Feed Stuffs 326
      Utilization of Alcohol Distillation Wastes as Feed Stuffs 326
      Utilization of Brewery Wastes as Feed Stuffs 326
      Utilization of Microbial Cells Grown on Waste Waters 326
   Utilization as Fertilizers 327
      Production of Organic Fertilizers from Fermentation Wastes 327
      Other Utilization of Fermentation Wastes as Fertilizers 329

Practice of Fermentation Waste Treatment and Its Problems 330
   Behavior of BOD and COD in the Biological Treatment 330
   Rate of Oxygen Consumption During the Biological Treatment and Treatment of High Loads of Waste . . 332
   Characteristics of Microflora in the Biological Treatments 332
   Residual COD Components and Their Removal by Microorganisms 334
   Combination of Biological Treatment with Physicochemical Treatments 336
   Reverse Osmosis as a Promising Tool for Fermentation Waste Treatment 337

Conclusions 339

References 340

INTRODUCTION

Fermentation industries have progressed rapidly in Japan since the last World War. Especially, the production of antibiotics such as penicillin and streptomycin, which were introduced from the United States, and the subsequent domestic inventions of amino acid and nucleotide fermentations, in which genetically designed, artificial mutants of microorganisms were utilized, have very much accelerated the growth of the Japanese fermentation industries.

At present, the fermentation industries have a very important role in the Japanese chemical industry, and supply a variety of products such as industrial alcohol, alcoholic beverages, fermented foods, baker's yeast, amino acids, nucleotides, organic acids, saccharides, enzymes, coenzymes, antibiotics and other pharmaceuticals, agricultural chemicals, feed additives, and so forth.

On the other hand, the pollution problems caused by industrial wastes have become more and more serious in Japan owing to the rapid expansion of various industries. The fermentation industries, of course, could not avoid these problems. Therefore, the social opposition and governmental regulations against such pollution are becoming more severe each year. Accordingly, solutions to such pollution problems are now the most urgent target of each industry in Japan.

Fermentation industries usually use a vast quantity of water. For example, in the alcohol fermentation industry, about 60 to 120 times the volume of water is used in the process per unit volume of the product alcohol. Even the volume of the residual liquor from distillation columns is about 8 to 9 times larger than the volume of the product alcohol. However, in fermentation industries, the main raw materials are usually of natural origin such as cane molasses, starch, sugars, etc., and generally do not contain materials hazardous to living organisms. Due to this, biological procedures such as the activated sludge process and methane fermentation have long been employed for the treatment of fermentation waste waters. However, secondary and tertiary treatments for such fermentation waste waters have not been reported until recently.

Presently, highly purified effluents must be discharged from waste treatment plants. The legislative limits for such waste effluent parameters as BOD (biochemical oxygen demand), COD (chemical oxygen demand), SS (suspended solids), and so on, are being lowered every year. Moreover, other parameters, such as colored contaminants and some inorganic substances, are also being considered as the subjects for legal regulation. In recent years, there has also been a governmental movement to regulate the total mass of waste discharged in place of the strength of waste effluent.

Under such circumstances, much more refined and novel treatment processes, including secondary and tertiary processes, are needed for each industry. Of course, such refined treatment processes often are an economic burden to industry. Therefore, in order to overcome such economic pressures, more efficient treatment processes, which include reuse of waste materials and treated water, must be considered in the future.

In this chapter, the authors wish to present certain problems and practices in the treatments of fermentation waste waters, mostly in Japan.

CHARACTERISTICS OF FERMENTATION WASTE WATERS

The characteristics of fermentation waste waters vary widely with the difference of products, raw materials, and manufacturing processes. In recent years, various new materials such as n-paraffin, acetic acid, ethanol, methanol, and gaseous hydrocarbons have been investigated as possible fermentation raw materials. Nevertheless, natural carbohydrate materials such as cane molasses, starch, and raw sugars are the principal raw materials in many fermentation industries. Thus, the general charcteristics of fermentation waste waters derived from such carbohydrate materials can be described as follows:

1. Concentrations of organic materials in the waste waters are usually high, and consequently BOD, COD, and SS are also high.
2. In many cases, colored impurities derived mainly from raw materials are included.
3. Certain inorganic substances such as phosphate, ammonia nitrogen, and other minerals are contained as the nutrients for microorganisms.
4. The pH values of waste waters are neutral or weakly acidic.
5. Hazardous substances like heavy metals, cyano compounds, and so forth are not contained.

Numerical characteristics of various fermentation waste waters are shown in Tables 1 to 3.

TABLE 1

Characteristics of Distillation Residual Wastes in Alcohol Fermentation

Parameters		Waste from sweet potato	Waste from cane molasses
Specific gravity		1.01–1.02	1.01–1.04
pH		4.0–5.0	4.5–5.5
Total solids	(% w/v)	1.0–3.0	5.0–9.0
Total organic matter	(% w/v)	1.0–3.0	3.0–7.0
Total sugars	(% w/v)	0.2–1.0	1.0–1.5
Total nitrogen	(% w/v)	0.01–0.10	0.05–0.15
BOD	(ppm)	8,000–20,000	20,000–30,000
COD	(ppm)	8,000–25,000	20,000–40,000
Suspended solids	(ppm)	10,000–15,000	2,000–6,000

From Ono, H., *Kogai Boshi Sangyo* (translated from Japanese), 2(7), 1, 1972. With permission.

TABLE 2

Characteristics of Waste Waters from Antibiotic Fermentations

Parameter		Antibiotic		
		Streptomycin	Kanamycin	Penicillin
pH		6.1	5.0	7.6
Total solids	(ppm)	9,220	27,760	19,560
Suspended solids	(ppm)	2,964	60	380
BOD	(ppm)	12,600	22,200	10,000
COD	(ppm)	3,260	15,600	5,330
Total nitrogen	(ppm)	616	790	578
Ash	(ppm)	3,168	12,092	10,324

TABLE 3

Characteristics of Waste Waters from Beer and Yeast Manufacturing

Parameters		Beer	Yeast
pH		3.7	5.4
Total solids	(ppm)	1,416	35,150
Ash	(ppm)	248	11,200
Total organic matter	(ppm)	1,162	23,950
Suspended solids	(ppm)	396	2,260
COD	(ppm)	550	21,120
BOD	(ppm)	620	16,300
Total nitrogen	(ppm)	24	594

TECHNOLOGY FOR TREATMENT OF FERMENTATION WASTE WATERS

General

In the practice of waste water treatment, it is important to know well the types, quantities, and characteristics of the waste waters which must be treated. Also, in order to design the whole waste treatment process, it is necessary to consider what types of impurities and what fraction of the impurities should be eliminated from the waste waters to conform to governmental regulations. Selection and combination of treatment procedures must also be evaluated on an economic basis.

In the treatment of fermentation waste waters, a microbial treatment process is usually selected as the primary treatment tool because of the nonhazardous nature of such waste waters and the low cost of such processes. Certain chemical and physical processes are also selected as secondary and tertiary treatment processes. Of course, such a combination of processes is not always fixed, but can be altered case by case. Sometimes, special treatment processes such as ammonia stripping and so on are added in the series.

Preceding Measures for Treatment

Before actual design of the whole treatment

process, it is important to take certain measures in order to minimize the total treatment costs. The following items are important.

Changing Manufacturing Processes

The raw materials and unit operations in the manufacturing processes must be reinvestigated from the standpoint of the ease of the subsequent waste treatment. Reduction of waste volume and strength must be always considered. For example, total or partial changes of raw materials are often very effective in decreasing waste strength, and alteration of extraction and recovery processes can often reduce the waste volume to a great extent.

Classification of Waste Waters

By classifying waste waters, the volume of waste waters which require intensive treatment can be reduced considerably, and each classified waste water can be treated in a more rational way. Waste waters may be classified by the similarity of waste characteristics or by the strength of the waste waters. In the fermentation industries, most waste waters are similar in their characteristics, as mentioned before. Therefore, it is more common to classify waste waters by their strength. For example, in a fermentation plant of the Kyowa Hakko Kogyo Co., Ltd., in Japan, fermentation waste waters were classified as follows: (a) high-strength waste water, (b) lower-strength waste water, and (c) discharged cooling water.

The high-strength waste water is mainly composed of residual fermentation broths from which fermentation products have been recovered, but still containing high concentrations of organic and inorganic solids. The lower-strength waste water consists mainly of washing waters and a part of the process waters, which contain lower concentrations of solids.

Such classification of waste waters is quite important from the standpoints of reuse of waste water for raw water supplies and utilization of useful waste components, and finally can reduce treatment costs to a great extent.

Survey of the Seasonal and Daily Changes of Waste Flows and Equalization of Wastes

In order to keep operations in the waste treatment plant stable and the quality of waste effluent at a certain level, intensive surveys of seasonal and daily changes of waste flows and planning for equalization of waste feed are important.

Consideration of the "Closed" System

It is most desirable that a "closed" system be established where production and utilization of wastes are in balance. In the case of fermentation industries, waste waters are principally nonhazardous to living organisms and moreover contain useful materials for feed stuffs and fertilizers (i.e., microbial cells, minerals of plant origin, and so forth). Therefore, considering the fermentation industries together with agricultural industry, an example of a "closed" system as shown in Figure 1 is proposed.[1]

In Figure 1, recovered and regenerated water can be reused as make-up water of fermentation media and other process water. Also, some parts of fermentation wastes such as microbial cells and high-strength waste waters can be utilized as feeds, fertilizers, and ingredients for fermentation media. Therefore, investigation of utilization technologies of waste components is another important subject in the treatment of fermentation waste waters.

Biological Treatment

Fermentation waste waters usually contain high concentrations of organic matter and, at the same time, certain amounts of nitrogen, phosphate, and minerals which are necessary for microbial nutrition. Also, pH values of such waste waters are usually neutral or weakly acidic. Therefore, many kinds of microorganisms can grow easily in such waste water without special pretreatment, and can reduce the organic matter very efficiently. This is the reason why the biological treatment is usually selected as a primary treatment process for fermentation waste waters.

There are two major types of biological treatment, the aerobic type process and the anaerobic type. Activated sludge treatment, trickling filtration, lagooning in oxidation ponds, and so forth have been well known as the representatives of aerobic type processes, and methane fermentation (anaerobic digestion) and anaerobic lagooning are popular as representatives of the anaerobic type processes.

General theroies and practices for biological treatment of industrial wastes are summarized in several monographs.[2-5]

Activated Sludge Treatment

Activated sludge treatment is the most popular method of aerobic biological waste treatment. The principle of the treatment process is schematically shown in Figure 2.

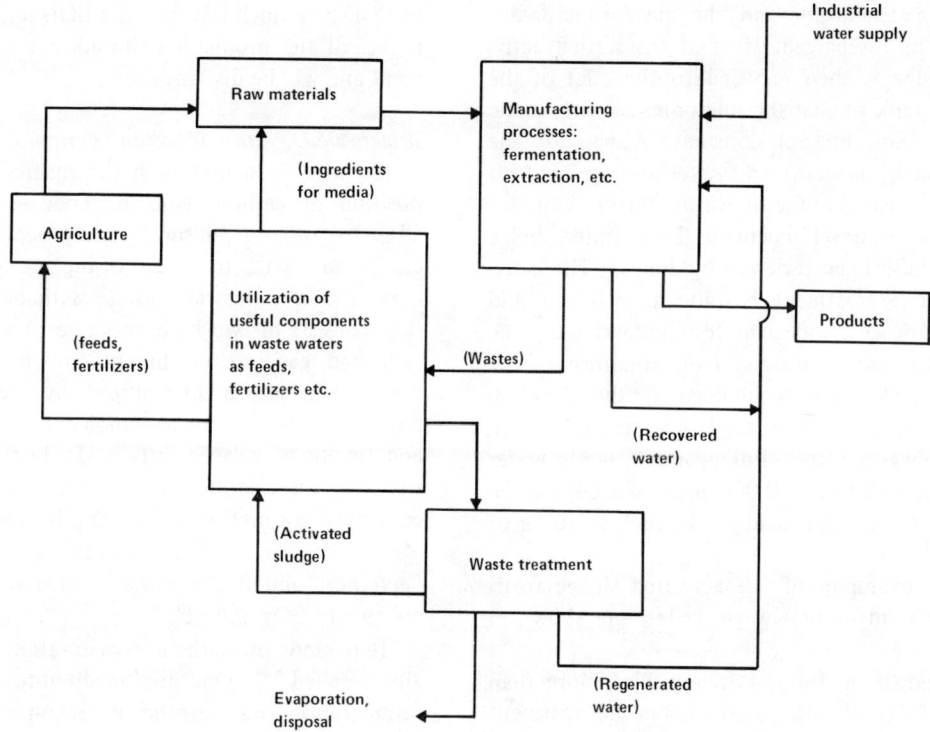

FIGURE 1. A "closed" system in the fermentation industries. (From Koike, T., *Ferment. Ind.* [translated from Japanese], 34, 163, 1976. With permission.)

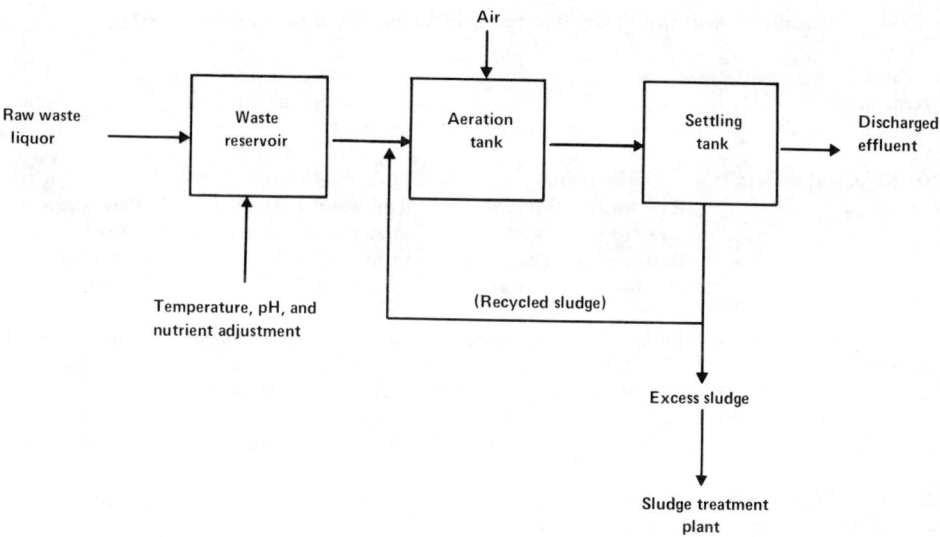

FIGURE 2. The principle of activated sludge treatment.

As shown in Figure 2, raw waste waters are introduced in the waste reservoir where the temperature, pH, and nutrients of the waste water are adjusted to a certain level. The resultant equalized waste water is pumped to the aeration tank, where the waste water is aerated vigorously together with a suspension of microorganisms (activated sludge) for a sufficient period of time for the microorganisms to oxidize the polluting constituents to the required extent. The treated waste water together with the activated sludge is then sent to the settling tank where the activated sludge is settled mechanically and the clear supernatant solution is discharged. Most of the settled activated sludge is then recycled to the inlet of the aeration tank so that the microbial activity in the aeration tank is kept constant. A part of the settled sludge is recovered as excess sludge and this is further treated at the sludge treatment plant.

The recent developments in the activated sludge process have been reviewed by Jones.[6] The activated sludge treatment is quite an efficient and compact process, especially for eliminating organic matter from waste waters, if the condition of the aeration tank is well maintained and the microbial floc (activated sludge) is adapted to fairly high-strength waste water. For example, waste water containing 5,000 to 10,000 ppm of COD can be treated at very high load levels, such as 10 kg of $COD/m^3/day$.

Some examples of the activated sludge treatment of fermentation waste waters are shown in Table 4.[7]

As shown in Table 4, removal of more than 90% of BOD was always attained in the treatment of fermentation wastes. However, percent removal of COD was much less than the BOD removal. This is one of the problems of biological waste treatment and will be discussed later.

Anaerobic Digestion (Methane Fermentation)

Methane fermentation is the anaerobic decomposition of carbonaceous material to CO_2 and CH_4 by microorganisms. This process is often employed for the treatment of organic sludges and concentrated industrial wastes. Methane fermentation consists of two basic processes, i.e., liquefaction and gasification. In the liquefaction step, organic matter is hydrolyzed by extracellular microbial enzymes to low molecular compounds and finally to volatile fatty acids. In the gasification step, the volatile fatty acids thus produced are consumed by methane-producing bacteria such as *Methanococcus*, *Methanosarcina*, and *Methanobacterium*, and finally are converted to a gaseous mixture of CO_2 and CH_4.

Two kinds of methane fermentation are generally applied.[8,9] One is "medium-temperature" fermentation and the other is "high-temperature" fermentation; the optimum temperature range for

TABLE 4

Results of Activated Sludge Treatment of Various Fermentation Waste Waters

Sort of fermentation		Yeast		Streptomycin		Penicillin	
Waste load [BOD(COD)·Kg/m³·d]		2.5 (BOD)		1.8 (COD)		1.6 (COD)	
Sort of waste water		Raw waste water	Treated water	Raw waste water	Treated water	Raw waste water	Treated water
Color		Dark brown	Yellowish brown	Light yellowish grey	Yellow	Yellowish brown	Brown
Turbidity		Turbid	Transparent	Turbid	Transparent	Turbid	Transparent
pH		5.4	8.1	6.1	8.4	7.6	7.8
Total solids	(ppm)	35,150		9,220	2,988	19,560	5,060
Ash	(ppm)	11,200		3,168		10,342	
Total organic matter	(ppm)	23,950		6,052	624	9,236	856
Suspended solids	(ppm)	2,260		2,964	28	380	13
COD	(ppm)	21,120	850	3,260	345	5,330	580
COD removal	(%)		59.6		84.6		70.6
BOD	(ppm)	16,300	163	12,600	27	10,000	12
BOD removal			90.0		99.7		99.7
Total nitrogen	(ppm)	594		616	172	578	147

Note: Dilution rate: yeast × 10, streptomycin × 1.5, penicillin × 2.7.

From Ono, H., *Kogai Boshi Sangyo* (translated from Japanese), 2(7), 1, 1972. With permission.

the former is 36 to 38°C, and for the latter is 51 to 53°C as shown in Figure 3.[10] The "high-temperature" fermentation can treat 2 to 2.5 times larger amounts of waste load per unit volume of fermentor than the "medium-temperature" fermentation. Therefore, if some heat source is available to keep the temperature of the fermentation liquor high enough, the "high-temperature" fermentation is much more desirable for treating larger quantities of wastes using the same size fermentor. However, in the "high-temperature" fermentation, the optimum temperature range is much narrower than that of "medium-temperature" fermentation. Therefore, the temperature of the fermented liquor must be more carefully controlled in "high-temperature" fermentation.

Some results of fermentation waste treatment by the methane fermentation are shown in Table 5.[10]

The advantages of the methane fermentation over the aerobic treatment are its capability to treat high-strength waste waters directly and to produce a large quantity of methane gas which is useful as fuel. However, the relatively lower BOD removal of the methane fermentation is often considered as a drawback of this process. Therefore, methane fermentation is sometimes used as a pretreatment process for a subsequent activated sludge treatment. Such an example of the combination of these processes applied to the treatment of distillation residues in the alcohol industry is shown in Table 6.[7]

In order to overcome the above-mentioned drawback of methane fermentation, Steffen,[11] Schropfer,[12] and Dietz[13] have developed anaerobic contact processes in which sludges are recovered from the treated effluents and recycled to the fermentor to maintain higher concentrations of sludge in the fermentor. A similar study was made by Sonoda and Tanaka[14] using a waste water discharged from a domestic food industry.

Recently, Scammell[15] reviewed the recent development of anaerobic treatment of industrial wastes and emphasized the economic advantages of the anaerobic treatment over the aerobic process. Also, Taylor[16] reviewed the properties of methanogenic bacteria in relation to the properties of the overall process of anaerobic digestion.

Miscellaneous Treatments

Besides activated sludge treatment and anaerobic digestion, other biological waste treatments using *Chlorella*,[17] photosynthetic bacteria,[18] yeast,[19] and fungi[20] have been reported. Waste treatment using *Chlorella* in combination with other measures such as activated sludge treatment[21] and anaerobic digestion[22] is described in the patent literature.

However, these processes are usually not as effective as activated sludge treatment and anaerobic digestion because of longer retention time and lower removal of BOD and COD. Therefore, they have not been developed for waste treatment purposes.

Among new biological waste treatments, microbial denitrification is one of the promising biological treatment processes. In this process, organic and ammonia nitrogen is first oxidized microbiologically to nitrate, and the nitrate thus produced is then reduced microbiologically to nitrogen gas.

Chemical and Physical Treatments

As mentioned before, biological treatments are quite efficient for treating fermentation waste

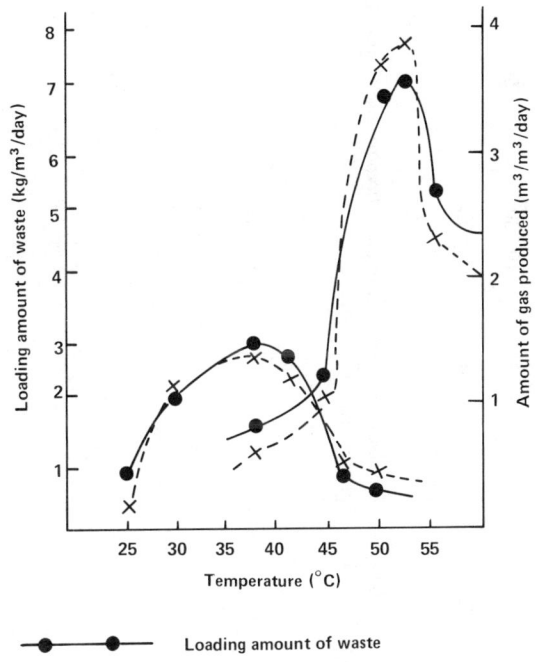

FIGURE 3. Comparison of "medium temperature" and "high temperature" methane fermentation. (From Misono, T., *Hakko To Biseibutsu* [*Fermentation and Microorganisms*, translated from Japanese], Vol. 3, Uemura, T. and Aida, H., Eds., Asakura Shoten, Tokyo, 1970, 167. With permission.)

TABLE 5
Treatment of Various Fermentation Waste Liquors by Methane Fermentation

Sort of waste liquor	Raw waste liquor		Waste load			Gas produced			BOD Removal (%)
	Total organic matter (%)	BOD (ppm)	Waste liquor loaded (e/m³/day)	Organic matter loaded (Kg/m³/day)		Total gas produced (l/m³/day)	Gas produced per unit amount of organic matter (l/kg)	CH_4 content (%)	
Distillation residue from alcohol production (high temp. treatment)	3.0–6.0	20,000–50,000	150–250	5.5–6.5		3,000–4,000	500–700	50–60	80–90
Distillation residue from alcohol production (medium temp. treatment)	3.0–6.0	20,000–50,000	80–125	2.0–3.0		1,000–1,800	500–700	50–60	95
Penicillin fermentation waste liquor	0.92	10,000	250	2.3		480	208	50	86
Streptomycin fermentation waste liquor	0.61	12,600	300	1.8		920	506	53	63 (COD)
Kanamycin fermentation waste liquor	1.57	22,200	300	4.7		1,990	423	51	91
Yeast production waste liquor (high temp. treatment)	2.89	10,280	350	10.2		2,910	287	58	52
Yeast production waste liquor (medium temp. treatment)	2.89	10,280	125	3.6		950	262	61	72

From Misono, T., *Hakko To Biseibutsu* (*Fermentation and Microorganisms*, translated from Japanese), Vol. 3, Uemura, T. and Aida, H., Eds., Asakura Shoten, Tokyo, 1970, 167. With permission.

TABLE 6

Treatment of Distillation Residue of Alcohol Production

Parameter		Raw waste liquor	Methane fermentation (primary treatment)	Activated sludge process (secondary treatment)	Chemical agglomeration and settlement (tertiary treatment)
Volume treated	(m³/day)	500	500	5,000	5,000
BOD	(ppm)	40,000	8,000	56	20
BOD removal	(%)		80	93	65
COD	(ppm)	45,000	27,000	1,220	400
COD removal	(%)		60	55	67
Suspended solids	(ppm)	4,000		150	150
Running cost	(yen/m³)*		2	178	1,280
	($/m³)		0.07	0.59	4.27

Note: Waste liquor was diluted by 10 times after the methane fermentation.

*Running cost is calculated as 1 dollar equal to 300 yen.

From Ono, H., *Kogai Boshi Sangyo* (translated from Japanese), 2(7), 1, 1972. With permission.

waters. Nevertheless, biological treatments alone cannot remove all the impurities from the wastes. For example, some parts of COD components, colored impurities, and most of inorganic substances are hardly removed by biological treatments. Therefore, in order to satisfy the more severe regulations for the waste water quality, certain chemical and physical treatment processes must be employed together with biological treatments. In the case of fermentation waste treatment, such chemical and physical processes are usually applied as secondary and tertiary treatments.

Chemical and physical treatment processes and waste components removed by these treatment processes are listed in Table 7. Some of the chemical and physical treatment processes are described in the following sections.

Chemical Coagulation and Sedimentation

Fermentation waste waters derived from cane molasses contain fairly large amounts of melanoidin pigments which have been produced by the amino-carbonyl reaction between reducing sugars and various amino compounds. Those melanoidin pigments are the main portion of biologically nonassimilable COD and colored impurities in the fermentation waste waters.

Chemical coagulation and sedimentation are often used effectively for removing such melanoidin pigments as well as other colloidal impurities. There are two main types of coagulant, inorganic and organic.

Inorganic coagulants most commonly used are alum, $Al_2(SO_4)_3 \cdot 18H_2O$; ferrous sulfate, $FeSO_4 \cdot 7H_2O$; ferric sulfate, $Fe_2(SO_4)_3$; ferric chloride, $FeCl_3$; and so forth. Aluminum sulfate (alum) appears to be more effective in coagulating carbonaceous waste components, and iron sulfate or chloride is more effective when a considerable amount of proteins are present in the waste.

Chemical coagulation and decolorization of

TABLE 7

Chemical and Physical Treatment Processes

Treatment process	Main waste components removed
High concentration waste liquor	
1. Thermal condensation	All solid matters
2. Freezing	Suspended solids
3. Wet combustion	Organic matters, COD, colored impurities
4. Incineration	Organic matters
Low concentration waste liquor or waste liquor after biological treatment	
5. Neutralization	pH (acidity or alkalinity)
6. Coagulation and sedimentation	Suspended solids; Part of BOD, COD, colored impurities, phosphate, etc.
7. Flocculation	Suspended solids; part of BOD, COD, colored impurities, phosphate, etc.
8. Oxidation with O_3, Cl_2, etc.	BOD, COD, colored impurities, NH_3 N, odor
9. Gas stripping	NH_3-N
10. Ion exchange resin	Inorganic ions, colored impurities
11. Electro dialysis	Inorganic ions
12. Reverse osmosis	Suspended solids, BOD, COD, colored impurities, inorganic ions
13. Ultrafiltration	Suspended solids, high molecular organic matters
14. Mechanical filtration	Suspended solids; Part of COD and BOD
15. Activated carbon adsorption	BOD, COD, colored impurities, suspended solids, odor
16. Zeolite adsorption	NH_3-N
17. Electrolysis	Suspended solids; BOD, COD, colored impurities, phosphate
18. Ultrasonication	Suspended solids

alcohol fermentation waste waters (derived from cane molasses) was investigated by Dazai et al.[2,3] In their experiments, the waste water, just after the activated sludge treatment, was further treated with alum, $Al_2(SO_4)_3 \cdot 18H_2O$, and ferric chloride, $FeCl_3 \cdot 6H_2O$. The results are shown in Tables 8 and 9.

Synthetic organic polyelectrolytes such as polyacrylates, polyacrylamides, polyethyleneimines, polyoxyethylenes, and so forth are also effective coagulants. They are usually used in combination with inorganic coagulants. In such combined uses of inorganic and organic coagulants, the inorganic coagulants are first added to the waste water with vigorous agitation and then the organic polyelectrolytes are added.

Colloidal particles in the waste waters usually have negative electrical charges, and addition of inorganic coagulants neutralizes such electrical charges and lets the colloidal particles coagulate. Subsequent addition of organic polyelectrolytes further agglomerates the coagulated colloidal particles by their adsorption and crosslinking activities and produces very large flocs. By such com-

TABLE 8

Chemical Treatment of Alcohol Fermentation Waste Water* with Alum, $Al_2(SO_4)_3 \cdot 18H_2O$

Amount of alum added (mg/ml)	pH	Weight of sludge produced (%)	Volume of sludge produced (v/v %)	Transparency (%)	Color	COD** (ppm)	BOD** (ppm)
0	7.8	—	—	59	Red brown	840	110
500	6.8	0.09	68	61	Yellow brown	—	—
1,000	6.7	0.17	35	81	Yellow	—	—
2,000	5.5	0.26	83	100	Light yellow	470 (44.0)	10 (90.9)
3,000	4.1	0.26	89	100	Light yellow	—	—
4,000	3.7	0.26	89	100	Light yellow	—	—

*Waste water is derived from cane molasses medium and treated primarily by activated sludge process.
**Numbers in parentheses show COD or BOD removal percent.

From Dazai, M., Yoshida, Y., Ogawa, M., and Ono, H., *Report of the Fermentation Research Institute,* No. 26, 1964, p. 119. With permission.

TABLE 9

Chemical Treatment of Alcohol Fermentation Waste Water* with Ferric Chloride, $FeCl_3 \cdot 6H_2O$

Amount of $FeCl_3 \cdot 6H_2O$ added (mg/ml)	pH	Weight of sludge produced (%)	Volume of sludge produced (v/v, %)	Transparency (%)	Color	COD** (ppm)	BOD** (ppm)
0	7.8	—	—	59	Red-brown	840	110
500	6.5	0.09	20	67	Yellow-brown	—	—
1,000	5.7	0.15	46	85	Yellow	—	—
1,500	4.8	0.22	92	100	Almost colorless	275 (67.3)	18 (83.6)
2,000	2.6	0.11	70	70	Yellow	—	—
4,000	—	—	—	60	Red-brown	—	—

*Waste water is derived from cane molasses medium and treated primarily by the activated sludge process.
**Numbers in parentheses show COD or BOD removal percent.

From Dazai, M., Yoshida, Y., Ogawa, M., and Ono, H., *Report of the Fermentation Research Institute,* No. 26, 1964, p. 119. With permission.

bined uses of inorganic and organic coagulants, the following synergistic effects are expected: (1) acceleration of clarification of waste water, (2) increase of the size and stability of coagulated flocs, (3) increase of crosslinking activity of organic polyelectrolytes, (4) increase in treatment capacity according to the increase in sedimentation and filtration rates, (5) decrease of sediment volume, (6) reduction of costs required for coagulants, and (7) increase in operational stability.

Recently, natural polyelectrolytes such as chitosan have become popular as effective coagulants for suspended solids in food processing wastes and activated sludge.[24] Chitosan is a high molecular aminoglycoside polymer manufactured by deacetylation of chitin which is obtained as shrimp and crab wastes. Chitosan is particularly effective in removing proteins and biomass from waste waters. Therefore, it can be successfully applied to recovery of excess sludge in activated sludge treatment. The sludge recovered by chitosan treatment can be safely utilized as raw materials for feed stuffs and fertilizers.

Activated Carbon Adsorption

Activated carbon can adsorb a variety of organic matter from aqueous solutions. Wang et al.[25] studied the activated carbon adsorption of various organic compounds which might occur in industrial effluents. The technology of activated carbon treatment in waste disposal is now quite popular and the interested reader can refer to several monographs.[26,27]

Activated carbon adsorption is especially effective to remove organic contaminants from waste effluent when the waste effluent contains rather small quantities of such organic material. The maximum concentration of organic matter in the effluent that can be treated economically by activated carbon will depend on the particular case. However, as a rule, when the concentration of organic matter is 200 ppm or more, pretreatment to remove the bulk of the organic matter prior to carbon treatment is desirable. Therefore, in the treatment of fermentation waste waters, application of activated carbon adsorption is recommended as the secondary or tertiary treatment. Zuckerman et al.[28] mentioned in their report on the removal of organic matter from sewage that organic matter having molecular weights less than 400 was easily removed biologically but organic matter having molecular weights more than 1,200 was difficult to remove by biological treatment. A similar tendency was observed in the removal of organic material by activated carbon treatment. Therefore, they recommended the use of carbon treatment after the removal of high-molecular-weight organics by chemical coagulation and sedimentation.

There are two types of carbon treatment, batchwise treatment using pulverized carbon, and the countercurrent continuous treatment using granular carbon columns. In most cases of waste treatment, granular carbon in a column system is preferred for the following reasons: (1) More efficient use of the total adsorption capacity of the carbon can be achieved. (2) Continuous operation is available, and consequently operation cost is lowered and flexibility against load fluctuations is increased. (3) Granular carbon can be thermally reactivated for reuse. In case of pulverized carbon, it can be reactivated experimentally but is difficult to reactivate on a large scale. On the other hand, granular carbon is easily reactivated thermally with about 5% or less burning and attrition loss.

Activated carbon adsorption is a very effective way to obtain high-quality waste effluent. However, the elimination of suspended solids prior to the column feeding and the reactivation of carbon makes this process rather expensive.

Chemical Oxidation

Chemical oxidation using ozone (O_3), chlorine gas (Cl_2), hydrogen peroxide (H_2O_2), sodium hypochlorite (NaOCl), and so on is effective in removing BOD, COD, colored impurities, and odors from the waste liquors.

In experiments by the author, residual waste water from a distillation column in the alcohol industry was directly treated with chlorine gas. Removal of COD and coloration from the waste water are shown in Figures 4 and 5, respectively.

A large decrease of COD and coloration in the waste water was observed in accordance with the formation of a large quantity of precipitates. However rather large consumption of chlorine gas per unit volume of the waste water makes this treatment expensive. Therefore, it is suggested that such chemical oxidation must be applied as the secondary or tertiary treatment after the biological treatment process. Also, in the case of chemical oxidation using chlorine gas, the toxicity of the treated water must be further investigated.

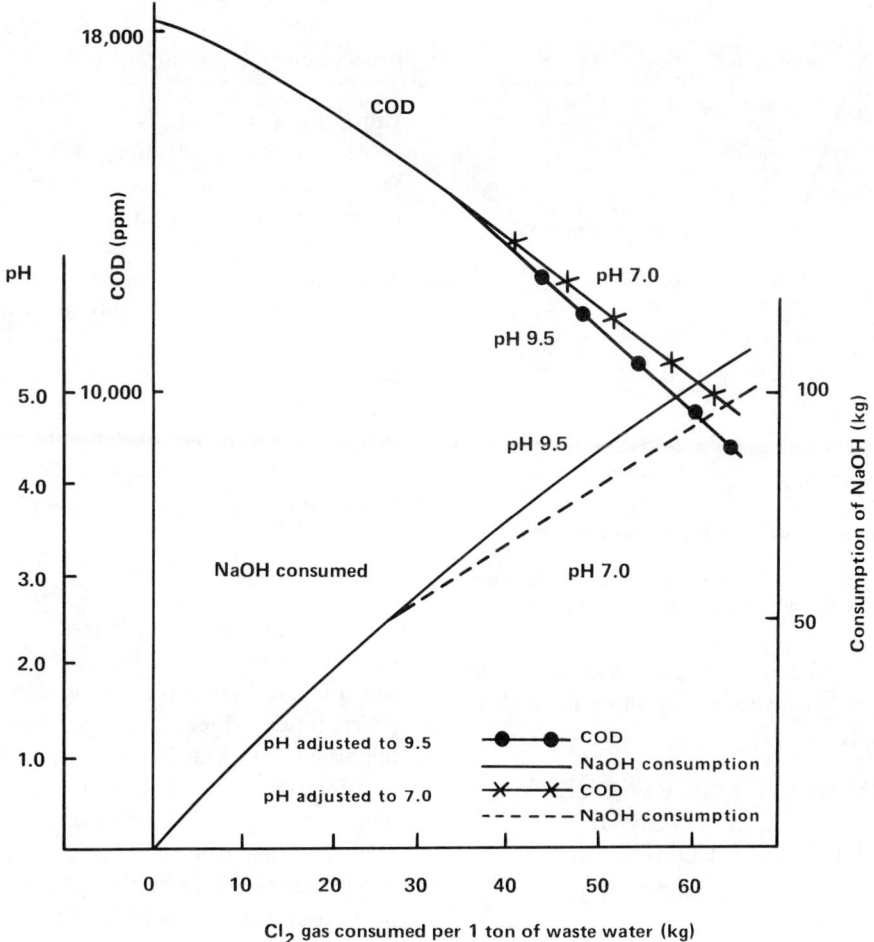

FIGURE 4. Changes of COD and NaOH consumption during Cl_2 treatment of alcohol distillation waste water.

Ammonia Stripping

In the fermentation industries, ammonia or ammonium salts are usually used as a raw material for microbial production and also aqueous ammonia is often used in extraction and purification processes like the ion exchange process. Therefore, the ammonium content of fermentation waste waters becomes fairly high in some cases.

A part of the ammonia in such waste waters can be removed by biological treatment, but it is desirable to remove the ammonia first if the ammonium content of the waste water is too high. The ammonia stripping method can be effectively applied for such purposes.

Miscellaneous Treatments

Concentration of any waste effluent by evaporation, especially high-strength aqueous waste, is often a desirable process in order to recover and utilize useful waste components. Various types of evaporators such as multiple-effect, flush, etc., are used for this purpose.

Electrolysis using aluminum and iron electrodes in the waste water is sometimes effective for the removal of COD and colored impurities. In this case, Al^{+3} and Fe^{+3} ions produced by electrolysis coagulate such impurities. Electrolytic treatment of a yeast fermentation waste before use of the activated sludge process has been studied. However, scaling on the surface of the electrodes and relatively high cost of the operation make the actual practice of this process difficult.

Application of various membrane processes for water treatment is becoming popular especially for the desalination of sea water and waste treatment in other industries. However, in the case of fermentation waste water treatment, the high

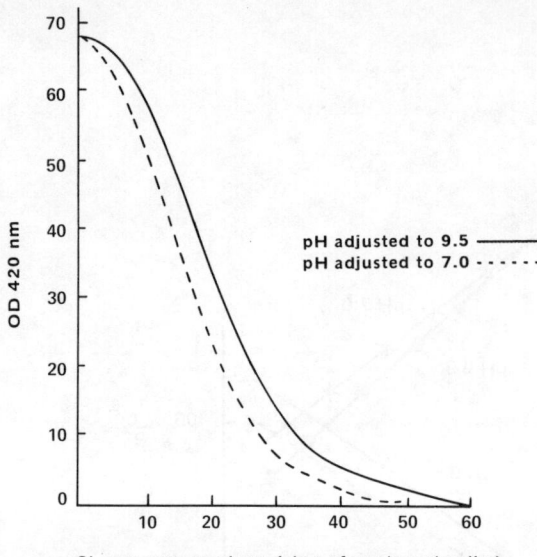

FIGURE 5. Changes of color (OD at 420 nm) during Cl_2 treatment of alcohol distillation waste water.

content of suspended solids and colloidal matter are still big problems for the commercial application of this process.

TECHNOLOGY FOR RECOVERY AND UTILIZATION OF USEFUL WASTE COMPONENTS

General

Due to governmental regulations concerning environmental protection, more effective treatments of industrial wastes are being adopted by every industry. Needless to say, this tendency makes the cost of waste treatment increase year after year. On the other hand, many industrial wastes still contain useful components if they are properly utilized. Therefore, if such useful components of the industrial wastes can be reused properly, it is obviously beneficial both for society and industry.

In the case of fermentation wastes, there are many useful components such as sugars, crude proteins, amino acids, vitamins, minerals, unidentified growth factors (UGF), and so forth. Accordingly, investigations on the utilization of such useful components have been carried out. Actually, utilization of such waste components as feed stuffs and fertilizers has already been industrialized. If such utilization of useful waste components is properly combined with the waste treatment process, including primary, secondary, and tertiary treatments, it is possible to decrease the waste treatment cost to a great extent and also to establish an ideal closed system including production and waste disposal.

Utilization as Feed Stuffs

As for the production of feed stuffs from fermentation waste, the following three main procedures can be expected: (1) direct concentration and drying of the waste water, (2) mixing the waste water or its concentrate with raw feed materials like rice bran, soy bean meals, and so on, and subsequently drying the mixtures, and (3) cultivation of useful microorganisms using the waste water as fermentation media and utilization of microbial cells and metabolites thus produced as feed stuffs. Examples of such utilization are given below.

Utilization of Alcohol Distillation Waste as Feed Stuffs

In the United States, domestic corn is used as the main raw material for alcohol fermentation, and the alcohol distillation wastes are directly concentrated, dried, and then marketed as an important feed ingredient. In the Japanese alcohol fermentation industry, however, cane molasses is now the biggest raw material because of the relatively inexpensive supply. Therefore, the distillation waste from the Japanese alcohol fermentation industry contains rather high concentrations of ash, especially potassium, as shown in Table 10.

The high content of potassium causes diarrhea when a large amount of such concentrate or dried powder is fed to animals. Therefore, only a small amount of such material can be added in feeds as a flavor enhancer or vitamin and UGF source.[29,30] Presently, the removal of potassium from the alcohol distillation waste is being investigated.

Utilization of Brewery Wastes as Feed Stuffs

Wastes from brewery industries such as beer, whiskey, and domestic fermented foods industries are also utilized as feed ingredients. For example, whiskey distillation waste is successfully applied as a feed ingredient and its composition is shown in Table 11.

Utilization of Microbial Cells Grown on Waste Waters

Fermentation waste waters still contain enough nutrients for microbial growth. Therefore, growth studies were made of certain unicellular algae,

TABLE 10

Components of Alcohol Distillation Waste Water and Its Concentrate and Dried Powder Derived from Cane Molasses

Component		Raw waste water	Concentrate	Dried powder
General				
Water	(%)	93.40	52.93	8.38
Crude protein	(%)	0.83	4.99	6.74
Crude fat	(%)	0	0.82	0.78
Crude fiber	(%)	0	0.27	0.79
Ash	(%)	1.40	11.40	26.11
Soluble non-nitrogenous substances	(%)	4.37	29.50	57.20
Minerals				
Ca	(%)	—	0.69	2.05
P	(%)	—	0.26	0.45
K	(%)	—	4.06	8.82
Mg	(%)	—	0.33	0.51
Fe	(%)	—	0.06	0.56
Mn	(mg%)	—	11.02	18.03
Zn	(mg%)	—	0.75	5.38
Cu	(mg%)	—	2.64	10.10
Co	(mcg%)	—	110.46	237.86
Vitamins				
Thiamine	(mg%)	—	0.09	0.01
Riboflavin	(mg%)	—	0.06	0.10
Pyridoxine	(mg%)	—	2.48	0.20
Niacin	(mg%)	—	—	14.0
Choline	(mg%)	—	58.8	40.5
Biotin	(mg%)	—	—	0.32
B_{12}	(mcg%)	—	1.69	2.52

photosynthetic bacteria, and fungi in fermentation waste waters, and then these microbial cells were utilized as feed ingredients.[17,31,32] Successive cultivation of yeast and *Chlorella* in the alcohol fermentation waste and the utilization of these microbial cells as feed stuffs has also been investigated.[33]

Activated sludge itself is a kind of biomass and contains 20 to 55% protein. Therefore, the use of activated sludge as a feed ingredient was also investigated.[34,35] However, in this case the rather high ash content in activated sludge is a problem.

Utilization as Fertilizers

Among various utilizations of fermentation wastes, the technology for fertilizers is most advanced. As shown in Tables 12[36] and 13,[36] fermentation wastes contain useful fertilizer components.

Production of Organic Fertilizer from Fermentation Wastes

Production of organic fertilizer from alcohol and amino acid fermentation wastes has been industrialized since 1964 at the Hofu Plant of the Kyowa Hakko Kogyo Co., Ltd., in Japan. One significance of this technology is the establishment of a closed system for the fermentation industry by recycling useful plant nutrients in the fermentation wastes into soils as valuable organic fertilizers. Another significance of this technology is the economic credit for decreasing waste treatment cost by converting waste components to organic fertilizers having economic value.

The process flowsheet for production of organic fertilizer from fermentation wastes is shown in Figure 6.[36]

In general practice, concentration of fermentation waste waters up to 40 to 50% of solid content

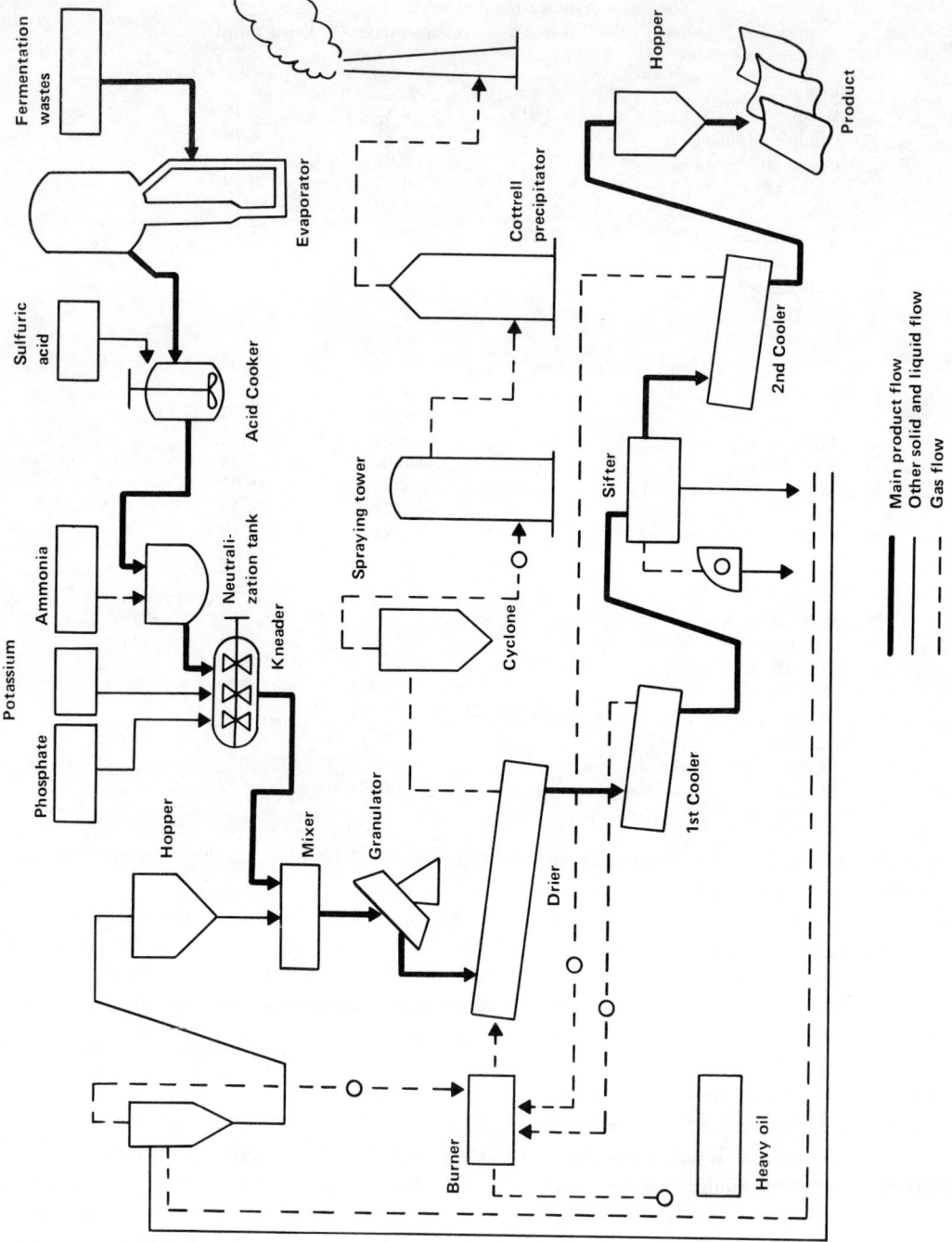

FIGURE 6. Manufacturing flow diagram for production of organic fertilizer from fermentation wastes. (From Tsuchiya, Y., *Kankyo Sozo* [translated from Japanese], 1973, p. 38. With permission.)

TABLE 11

Components of Whiskey Distillation Waste (Dried Solid)

Component	Content	
General		
Water	5	(%)
Crude protein	27	(%)
Crude fat	0.16	(%)
Crude fiber	0.17	(%)
Ash	17	(%)
Carbohydrate	51	(%)
Amino acid		
L-Lysine	1.19	(%)
L-Methionine	0.37	(%)
L-Arginine	0.45	(%)
L-Cystine	0.08	(%)
L-Histidine	0.43	(%)
L-Phenylalanine	0.70	(%)
L-Tryptophane	0.07	(%)
Vitamin		
Vitamin B_1	1.5	(ppm)
Niacin	510	(ppm)
Pantothenic acid	67	(ppm)
Riboflavin	21	(ppm)
Choline	2,000	(ppm)
Pyridoxin	19	(ppm)
Biotin	0.9	(ppm)
Inositol	10,000	(ppm)

TABLE 12

Composition of Fermentation Wastes (g/100 g of dry matter)

	Alcohol fermentation waste	Glutamic acid fermentation waste
Water	4.51	3.54
Crude protein	11.43	17.11
Ash	27.51	12.92
Crude fat	0.44	0.52
Crude fiber	0.40	0.02
Soluble nonnitrogenous Substance	55.71	55.17

From Tsuchiya, Y., *Kankyo Sozo* (translated from Japanese), 1973, p. 38. With permission.

is rather difficult because of the tremendous increase in viscosity. Even if the waste waters are dried to the solid state by other means, such dried matter is strongly hygroscopic and cannot be stored for a long time. Therefore, in order to convert fermentation wastes to valuable organic fertilizers, some measures must be taken to improve such physicochemical defects of waste

TABLE 13

Contents of Plant Nutrition Elements in Fermentation Wastes (g/100 g of dry matter)

	Alcohol fermentation waste	Glutamic acid fermentation waste
Total N	2.0	10.5
NH_3-N	—	6.5
Organic nitrogen	2.0	4.0
Soluble P_2O_5	0.1	0.5
Soluble K_2O	11.0	4.5

From Tsuchiya, Y., *Kankyo Sozo* (translated from Japanese), 1973, p. 38. With permission.

organics; also, the composition of nutritional elements must be balanced.

In order to solve these problems, the concentrate of fermentation wastes was cooked with mineral acids like sulfuric acid and the organics in the wastes were converted to a humus-like substance,[37,38] and then certain plant nutrition elements such as ammonia, potassium, and phosphate were added to the humus-like mash for balancing the composition as fertilizers. The resulting mixture was granulated mechanically, dried, sieved, and then packaged for sale. The effect of the organic fertilizers thus produced has been examined in agricultural fields for many years and the product has been widely adopted for many kinds of crops because of its high fertilization effect and capability for improvement of physicochemical characteristics of soils.

In this manufacturing process, not only fermentation wastes but also waste acids from amino acid extraction processes can be effectively utilized as a part of the raw material.[39,40] Also, excess sludges from the activated sludge treatment process can be easily utilized as a part of the raw material in this process.

Other Utilization of Fermentation Wastes as Fertilizers

Fermentation waste waters, especially the ones derived from cane molasses, contain minerals necessary for plant growth. Therefore, such fermentation wastes or their concentrates can be used as raw materials for liquid fertilizers if the transportation costs are acceptable.[41,42] The effect of such liquid fertilizers derived from fermentation waste concentrates is being investigated in the cane fields of a southern island of Japan with successful results.

Combined use of fermentation wastes with waste materials from other industries is also being investigated. For example, bark and sawdust from wood processing industries can be easily fermented in the solid state together with fermentation waste waters and converted to excellent humus for agricultural uses within a short time, namely 3 or 4 months.

PRACTICE OF FERMENTATION WASTE TREATMENT AND ITS PROBLEMS

Fermentation waste waters generally contain rather high concentrations of organic matter and necessary minerals for microbial growth; generally they do not contain materials destructive to microorganisms. Due to these reasons, biological treatment using microbes seems to be one of the most suitable means of treatment of fermentation waste waters, and at present the biological treatment is widely adopted for such purposes. However, biological treatment alone cannot remove all impurities to the required extent.

Limitations and problems in biological treatment, especially in the treatment of fermentation waste waters derived from cane molasses by the activated sludge process, are described below.

Behavior of BOD and COD in Biological Treatment

Fermentation wastes derived from cane molasses usually contain biologically nonassimilable organic matter like melanoidin pigments, nonassimilable sugars, and so forth. Therefore, if such wastes are treated biologically, the COD removal percentage is usually much lower than the BOD removal percentage. Changes of the relation between BOD and COD during the activated sludge treatment of amino acid fermentation waste waters were determined and the results are shown in Figure 7.[43]

Organic matter in a waste water can be expressed by different measurements such as BOD, COD, and TOD (Total Oxygen Demand).[44] For example, oxygen demand of waste waters before and after the activated sludge treatment was determined by different measurements, and the results are shown graphically in Figure 8.[1]

As shown in Figure 8, BOD of the fermentation waste water decreased significantly (ΔBOD =

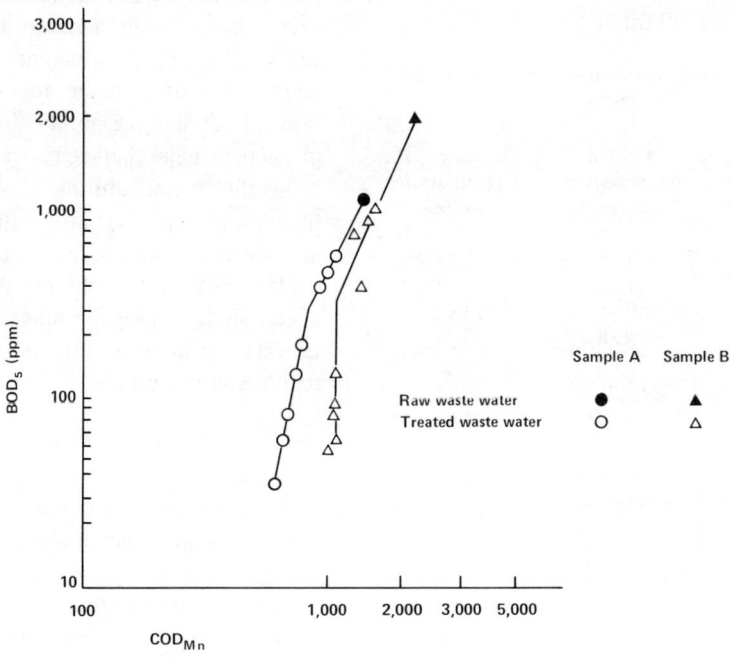

FIGURE 7. Relationship between BOD and COD during activated sludge treatment of amino acid fermentation waste waters. (From Koike, T., *Sangyo Kogai* [translated from Japanese], 10, 1611, 1974. With permission.)

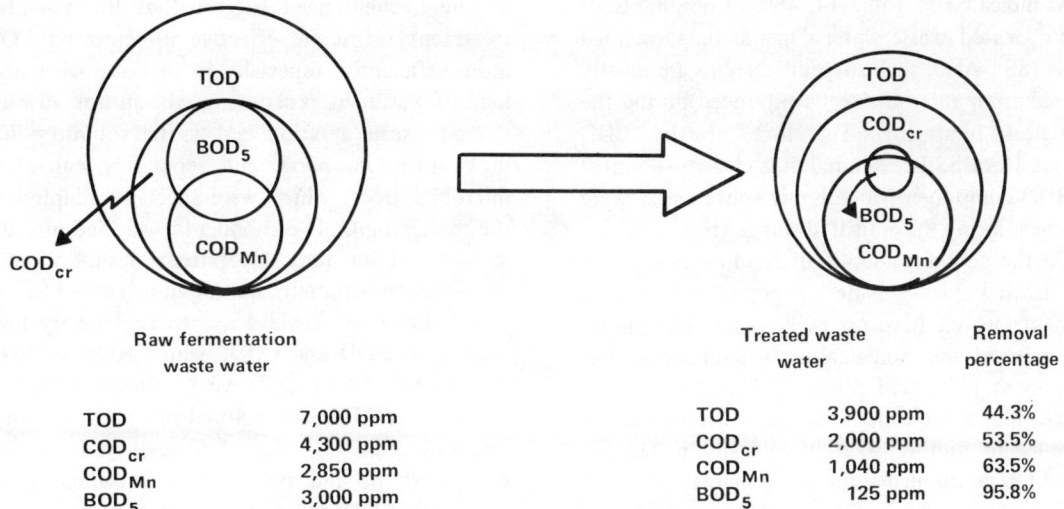

FIGURE 8. Behavior of BOD, COD, and TOD in the activated sludge treatment of fermentation waste water. TOD = total oxygen demand. COD_{cr} = COD expressed by potassium dichromate oxidation. COD_{Mn} = COD expressed by potassium permanganate oxidation. BOD_5 = BOD after 5 days' incubation at 20°C. (From Koike, T., *Ferment. Ind.* [translated from Japanese], 34, 163, 1976. With permission.)

TABLE 14

Distributions of BOD and COD among the Components of Waste Waters before and after the Activated Sludge Treatment[1]

	Before treatment				After treatment			
	BOD_5		COD_{Mn}		BOD_5		COD_{Mn}	
Component	ppm	Distribution %	ppm	Distribution %	ppm	Distribution %	ppm	Distribution %
Colored pigments	100	3.3	740	26.0	trace	—	930	89.4
SS	700	23.3	640	22.5	60	48.0	40	3.8
Organic acids	400	13.3	170	6.0	5	4.0	—	—
Lipids	840	28.0	730	25.6	20	16.0	30	3.0
Amino acids	280	9.3	70	2.5	15	12.0	10	0.9
Sugars	320	9.8	220	7.6	5	4.0	10	0.9
Others	360	12.0	280	9.8	20	16.0	20	2.0
Total	3,000	100	2,850	100	125	100	1,040	100

Note: Fractionation of components:
Sample → Centrifugation → Sephadex G 25 → Ether extraction
        (SS)   (Pigment)   (Organic acids and lipids)
→ Ion exchange (Dowex 50WX16) → Activated carbon → Effluent
  (Amino acids)   (Sugars)   (Others)

From Koike, T., *Ferment. Ind.* (translated from Japanese), 34, 163, 1976. With permission.

95.8%), but removal of COD_{Mn} was much lower (ΔCOD_{Mn} = 63.5%). These results show the limit of the biological treatment when the organic matter in the waste water is determined by COD value.

The same samples used in Figure 8 were fractionated to several components such as colored pigments, suspended solids, organic acids, lipids, amino acids, sugars, and other substances including nucleic acids. Then, the distribution of BOD and COD among those components was determined. These results are shown in Table 14.[1]

As indicated in Table 14, 48% of residual BOD in the treated waste water depends on suspended solids (SS). Also, such suspended solids are mostly derived from microbial cells produced during the biological treatment. Therefore, if the BOD derived from SS is excluded, it is clear that almost all BOD components in the raw waste water were removed during the activated sludge treatment.

On the contrary, 89.4% of residual COD_{Mn} in the treated waste water depends on colored pigments which have no BOD value. This shows that most of the biologically nonassimilable COD component is colored pigment. Therefore, it can be said that some other treatments must be taken in order to remove such biologically nonassimilable COD components, including pigments.

Rate of Oxygen Consumption During the Biological Treatment and Treatment of High Loads of Waste

The rate of oxygen comsumption per unit amount of suspended solids during the biological treatment was measured, and the results are shown in Figure 9.[1]

As shown in Figure 9, the rate of oxygen consumption changed significantly after about 10 hr of operation. Active consumption of oxygen was accompanied by the significant decrease of COD in the first 10 hr but after that the oxygen consumption rate dropped rapidly and COD did not decrease any more.

These phenomena suggest that the two-step treatment might be effective in removing COD more efficiently, especially in the case of a high load of waste. In experiments, the authors divided the pilot-scale aeration tank (total volume = 700 m^3) into two parts, and separately cultivated microbial flocs, which were specially adapted to the remaining COD components, were fed into the second aeration tank. Operational results in such pilot-scale experiments are shown in Table 15.[1]

As shown in Table 15, even under very high loading of BOD and COD, nearly 100% of BOD and 60 to 65% of COD were removed from the waste water. However, colored materials were not removed and in fact increased slightly. From this result and the data of Table 14, the majority of remaining COD components are expected to be colored pigments.

Formation of microbial flocs was normally maintained during the experiments even under the high-load conditions.

Characteristics of Microflora in the Biological Treatment

Investigations of the microflora in the first and second aeration tanks were carried out. Isolated microbial strains from both tanks were cultivated separately in a pure state using the raw fermentation waste water as a common medium, and the capabilities of respective strains to remove COD and color (OD at 420 nm) were determined. Those

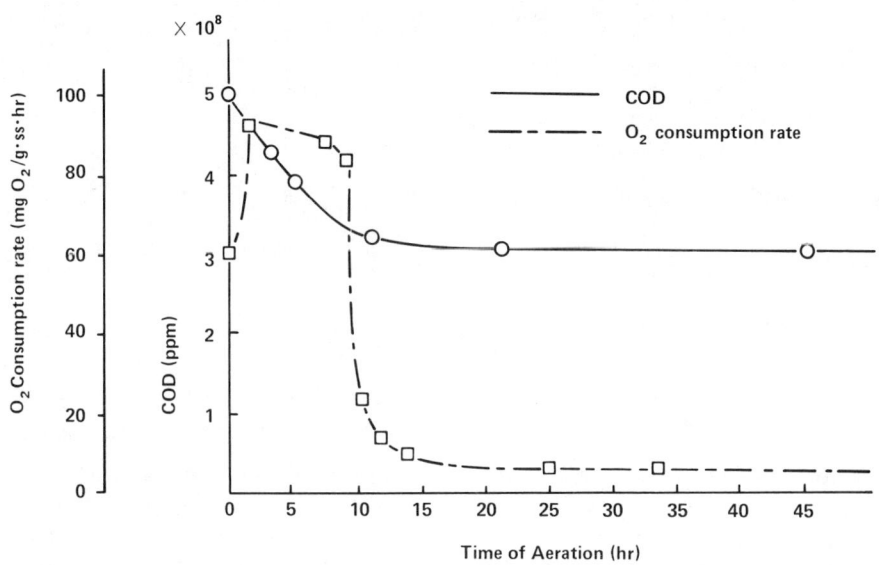

FIGURE 9. Changes of oxygen consumption rate and COD during biological treatment. (From Koike, T., *Ferment. Ind.* [translated from Japanese], 34, 163, 1976. With permission.)

TABLE 15
Operational Results of the Two-step Biological Treatment in the Pilot-scale Aeration Tank (700 m^3)[1]

No. of experiment	Operational condition			Parameter of water quality								Removal of COD (%)	Removal of BOD (%)
	Flow rate (m^3/hr)	COD loaded (kg COD/m^3/day)	BOD loaded (kg BOD/m^3/day)	Treated waste water				Raw waste water					
				COD (ppm)	BOD (ppm)	Color (OD) 420 nm	pH	COD (ppm)	BOD (ppm)	Color (OD) 420 nm	pH		
1	18	2.0	2.0	1,250	130	4.0	6.5	3,300	3,200	3.8	4.3	62.1	96.0
2	10	2.1	2.5	2,100	150	6.2	6.3	6,000	7,200	5.6	4.5	65.0	98.0
3	30	3.4	2.9	1,200	30	3.6	6.7	3,300	2,840	3.4	6.0	64.0	99.0
4	30	6.5	6.3	2,470	200	7.0	8.2	6,500	6,140	5.3	5.9	62.0	97.0
5	30	10.4	11.3	3,800	800	9.2	8.3	9,200	10,000	8.2	5.9	59.0	92.0

From Koike, T., *Ferment. Ind.* (translated from Japanese), 34, 163, 1976. With permission.

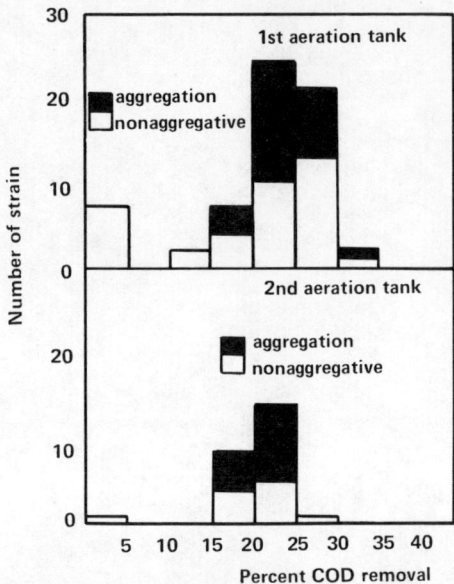

FIGURE 10. Distribution of COD removing capabilities of isolated microbial strains. (From Koike, T., *Ferment. Ind.* [translated from Japanese], 34, 163, 1976. With permission.)

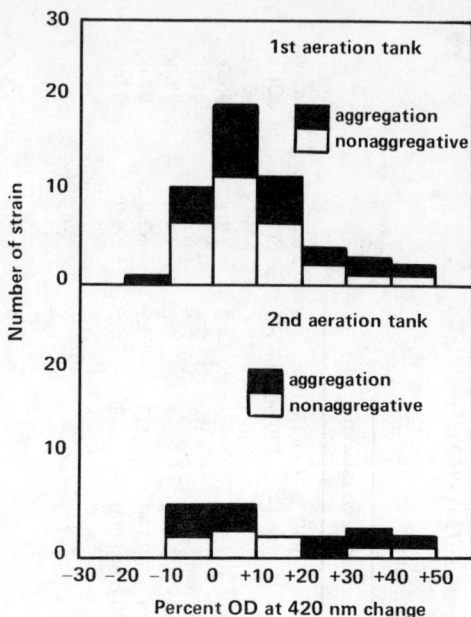

FIGURE 11. Color change by isolated microbial strains. (From Koike, T., *Ferment. Ind.* [translated from Japanese], 34, 163, 1976. With permission.)

results are graphically summarized in Figures 10[1] and 11.[1]

The majority of the microflora isolated from the second aeration tank were relatively simple and most of the strains were similar to the strains isolated from the first aeration tank. In Figure 10, the microflora isolated from the second tank show unexpectedly lower activity in COD removal under the experimental conditions used. However, such microflora can still be expected to have a significant role in COD removal under high-load conditions.

As shown in Figure 11, color-increasing strains were dominant in comparison with color-decreasing strains in both tanks, especially in the second tank.

Among the strains isolated from the aeration tanks, *Pseudomonas* sp., *Flavobacterium* sp., *Achromobacter* sp., and *Klebsiella* sp. were dominant. COD removal rates showed by such representative strains are indicated in Figure 12.[1]

Among the strains indicated in Figure 12, nonaggregative strains such as *Pseudomonas* sp. No. 471 and *Flavobacterium* sp. No. 481 contributed to the removal of COD to a great extent. Such nonaggregative strains are often isolated from microbial flocs in the aeration tanks and are assumed to be included in microbial flocs when they are cultivated together with aggregative microorganisms in waste waters.

Mixed cultures of isolated strains in the fermentation waste water were also carried out; the results are shown in Table 16.[1]

As shown in Table 16, mixed cultures show higher percentages of COD removal than single cultures. And also, it is evident that nonaggregative strains such as *Pseudomonas* sp. No. 471 and *Flavobacterium* sp. No. 481 are predominant in the reconstructed microbial flocs when they are cultured with other aggregative strains. Those phenomena suggest more complicated mutual interactions among different microorganisms in the treatment of waste water. Therefore, more ecological and biochemical studies on such synergistic effects of microorganisms are desirable in the future.

Residual COD Components and Their Removal by Microorganisms

As already mentioned, it can be said that most residual COD components after biological treatment are colored substances. Molecular weights of these colored substances are assumed to be 10^3 to 10^4 according to the results of Sephadex® filtration of such colored substances.

Waste waters before and after biological treatment were chromatographed with using Sephadex G25 columns, and the chromatographic patterns were traced by optical density at 260 nm. These results are shown in Figure 13.[1]

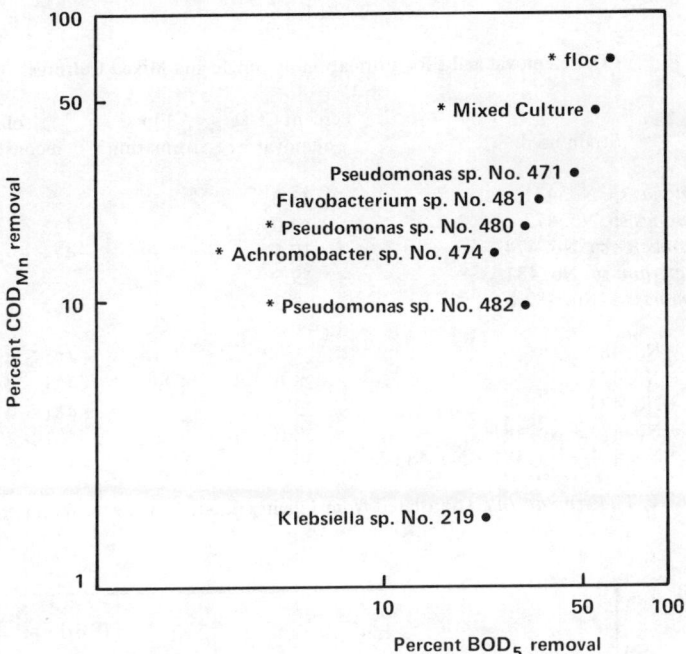

FIGURE 12. Relation of COD and BOD removal rates expressed by isolated strains. Asterisk indicates flocculent growth on waste medium. (From Koike, T., *Ferment. Ind.* [translated from Japanese], 34, 163, 1976. With permission.)

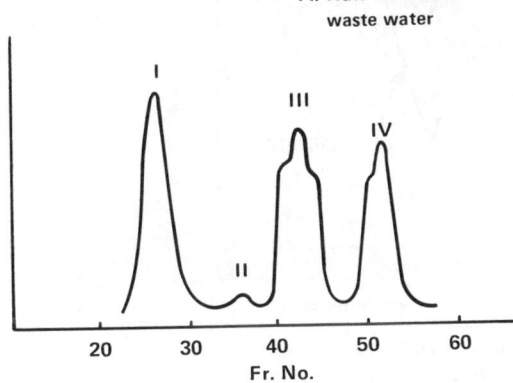

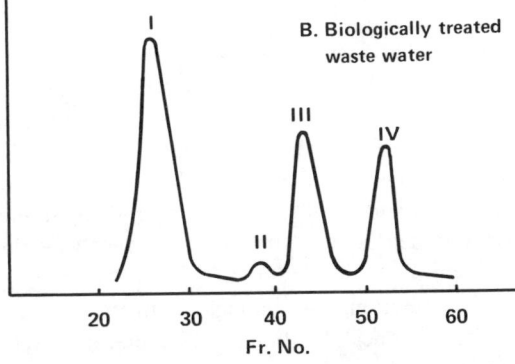

FIGURE 13. Sephadex G25 chromatographic patterns of waste waters traced by OD 260 nm. (From Koike, T., *Ferment. Ind.* [translated from Japanese], 34, 163, 1976. With permission.)

As shown in Figure 13, four relatively simple peaks were observed for the biologically treated waste water. On the other hand, the chromatographic pattern of raw fermentation waste water was a little more complicated. A similar chromatographic pattern was obtained when cane molasses itself was examined in the same way. Therefore, it can be said that the residual substances in the treated waste water are mostly derived from cane molasses. Further experiments showed that Fraction I occupied 50% of the residual COD components. Fraction I is a high molecular polymer which is easily coagulated by the change of electric charges. Fraction III shows a yellow color and strong ultraviolet absorption.

The infrared spectrum of Fraction I was examined. The spectrum is shown in Figure 14[1] together with the spectrum of the browning substance from maple syrup measured by Underwood[4,5] as well as the spectrum of humic acid.

The coloration of waste water usually increases 10 to 20% during biological treatment. This phenomenon can be explained by the dominance of color-increasing microorganisms in aeration tanks (see Figure 11).

In order to remove colored substances from waste waters, changes of coloration were examined by cultivating various bacteria, fungi, and *Basidio-*

TABLE 16

COD Removal and Floc Formation by Single and Mixed Cultures[1]

Strain used	Percent COD removal	Floc formation	Isolates from reconstructed floc
Pseudomonas sp. No. 471	23.0	−	−
Pseudomonas sp. No. 473	14.0	+	473
Achromobacter sp. No. 474	10.5	±	474
Flavobacterium sp. No. 481	19.5	−	−
Pseudomonas sp. No. 482	8.3	+	482
No. 471 + No. 482	25.0	±	471 > 482
No. 471 + No. 474	26.0	±	471 > 474
No. 481 + No. 474	23.0	±	481 > 474
No. 473 + No. 474 + No. 481	28.0	±	481 > 773, 474
No. 471 + No. 474 + No. 481 + No. 482	30.5	+	471, 481 > 474, 482

From Koike, T., *Ferment. Ind.* (translated from Japanese), 34, 163, 1976. With permission.

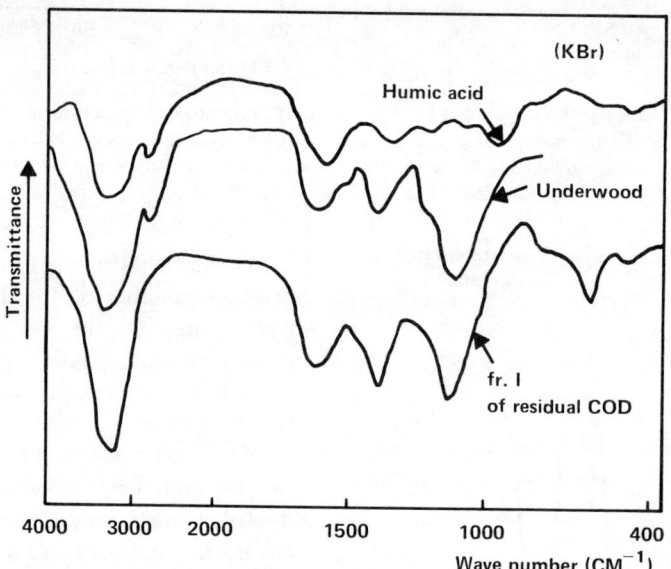

FIGURE 14. Infrared spectra of residual COD. (From Koike, T., *Ferment. Ind.* [translated from Japanese], 34, 163, 1976. With permission.)

mycetes in the fermentation waste waters containing glucose. As shown in Figure 15, some *Bacidiomycetes* can decrease coloration significantly.[1]

However, at present, practical microorganisms for decoloration of fermentation waste water are not available.

Combination of Biological Treatment with Physicochemical Treatments

As mentioned before, biological treatment can remove nearly 100% of BOD from fermentation waste water, but only about 60% of COD can be removed by the same treatment. Moreover, coloration of the waste water is usually increased to some extent during biological treatment. These are thought to be the limitations of the biological treatment of fermentation waste water. Under circumstances in which the legislative regulation of waste effluent quality is becoming more severe, some measures must be taken for resolving this problem.

Combination of biological treatment with physicochemical treatment was therefore investigated. There are many possible combinations. At the Kyowa Hakko Kogyo Co., Ltd., the combination of treatments shown in Figure 16 was tried.[1]

In this process, a biological treatment, namely

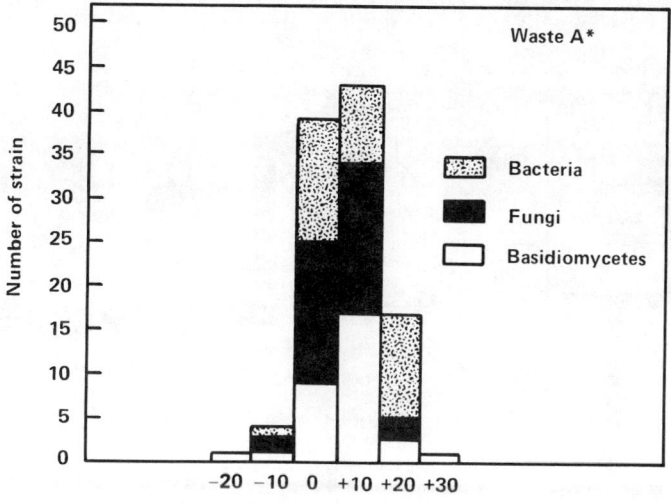

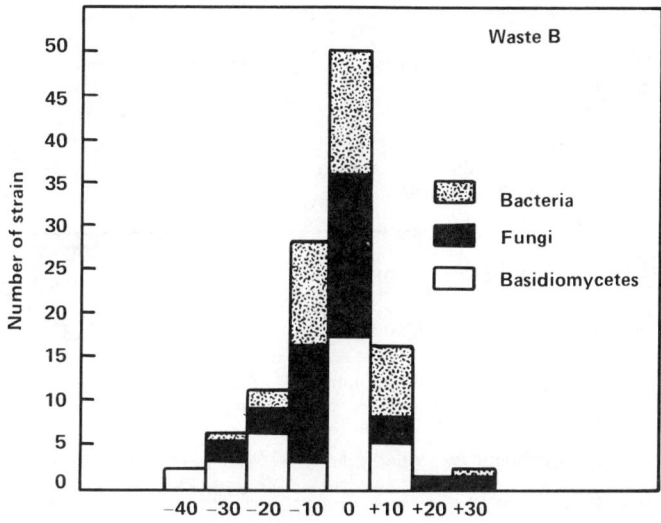

FIGURE 15. Distribution of decolorizing microorganisms. (From Koike, T., *Ferment. Ind.* [translated from Japanese], 34, 163, 1976. With permission.)

an activated sludge process which was specially adapted to high loads of waste, was used as the primary treatment, and a chemical treatment, namely coagulation with ferric chloride, was used as the secondary treatment. As the tertiary treatment, a physical treatment, namely electrolysis using iron and aluminum electrodes, was employed. The flow sheet is shown in Figure 16.

An example of the results of such combined treatment of fermentation waste water is shown in Table 17.[1]

As shown in Table 17, most of BOD, COD, SS and color (OD) were removed.

Reverse Osmosis as a Promising Tool of Fermentation Waste Treatment

Reverse osmosis is becoming popular as a water treatment process[46] and is now being applied to the treatment of sewage.[47] There has been no report, however, on the application of this technique to the treatment of waste water from the industrial fermentation plant. If the waste water contains valuable materials, the R.O. (Reverse Osmosis) system can be applied satisfactorily for their recovery on an economical basis. Though the fermentation waste water used here did not contain such materials, a preliminary examination

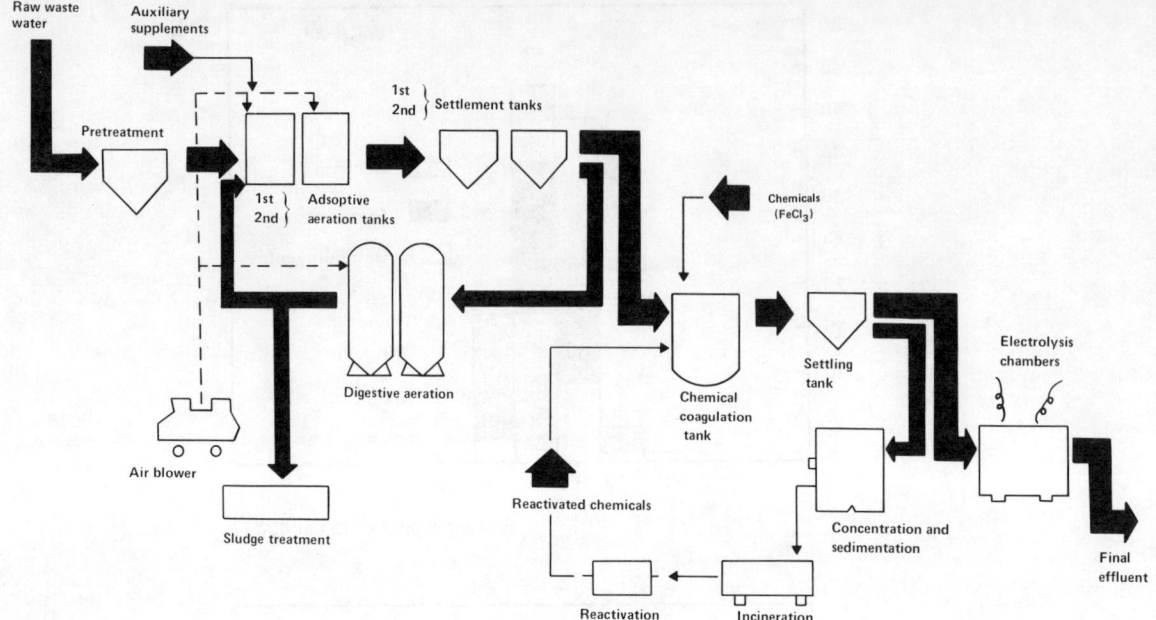

FIGURE 16. Combined treatment of fermentation waste water. (From Koike, T., *Ferment. Ind.* [translated from Japanese], 34, 163, 1976. With permission.)

TABLE 17
Result of Combined Treatment of Fermentation Waste Water[1]

Treatment	Parameter			
	COD_{Mn}	BOD_5	OD at 420 nm	SS
A. Raw fermentation waste water	6,500 ppm	6,140 ppm	5.3	1,400 ppm
B. Biological treatment	2,470 ppm	200 ppm	7.0	350 ppm
Removal % A-B/B × 100	62%	97%	−32%	71%
C. Chemical ($FeCl_3$) treatment	640 ppm	130 ppm	0.84	100 ppm
Removal % B-C/C × 100	74%	35%	88%	70%
Removal % A-C/C × 100	90%	98%	84%	93%
D. Physical treatment	110 ppm	50 ppm	0.42	100 ppm
Removal % C-D/D × 100	83%	62%	50%	0%
Removal % A-D/D × 100	93.3%	99.2%	92.0%	92.9%

From Koike, T., *Ferment. Ind.* (translated from Japanese), 34, 163, 1976. With permission.

of the R.O. technique revealed the successful removal of COD, inorganic phosphate, total nitrogen, and colored substances from the waste water.

The raw waste water used here was the effluent from the biological treatment process. It contained some suspended solids which significantly decreased the flux of permeate. Because of this the feed water to the R.O. system was previously filtered using Filter-aids®. The content of solutes in this filtrate is shown in Table 18.

The R.O. system used is schematically presented in Figure 17. The diameter of the membrane was 76 mm and the effective membrane area was 31 cm². The chamber was filled with 180 ml of the feed waste stirred with a magnetic stirrer and pressurized constantly to 50 kg/cm² with nitrogen gas until the volume of permeate reached 100 ml, then the permeate was mixed well and analyzed.

From these analytical data, rejection percentages of various solutes were calculated and are shown in Table 18. The application of the R.O.

TABLE 18

Analysis of Permeate and Rejection Percentage of Solutes

Solute		Feed	KP-98*		KP-90*		KP-00*	
					Membrane type			
COD	ppm	709	22	(97)**	16	(98)	85	(88)
Total nitrogen	ppm	710	67	(91)	100	(86)	320	(55)
Inorganic phosphate	ppm	22	0	(100)	1	(96)	7	(66)
Total cation	eq/l	0.11	0.005	(96)	0.011	(90)	0.045	(60)
OD 470 nm		1.23	0.023	(98)	0.014	(99)	0.062	(95)
OD 260 nm		12.0	0.16	(99)	0.20	(98)	1.29	(89)

*All membranes were the products manufactured by Eastman Kodak Co.
**The numbers in parentheses show rejection percentage of solutes.

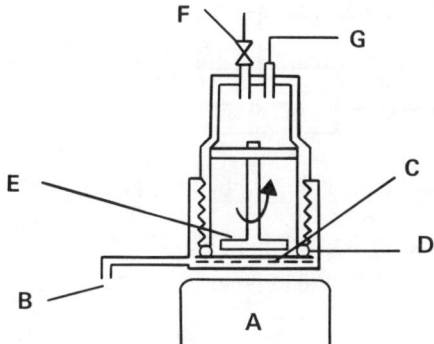

FIGURE 17. Diagram of R.O. device. (A) stirrer; (B) permeate outlet; (C) membrane, cellulose acetate film type (ϕ = 76 mm); (D) O-ring; (E) stirrer piece; (F) safety valve; (G) N_2 gas inlet.

process to the treatment of fermentation waste water seems promising from these data. However, it is needless to say that, before actual application of this process, there must be many studies which deal with the best pretreatment procedures, the selection of the best membrane and other suitable equipment, flux stability, membrane life, product quality, and, above all, economics of this process. But, if everything goes well, the problems in reducing COD, nitrogen, phosphate, and colored substances in fermentation waste water can be solved by this one process and the resulting permeate be reused as a raw water supply. Thus, it seems worthwhile to investigate further the possibility of actual application of this technique to the treatment of fermentation waste water.

CONCLUSIONS

Technologies for waste treatment are progressing rapidly due to social needs, and treatment of fermentation waste water is not exceptional. However, in this field, there are still many fundamental problems which must be solved in the future. For example, in the biological treatment of waste waters, there are many unknown aspects of microflora in the treatment tanks. Therefore, there are many possibilities of discovering new abilities of microorganisms still in the future. On the other hand, new technologies are now coming out in the fields of chemical and physical treatments. Membrane processes, for example, would be very promising if economic and operational problems can be solved.

Technologies for converting waste materials to useful products are also important. These technologies, properly combined with waste treatment technologies, would be quite beneficial not only for keeping the environment clean but also for finding new resources for human beings.

At the closing of this chapter, a pattern of treatment of total fermentation wastes is shown in Figure 18.[1]

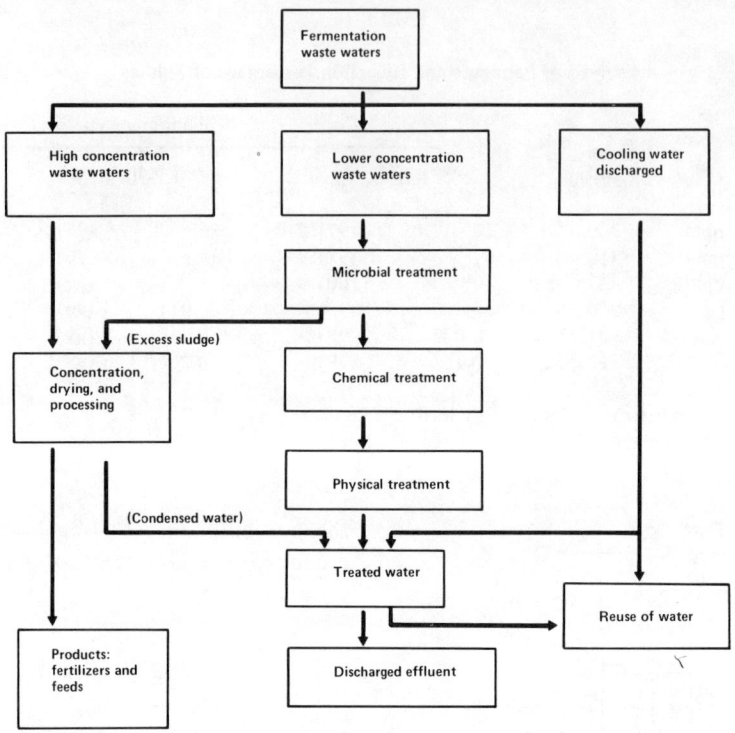

FIGURE 18. A pattern of treatment of total fermentation wastes. (From Koike, T., *Ferment. Ind.* [translated from Japanese], 34, 163, 1976. With permission.)

REFERENCES

1. **Koike, T.,** Waste water treatment in fermentation industries, *Ferment. Ind.* (translated from Japanese), 34, 163, 1976.
2. **Eckenfelder, W. W., Jr. and O'Conner, D. J.,** *Biological Waste Treatment,* Pergamon Press, Oxford, 1961.
3. **Nemerow, N. L.,** *Theories and Practices of Industrial Waste Treatment,* Addison-Wesley, Reading, Mass., 1963, chap. 12.
4. **Conway, R. A.,** Treatment of dilute wastewater, in *Industrial Waste Disposal,* Ross, R. D., Ed., Reinhold, New York, 1968, 101.
5. **Imhoff, K., Muller, W. J., and Thistlethwayte, D. K. B.,** *Disposal of Sewage and Other Water-born Waste,* Butterworths, London, 1971.
6. **Jones, G. L.,** Microbiology and activated sludge, *Process Biochem.,* p. 3, January/February 1976.
7. **Ono, H.,** Treatment of industrial waste waters by activated sludge process, *Kogai Boshi Sangyo* (translated from Japanese), 2(7), 1, 1972.
8. **Ono, H. and Tanaka, M.,** Studies on methane fermentation of alcohol distillation waste, No. 3, On the fermentation temperatures, *Hakko Kyokaishi* (translated from Japanese), 13, 538, 1955.
9. **Ono, H., Tanaka, M., Fukuoka, S., and Seiko, Y.,** Studies on the methane fermentation of alcoholic distillation slops, Part 10, On the thermophilic digestion, *Hakko Kyokaishi* (translated from Japanese), 15, 526, 1957.
10. **Misono, M.,** Methods of waste water treatment, in *Hakko To Biseibutsu (Fermentation and Microorganisms,* translated from Japanese), Vol. 3, Uemura, T. and Aida, H., Eds., Asakura-Shoten, Tokyo, 1970, 167.
11. **Steffen, A. J. and Baker, M.,** Operation of full-scale anaerobic contact treatment plant for meat packing wastes, *Proc. 16th Ind. Waste Conf.,* 106, 386, 1961.
12. **Schropfer, C. J. and Ziemke, N. R.,** Development of anaerobic contact process, *Sewage Ind. Wastes,* 31, 164, 1959.
13. **Dietz, J. C., Clinebell, P. W., and Strub, A. L.,** Design consideration for anaerobic contact system, *J. Water Pollut. Control Fed.,* 38, 517, 1966.
14. **Sonoda, Y. and Tanaka, M.,** Anaerobic digestion of low concentration wastes. Part 1, Continuous digestion tests for some industrial wastes, *J. Ferment. Technol.* (translated from Japanese), 46, 789, 1968.
15. **Scammell, G. W.,** Anaerobic treatment of industrial wastes, *Process Biochem.,* p. 34, October 1975.

16. Taylor, G. T., The formation of methane by bacteria, *Process Biochem.*, p. 29, October 1975.
17. Okubo, H., Tezuka, T., and Takao, K., Alcohol distillation wastes treatment by unicellular algae culture, Part 1, Cultural condition of unicellular algae in test tube, *Hakko Kyokaishi* (translated from Japanese), 25, 38, 1967.
18. Kobayashi, M. and Kitamura, H., Studies on purification and utilization of polluted water in food industry, *Hakko Kyokaishi* (translated from Japanese), 32, 158, 1974.
19. Yano, T., Horii, K., and Ozaki, A., Utilization of fermentation wastes, especially alcohol distillation wastes, as feeds, *Hakko Kyokaishi* (translated from Japanese), 27, 135, 1969.
20. Takiguchi, Y., Studies on treatment and application of Baker's yeast culture waste solution, Part 1, On screening of effective molds, *Hakko Kyokaishi* (translated from Japanese), 44, 711, 1966.
21. Kono, M., Method for Culturing Unicellular Algae by Oxidative Degradation of Wastes, Japanese Patent Sho 42-14030, 1967.
22. Yokouchi, H., Treatment of Anaerobic Fermentation Supernatant, Japanese Patent Sho 44-6618, 1969.
23. Dazai, M., Yoshida, Y., Ogawa, M., and Ono, H., Decolorization of alcohol fermentation waste water derived from cane molasses by coagulation-sedimentation method, *Report of the Fermentation Research Institute*, No. 26, 1964, p. 119.
24. Bough, W. A., Chitosan-A polymer from seafood waste, for use in treatment of food processing wastes and activated sludge, *Process Biochem.*, p. 13, January/February 1976.
25. Wang, L. K., Leonard, R. P., Wang, M. H., and Goupil, D. W., Adsorption of dissolved organics from industrial effluents on to activated carbon, *J. Appl. Chem. Biotechnol.*, 25, 491, 1975.
26. Joyce, R. S., Activated carbon adsorption, in *Industrial Waste Disposal*, Ross, R. D., Ed., Reinhold, New York, 1968, 144.
27. Miyazawa, S., Waste water treatment with reactivation by pulvarized activated carbon, *Sangyo Kogai* (translated from Japanese), 12, 220, 1976.
28. Zuckerman, M. M. and Molof, A. H., High quality reuse water by chemical-physical waste water treatment, *J. Water Pollut. Control Fed.*, 42, 437, 1970.
29. Amano, H., Utilization of alcohol distillation wastes as feed, *Hakko Kyokaishi* (translated from Japanese), 24, 531, 1966.
30. Yano, T., Horii, K., Ohara, H., and Ataka, K., Studies on digestibility of alcohol fermentation wastes concentrates in milk cow and sheep, *Hakko Kyokaishi* (translated from Japanese), 29, 240, 1971.
31. Tsuru, S., Yeast-like fungi relating to the production of protein for feed use, *Kagaku To Kogyo* (Tokyo), 27, 637, 1974.
32. Kumabe, K., Tomiyama, M., Hirao, M., and Mochizuki, K., Treatment of Fermentation Wastes using Fungi, Japanese Patent Sho 40-1152, 1965.
33. Nagase, T., Okubo, H., and Tezuka, T., Treatment of Alcohol Distillation Wastes, Japanese Patent Sho 45-9818, 1970.
34. Kuribayashi, M., Miyashita, T., Tanaka, M., and Ono, H., Treatment of alcohol distillation wastes and dehydration of activated sludge, *Hakko Kyokaishi* (translated from Japanese), 24, 177, 1966.
35. Ono, H. and Watanabe, K., Method for Producing Feed Additives from Activated Sludge, Japanese Patent Sho 46-13654, 1971.
36. Tsuchiya, Y., Concentration of fermentation waste waters and conversion of the concentrate to fertilizers by acid treatment, *Kankyo Sozo* (translated from Japanese), 38, 1973.
37. Hirano, K., Aramaki, G., and Sejima, T., Method for Producing Organic Fertilizers, Japanese Patent Sho 44-16332, 1969.
38. Hirano, K., Aramaki, G., and Sejima, T., Method for Producing Humic Acid Containing Substances, Japanese Patent Sho 44-28083, 1969.
39. Hirano, K., Aramaki, G., and Soda, A., Method for Producing Fertilizers Containing Nitrate-Nitrogen, Japanese Patent Sho 47-10004, 1972.
40. Kurihara, S., Kubota, Y., Noguchi, S., and Iwai, M., Treatment of Glutamic Acid Fermentation Wastes, Japanese Patent Sho 47-14883, 1972.
41. Matsuda, K. and Harada, M., Method for Producing Liquid Fertilizer from Whiskey Distillation Wastes, Japanese Patent Sho 45-11530, 1970.
42. Yoshino, Y. and Maeda, M., Method for Producing Liquid Fertilizer, Japanese Patent Sho 46-18562, 1971.
43. Koike, T., Treatment of fermentation waste waters, *Sangyo Kogai* (translated from Japanese), 10, 1611, 1974.
44. Testing Method for Industrial Waste Water, *Japan Industrial Standard (JIS) K-0102*, Nihon-Kikaku-Kyokai, Tokyo.
45. Underwood, J. C., Willits, C. O., and Lento, H. G., Browning of sugar solutions. VI. Isolation and characterization of the brown pigment in maple syrup, *J. Food Sci.*, 26, 397, 1961.
46. Anon., Reverse osmosis today, *Process Biochem.*, p. 32, January/February 1976.
47. Sammon, D. C. and Stringer, B., The application of membrane processes in the treatment of sewage, *Process Biochem.*, p. 4, March 1975.

DYNAMICALLY FORMED AND TRANSIENT MEMBRANES

E. Drioli

TABLE OF CONTENTS

Introduction 343

Early History 344

Recent Studies. 345

Applications 349

Conclusions 352

References 353

INTRODUCTION

Within the past 10 years, pressure-driven membrane processes such as hyperfiltration (HF) and ultrafiltration (UF) have found application in many areas, from seawater desalting and industrial waste treatment to the separation, purification, and concentration of liquid foods, proteins, and pharmaceuticals. Current interest in these advanced separation techniques is based on the discovery and development of new classes of selective membranes which combine high selectivity with satisfactory permeability.

Among these new classes, an interesting and promising group of HF and UF membranes are the so-called "dynamically formed membranes" pioneered by the group at Oak Ridge National Laboratory.[1,2] In general, these membranes are formed when solutions containing membrane-forming additives are circulated under pressure over finely porous bodies, or primary membranes, which serve as a support for the secondary membrane. Details of the mechanism of formation of the dynamic membranes and their role in HF and UF processes have not yet been thoroughly defined.

On the one hand, the dynamically formed membrane is a new kind of semipermeable membrane offering very high permeation rates and the possibility of easy *in situ* regeneration.[3,4] On the other hand, it may be speculated that the phenomenon of building up a thin, stable interfacial layer from a very low concentration of selected species present in the filtering solution is a general one. In some cases, this phenomenon may significantly affect or even dominate the overall HF or UF process. Even the asymmetric cellulose acetate membranes presently commercially used in these processes could be considered as suitable microporous supports for the formation of these thin interfacial films. Phenomena still currently under investigation, such as membrane compaction and the correlated flux decline or change in rejection, may well be influenced by dynamically formed secondary membranes.[5] Membrane fouling, caused by the build-up of deposits on the high-pressure side of a semipermeable membrane, could be considered as an uncontrolled dynamic membrane formation, with predominantly negative effects on the separation process. Clearly, fundamental but ill-defined aspects of membrane science such as solute-membrane and solute-solute interactions as well as concentration polarization are involved in dynamic membrane formation.

As noted, these membranes have been under investigation as HF and UF barriers. Recent studies have also illustrated the possibility of using this membrane formation technique to create an immobilized enzyme in gel form on a polymeric membrane. As pointed out below, certain advantages derive from the combination of selective mass transfer across the semipermeable membrane with chemical reactions occurring at the membrane surface or within its pores.[6] In this paper, we will review the early history as well as some of the recent innovations in this field.

EARLY HISTORY

In 1966, a research group at Oak Ridge National Laboratory (ORNL) found that membranes could be created by exposing a finely porous body to a pressurized and circulating solution containing a colloidal dispersion of a hydrous oxide.[1] Typical early experiments were performed by recirculating 0.02 M NaCl solutions containing a very low concentration (0.0005 M) of Fe(III) over the surface of a 0.8-μm diameter porous silver frit in a hyperfiltration apparatus. The initially very rapid permeation rate fell in an hour or so to a typical value of 0.2 cm/min at 35 atm applied pressure, as the rejection of NaCl rose from essentially zero to about 60%.

While a number of materials were found to produce dynamic membranes, the additives producing the most interesting results were those that would be expected to form ion exchange membranes. A partial list of the additives investigated at ORNL is reproduced in Table 1. A number of porous supports were studied, and while not all supporting structures were found to be equally satisfactory, the chemical nature of the porous support was not of primary importance. Among the materials examined were porous metals including silver, porcelain, porous carbons, sintered glass, and molecular filters of the Millipore® type.

The most thorough studies were made with membranes dynamically formed from Zr(IV) oxide on porous tubes. Salt rejection was measured as a function of salt type, concentration, and pH.[7] An ion-exclusion mechanism was found to offer at least a qualitative explanation of the results. Salt rejection was also found to agree qualitatively with the known ion exchange properties of the dynamically formed membranes.

The salt rejection exhibited by the early dynamically formed membranes was well below that attained with commercial cellulose acetate membranes. A significant improvement in performance was achieved, however, with the introduction of a new technique of preparation.[8] When a previously formed hydrous metal oxide membrane was exposed to a solution containing polyacrylic acid (PAA), a layer of PAA deposited on the first dynamically formed membrane. Salt rejection improved, particularly when the pH was raised thereby neutralizing the PAA. A number of questions remain unresolved: why the PAA attaches at neutral pH; why it remains attached instead of simply redissolving in the solution; and why the dependence of salt rejection on salt concentration is lower then expected. Nevertheless, these improved dynamically formed membranes appear promising for the desalination of low salinity waters by hyperfiltration.

The relatively low salt rejection exhibited by dynamically formed polyelectrolyte membranes has been attributed to a low fixed charge capacity of these membranes. Consistent with this, a good qualitative agreement was found between salt rejection observed with finely divided cation exchange particles (Dowex® 50W) dynamically formed on Millipore filters and a model based on Donnan's ion exclusion theory.[9] A small amount of salt leakage through the membrane was assumed. Tanny and Jagur-Grodzinsky assumed that the fixed charge density could be increased by excluding water from the vicinity of the polyelectrolyte, and they used partially cured Loeb-Sourirajan cellulose acetate (CA) membranes as supports for polyelectrolytes such as poly (2-vinyl pyridine) and polyvinylamine.[10] The size of the micropores in the surface of these membranes can be adjusted by annealing the membrane in water, and an apparently good fit between pore diameter and the size of the polyelectrolyte molecules could

TABLE 1

Typical Additives Tested at ORNL Which Can Form Salt-rejecting Membranes

Hydrous oxides [Al(III), Fe(III), Sl(IV), Zr(IV), Th(IV), U(VI)]
Finely ground low cross-linked ion-exchange resins
Clays (bentonite)
Humic acid
Poly(styrene sulfonic acid)
Poly(vinyl benzyl trimethyl ammonium chloride)
Cellulose acetate hydrogen phthalate
Cellulose acetate N,N-diethylaminoacetate
Poly(methyl vinyl ether/maleic anhydride) (Gantrez AN)
Poly(4-vinyl pyridine)
Poly(4-vinyl pyridinium butyl chloride)
Poly(vinylpyrrolidone)

TABLE 2

Effect of Colloidal Additive on Reverse Osmotic Properties of Cellulose Membrane Under 50 atm and at 25°C

Run	Feed (0.5N)	Annealing temp. °C	Additive	Salt rejection %		Additive rejection %	Water flux (ml/cm^2 hr)	
				No additive	Additive		No additive	Additive
1	NaCl	80	Fe	1.1	8.5	—	1.76	1.57
2	NaCl	80	Si	5.2	18.9	—	2.09	0.58
3	NaCl	80	Al	5.9	20.5	—	2.07	0.12
4	NaCl	80	Cu	19.8	31.1	—	0.78	0.54
5	NaCl	80	Na-polyacrylate	4.7	15.5	—	0.94	0.60
6	NaCl	80	Polyethyleneimine	7.9	27.4	—	1.23	0.36
7	NaCl	80	Poly (tri-) phosphoric acid	9.9	17.8	—	1.64	1.46
8	NaCl	80	Zr	3.8	32.2	—	0.78	0.40
9	NaCl	80	Zr	2.3	32.8	99	0.46	0.32
10*	NaCl	80	Zr	22.4	25.9	93	0.59	0.47
11*	NaCl	80	Zr	20.6	31.1	98	0.66	0.47
12*	NaCl	80	Zr	42.6	45.7	98	0.31	0.27
13*	NaCl	75	Zr	71.3	79.2	94	0.91	0.91
14*	NaCl	75	Zr	6.2	79.6	98	2.05	1.15
15*	NaCl	75	Zr	14.6	24.6	98	0.73	0.46
16*	NaCl	70	Zr	8.4	18.3	98	1.57	0.59
17*	NaCl	70	Zr	29.0	40.7	98	0.97	0.40
18*	NaCl	80	Th	5.7	21.8	—	1.91	1.01
19*	CaCl$_2$	80	Th	14.4	72.2	—	0.71	0.29

*pH of the food solution: 2.7 ~ 3.5.

be achieved. Salt rejection observed in hyperfiltration indicated that the improvement caused by PAA addition was a function of the initial salt rejection of the CA membrane. A maximum in salt rejection increase was observed with CA membranes annealed at 40 to 50°C. Salt rejection did not improve significantly when PAA was circulated over CA membranes annealed at temperatures above 60 to 70°C. The authors speculated that the behavior of these membranes was related to the existence of active groups distributed along the surface of the narrow pores which effectively trap the polyelectrolyte within the pore. It has been shown by Spiegler[11] and Pusch,[12] in fact, that CA membranes carry a small fixed charge capacity due to the existence of free carboxylic groups in the matrix. Pusch[12] has reported a charge density of 3.5 meq/dry gram of CA membrane.

Specific interactions, such as hydrogen bonding, between the supporting matrix and the additive may play an important role in dynamic membrane formation. This is suggested by work at Gulf General Atomic[13-16] in which PAA was dynamically formed on supports of mixed cellulose esters or porous polysulfone. Effects of pore size distribution in the support on membrane performance were also reported by Sachs and Lonsdale[13] in related studies of dynamically formed PAA membranes. Some data reflecting the effect of various colloidal additives on membrane performance are presented in Table 2.

RECENT STUDIES

Membranes dynamically formed on asymmetric cellulose acetate substrates have been studied in some detail recently at the Max Planck Institute for Biophysics in Frankfurt am Main. The results illustrate the transient nature of certain types of dynamically formed membranes. The measurements were carried out in a standard hyper-

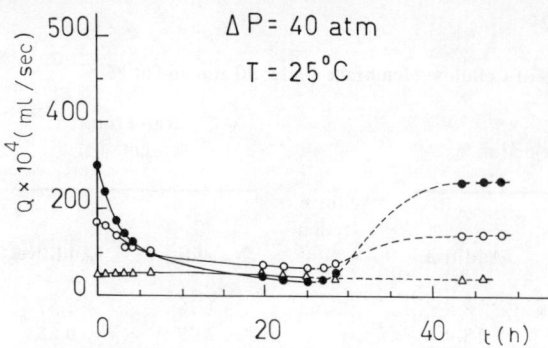

FIGURE 1. Volume flux, Q, vs. time for three asymmetric cellulose acetate membranes; ΔP = 40 atm, T = 25°C. (From Drioli, E., Lonsdale, H. K., and Pusch, W., *J. Colloid Interface Sci.*, 51, 355, 1975. With permission.)

filtration apparatus[15] with CA membranes made by the procedure of Manjikian, Loeb, and McCutchan.[16] The membranes were annealed in water at temperatures of 60 to 90°C prior to use. The experiments were performed at 25°C and at 40 atm applied pressure in a cell of area 12 cm², with a feed brine velocity sufficient to minimize boundary layer effects.[17]

The measurements were performed mainly with mixtures of $AlCl_3 \cdot 6H_2O$ and either LiCl or NaCl. Typical plots of volume flux, Q, through three membranes as a function of time are presented in Figure 1. The membranes are designated according to their annealing temperature. Thus, the "CA 60" membrane was annealed at 60°C. Prior to "time zero" in the figure, the volume flux was measured for several hours in the presence of a 10^{-4} M LiCl solution, and the flux for each membrane was quite stable at the initial value shown in the figure. At t = 0, the feed was switched to a mixture of 10^{-4} M LiCl and 3×10^{-4} M $AlCl_3$, at which point all the membranes exhibited a rapid and sharp decline in the rejection of Li^+ ion. This was followed by a slower but major decline in water flux through the CA 60 and CA 75 membranes and an increase in the Li^+ ion rejection to values considerably higher than exhibited by these membranes prior to the addition of Al^{+++} ion. The membrane properties stabilized after about a day at which time the concentration of the feed solution was increased to 10^{-2} M LiCl and 3×10^{-2} M $AlCl_3$. There was essentially a return to the initial flux values, as shown in Figure 1, and the rejection of Li^+ ion was again much lower than that observed prior to the addition of Al^{+++}. The rejection of Al^{+++} ion by all the membranes was practically complete throughout this series of experiments. This is consistent with the known ability of CA membranes to strongly reject multivalent ions.

The rejection of Li^+ ion during a similar series of tests is shown in Figure 2. Here, rejection has been plotted against the annealing temperature of the membranes. The initial performance, obtained with 10^{-3} M LiCl in the feed, is shown as Curve A. After steady state performance was reached, the feed was changed to 10^{-3} M LiCl and 3×10^{-3} M $AlCl_3$ and the results are shown as Curve B. The rejection of Li^+ ion fell almost immediately to about 0% and became stabilized there. When the feed was changed again to 10^{-4} M LiCl and 3×10^{-4} M $AlCl_3$ (Curve C), the Li^+ ion rejection of the CA 60 and CA 75 membranes increased over a period of about 20 hr to the values shown in the figure, while the volume flux steadily declined in the manner exemplified in Figure 1. Finally, the feed concentration was increased to 10^{-2} M LiCl and 3×10^{-2} M $AlCl_3$. The rejection again dropped precipitously for the CA 60 and CA 75 membranes (Curve D) and the volume flux for all the membranes returned essentially to their initial values.

A similar pattern is exhibited with these membranes when exposed to NaCl-$AlCl_3$ mixtures, as shown in Figure 3. Again, changes in performance were limited to the membrane annealed at 60°C and the performance of the membrane annealed at 85°C was essentially invariant throughout the experiment. An additional observation of note is that the membranes are strongly history dependent. For example, when a membrane substrate was initially exposed to a feed of 0.1 M NaCl and 3×10^{-4} M $AlCl_3$ (Curve C-D in Figure 3), there was no apparent change in performance, However, when a dynamically formed membrane was prepared at lower NaCl concentrations, it persisted when exposed to the same conditions of high NaCl concentration (Curve F-G).[18]

Several factors should be considered in discussing these results. First, CA membranes are known to possess a low fixed charge concentration of approximately 10^{-3} meq/dry gram because of residual carboxylic acid groups. Thus, these membranes will act as weak cation exchangers and tend to exclude coions (e.g., Cl^- ions) at concentrations of 10^{-3} M and lower. A simple solution-diffusion model of membrane transport will not be applicable. The ion-exclusion effect is

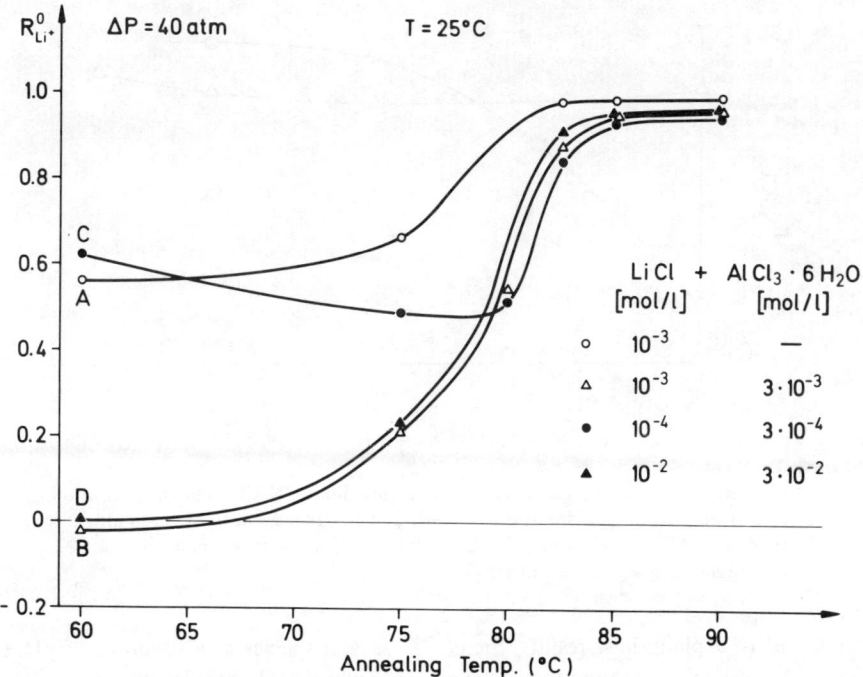

FIGURE 2. Steady-state rejection of Li^+ vs. annealing temperature of asymmetric cellulose acetate membranes for various feed solution compositions. (From Drioli, E., Lonsdale, H. K., and Pusch, W., *J. Colloid Interface Sci.*, 51, 355, 1975. With permission.)

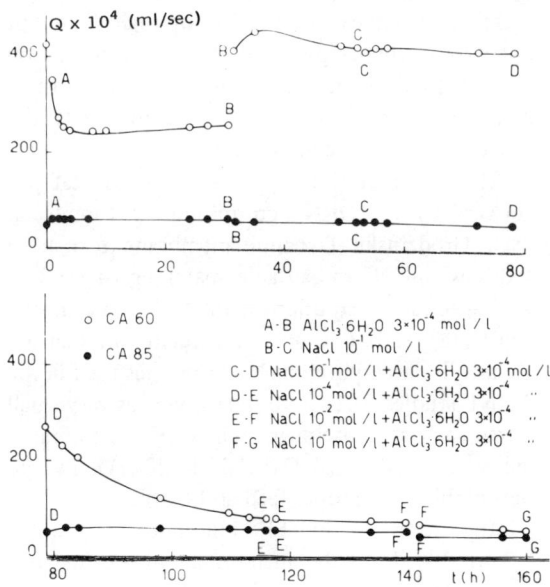

FIGURE 3. Volume flux, Q, vs. time for two asymmetric cellulose acetate membranes and different salt solutions; $\Delta P = 40$ atm, $T = 25°C$.

illustrated in the series of measurements shown in Figure 4. There, NaCl rejection is plotted against salt concentration with feed solutions of NaCl alone and solutions containing $AlCl_3$ as well. With only NaCl present, rejection increases with decreasing salt concentration, consistent with a Donnan exclusion mechanism. With $AlCl_3$ present, the NaCl rejection is nearly independent of concentration, remaining at the low value observed at high concentrations in the absence of $AlCl_3$. Donnan exclusion effects are inoperative at these higher concentrations because the electrolyte concentration markedly exceeds that of the fixed charges. One can conclude that in the presence of $AlCl_3$, the fixed charges of the membrane are neutralized, and this is consistent with the observations made at low concentration with $LiCl-AlCl_3$ mixtures. The neutralization could be due to (1) Al^{3+} ions taken up by the membranes in preference to monovalent cations; (2) H^+ ions resulting from the hydrolysis of $AlCl_3$ and being taken up by the carboxylic acid groups; or (3) the formation and uptake of complexes such as $Al_3(OH)_{3n}^{n+3}$, which will be discussed below. The uptake of these species onto the pore walls would affect both the ability of the membrane to reject salts[19,19a] as well as the mechanical permeability of the membrane due to the constriction of the pores.

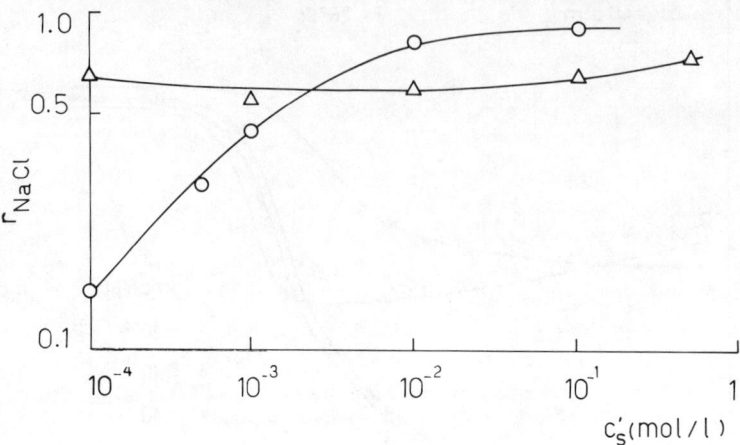

FIGURE 4. NaCl rejection, r_{NaCl}, as a function of NaCl brine concentration, C'_s, for pure NaCl feed solutions (○——○) and NaCl feed solutions containing 3×10^{-4} M/l $AlCl_3$ (△——△) at 50 atm and 25°C, using an asymmetric cellulose acetate membrane annealed at 60°C.

A plausible model to explain these results, then, is the following. A reversible, dynamic membrane is created in the micropores of the skin of the asymmetric membrane, caused by some hydrolysis product of aluminum present in the feed solution. Dynamic membrane formation appears to require both certain solution conditions (pH and concentration) and an appropriate finely porous substrate. This has been observed in previous work as well. The pH range in the experiments described here was 4.1 to 4.2 (at 3×10^{-3} M $AlCl_3$) to 3.5 (at 3×10^{-2} M $AlCl_3$). From these pH values and the complex aqueous solution chemistry of aluminum,[20,21] we hypothesize that a complex such as $[Al_3(OH)_8]_n^{n+3}$ is responsible for the formation of a dynamic membrane. Hydrolysis products of aluminum ion include several species:

$Al^{3+} + H_2O \rightleftharpoons AlOH^{2+} + H^+$

$Al^{3+} + 2H_2O \rightleftharpoons Al(OH)_2^+ + 2H^+$

$Al^{3+} + 3H_2O \rightleftharpoons Al(OH)_3 + 3H^+$

$Al^{3+} + 4H_2O \rightleftharpoons Al(OH)_4^- + 4H^+$

Several investigators have postulated that some of the above species polymerize, particularly the water-soluble species $Al(OH)^{2+}$ and $Al(OH)_2^+$, to yield polynuclear species. Singley and co-workers[22] have shown that, under nonequilibrium conditions, the predominant species may not be simply neutral aluminum hydroxide but a positively charged species. The same authors have presented a concentration distribution for the several species as a function of pH, as is shown in Figure 5. At the pHs of interest here (about 4 to 4.5) the two soluble, polymerizable species $AlOH^{2+}$ and $Al(OH)_2^+$ are present and it is these polynuclear species that may be responsible for dynamic membrane formation. The fact that membranes annealed at 80°C or higher do not appear to support dynamic membrane formation suggests that the micropores in the surface of these membranes are too small to accommodate the charged aluminum complex ions.

The interpretation is consistent with that put forward by the ORNL group and by Tanny and Jagur-Grodzinski. Dynamic membrane formation requires only (1) a suitable matching of sizes of the species in solution with the pores in the substrate, and (2) an attachment mechanism. Relatively large pores such as the ones used in the ORNL studies (0.1 to 10 µm) as well as very small pores such as those suggested for CA membranes annealed at 60 to 70°C (i.e. 10 to 20 Å) appear to be suitable with proper feed additives.

The decrease in Li^+ ion rejection, which occurred with all the membranes in the first few minutes after addition of aluminum ion and before dynamic membrane formation, requires comment. This is apparently the result of a "Donnan membrane effect" in which the rejection of a membrane-permeable ion can be reduced, even to negative values, by the presence of a membrane-impermeable coion. This was originally demonstrated by Lonsdale et al.[23] with $NaCl$-Na_2SO_4 mixtures and is the subject of a more recent work.[24] As noted, rejection of aluminum ion was

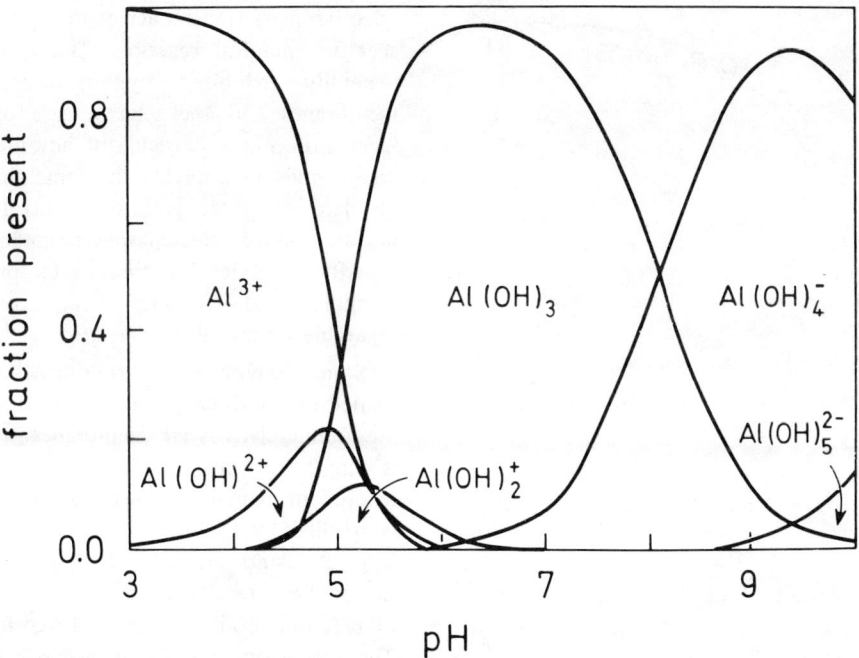

FIGURE 5. Species present vs. pH for 1×10^{-4} M aluminum perchlorate under nonequilibrium conditions. (From Committee report, State of the Art of Coagulation, presented at the Annual Conference of the *Journal of the American Water Works Association*, June 23, 1970.)

virtually quantitative in the work described above and the initial depression of Li$^+$ ion rejection on introduction of aluminum ion (Curves A and B in Figure 2) or when the dynamic membrane was destroyed (Curves C and D in Figure 2) is due to this phenomenon. The steady-state decrease in Li$^+$ ion rejection observed with the membranes annealed at the higher temperature, where no dyanmic membrane was apparent, is in reasonable agreement with a quantitative theory of this phenomenon recently proposed.[24] According to this treatment, the decrease in rejection of a membrane-permeable ion (Li$^+$ in this case) depends only on the ratio of the equivalent concentrations of impermeable to permeable ions. In the above experiments, this ratio was fixed at 9:1.

APPLICATIONS

The industrial potential of dynamically formed membranes has been extensively investigated, primarily by the group at ORNL. Interesting and promising results have been obtained in a variety of applications involving both HF and UF. Dynamic membranes have been successfully formed from species already present in the feed solution. For example, Perona et al.[25] demonstrated that the lignosulfonates present in spent sulfite liquors produced dynamic membranes on porous carbon and ceramic substrates, which then ultrafiltered colored matter and sugars from the solution. UF was carried out using a series of porous substrates to improve the quality of hardwood reducing sugars present in the permeate to the point where they could be used for fermentation.[26] An extensive study was performed by the ORNL group in cooperation with the International Paper Co. of the treatment of waste effluents generated in the Kraft pulping and bleaching processes. Dual layer hydrous Zr(IV) oxide-PAA membranes can be operated at the temperature of the bleaching process, and energy economies can be effected by immediate reuse of the hot water in the caustic extraction stage.[27]

Membranes dynamically formed from feed constituents have also been generated with sewage water and surface water feeds.[28] In addition, effluents from the dyeing and finishing of textiles have been treated with dynamically formed Zr-PAA membranes with encouraging results.[29] More than 80% of the organic carbon was removed from effluents generated in the dyeing of cotton-

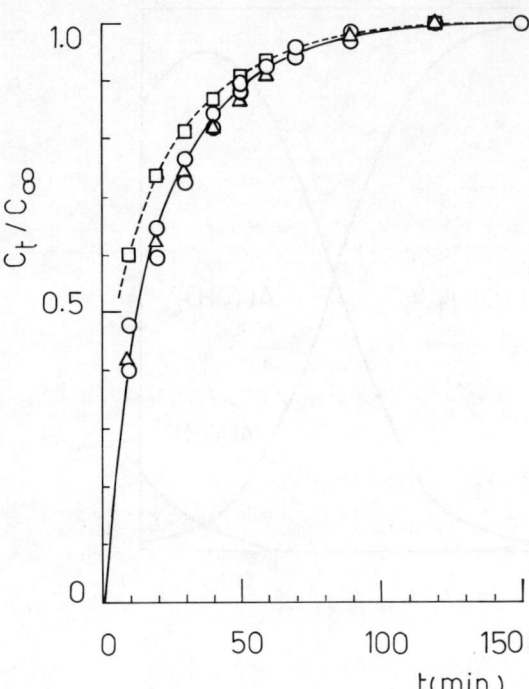

FIGURE 6. Behavior of the parameter C_t/C_∞ as a function of time: C_t = p-nitrophenol concentration at time t; C_∞ = p-nitrophenol at steady state. Initial substrate concentration: X = 30 μM, ○ = 60 μM; △ = 90 μM; □ = 120 μM. (From Drioli, E. and Scardi, V., *J. Membr. Sci.*, 1, 237, 1976. With permission.)

polyester fabrics, and 90% was removed from effluents resulting from dyeing of other synthetics. Water fluxes were high, 50 to 200 gal/ft²/day at 1000 psi, as is typical with dynamically formed membranes. These same membranes removed 98% of the organic carbon in effluents from commercial laundries. The filtrate appeared to be of sufficiently high quality for reuse in laundering.[30] Tests of waters simulating shower water to be generated during space missions were also promising.[31]

Dynamically formed UF and HF membranes appear promising for enzyme processing as well. Such membranes can be formed either by ultrafiltering the aqueous enzyme solution to form a gelled enzyme membrane on the porous support prior to adding substrate, or by recirculating a mixture of enzyme and substrate over the porous support so that the enzyme is continuously present in the substrate solution during and after the formation of the gelled enzyme membrane. These kind of dynamically formed gelled enzyme composite membranes may be particularly useful in processes where it is advantageous to combine a selective mass transfer across the membrane with a specific chemical reaction. The high flow rates generally exhibited by dynamically formed membranes and the relative ease of membrane regeneration are particularly advantageous. One could envision applying these membranes to the problem of water pollution control. Enzymatic degradation of selected species could be used to transform toxic into nontoxic products, for example, or to produce species that are easily separable via ultrafiltration.

Some preliminary experiments have been carried out with enzymes immobilized in gel form on commercial CA UF membranes.[32] The UF cell is thus tranformed into a heterogeneous enzyme reactor in which the distribution of enzyme is controlled by the hydrodynamics of the system. Typical results are presented in Figure 6, which shows the time dependence of the concentration of reaction product in an unstirred batch system. The supporting membrane was a CA membrane annealed at low temperature. The total amount of enzyme, acid phosphatase, gelled on the membrane was 11.5 $\mu g/cm^2$ and the thickness of the dynamically formed membrane can therefore be assumed to be approximately 100 Å. A non-steady state product concentration persisted for approximately 2 hr; this may be attributed to the time required for the development of a stationary concentration profile within the gelled enzyme layer. The curves of reduced concentration, C_t/C_∞, obtained with different substrate concentrations nearly superimpose, indicating that the transient behavior is not affected by substrate concentration.

The apparent Michaelis constant, V_m', calculated from these data is 1.35×10^{-4} M, a value seven-fold lower than that obtained with the same enzyme in free solution. This may be the result of the fact that the enzyme was present in a dense gel layer. However, the narrow substrate concentration range used in measuring K_m' was rather narrow, introducing considerable error in the calculation of K_m'. Another possibility is that the enzyme molecules have a favorable surface orientation in the gelled layer.

The steady-state concentration of p-nitrophenol in the permeate decreased with increasing pressure, i.e., with increasing flow rate, as shown in Figure 7. This is the result of the decreased contact time between enzyme and substrate with increasing flow rate. The asymptotic decline in enzyme activity to a definite value with

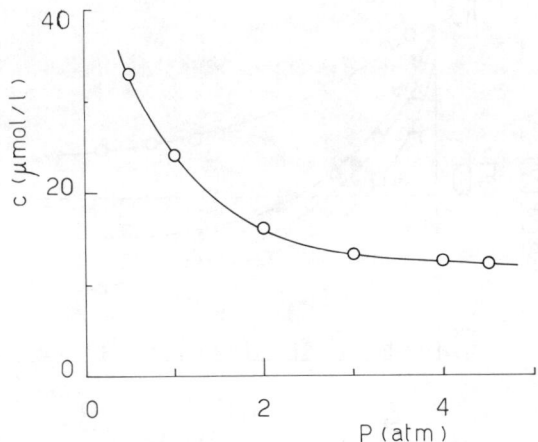

FIGURE 7. p-Nitrophenol concentration in the permeate at steady state as a function of applied pressure. Substrate concentration = 60 μM; enzyme loading = 6.8 $\mu g/cm^2$. (From Drioli, E. and Scardi, V., *J. Membr. Sci.*, 1, 237, 1976. With permission.)

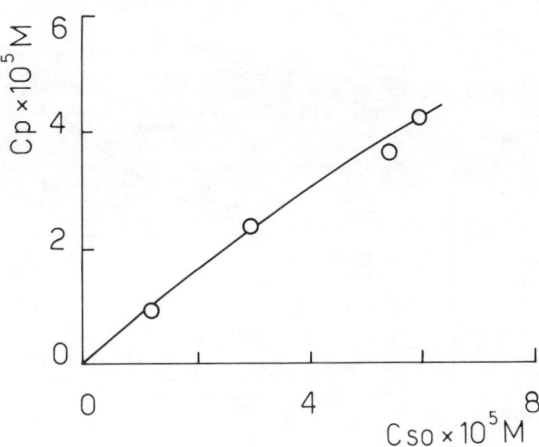

FIGURE 8. Behavior of the product concentration C_p in the permeate vs. the initial substrate concentration. o = experimental data; ——— analytical curve, Equation 4. (Capobianco, G., *Comportamento Cinetico di Enzimi Gelifica ti su Membrane Semipermeabili*, Ph.D. thesis in Chemical Engineering, University of Naples, 1976.)

increasing pressure is in agreement with the "gel model" theory,[33] which predicts that the permeate flow rate becomes pressure independent when the gel layer is completely formed.

This can be shown as follows: Assume that the reaction catalyzed by the dynamically formed, gelled enzyme layer follows Michaelis-Menten kinetics. The reaction rate is then given by Equation 1:

$$V = V'_{max} C_s/(K'_m + C_s) \quad (1)$$

A simple model can then be developed based on the following hypotheses:

1. Plug flow exists in the UF reactor through the gelled membrane.
2. The gelled enzyme layer on the membrane has a constant thickness δ_E.
3. The membrane does not reject either substrate or reaction products at all and there is no back diffusion of enzyme from the gel into the solution. Thus, the concentration profiles of substrate and enzyme upstream and downstream from the composite membranes are flat.

A mass balance across the gel layer can then be written as follows:

$$Q_T(dc/dx) = -\Omega V \quad (2)$$

where Q_T is the volumetric flow rate across the membrane, Ω is the membrane surface area, and x is the coordinate normal to the membrane surface. The boundary conditions are $C_s = C_{so}$ at x = 0 and $C_s = C'_s$ at $x = \delta_E$. Integration of Equation 2 then gives

$$K'_m \ln(C'_s/C_{so}) + C_{so} - C'_s = V'_{max}\Omega\delta_E/Q_T \quad (3)$$

The degree of conversion, $X = (C_{so} - C'_s)/C_{so}$, can now be introduced to give

$$-K'_m \ln(1 - \bar{X}) + C_{so}\bar{X} = V'_{max}\delta_E/\gamma_E\epsilon Q_T \quad (4)$$

where γ_E is the specific weight of the gelled enzyme and ϵ is its porosity. The terms V'_{max}, γ_E, and ϵ can be assumed constant to a first approximation. From Equation 4, K'_m can be obtained from the slope of a plot of $C_{so}X$ against $\ln(1 - X)$.

A comparison between theory and experiment is presented in Figure 8. Here, the concentration of product in the permeate is plotted against the initial feed concentration of substrate. Equation 4 was used to calculate K'_m as described above. Know values of the quantities on the right hand side of Equation 4 were then used to generate the theoretical plot of C_p vs. C_{so}. The agreement is clearly satisfactory showing that the model is self consistent. However, the value of apparent Michaelis constants calculated in this way do not agree with Michaelis constants measured in homogeneous systems, suggesting that these heterogeneous systems cannot be treated with a model derived for a homogeneous system.[34]

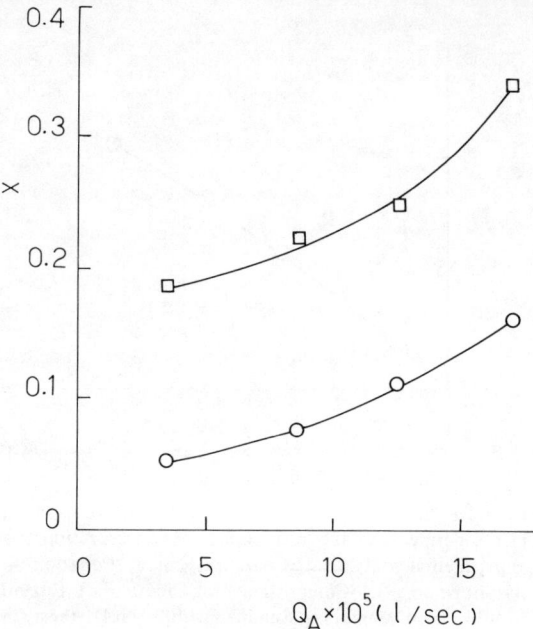

FIGURE 9. Behavior of X_1 and X_2 as a function of the axial flow rate Q_A. $C_{so} = 6 \times 10^{-5}$ M; $R = 1.5$ atm. (From Drioli, E. and Scardi, V., *J. Membr. Sci.*, 1, 237, 1976. With permission.)

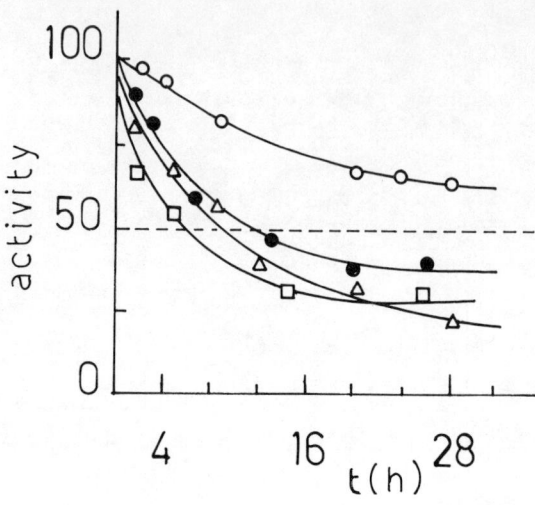

FIGURE 10A. Enzyme stability in free solution at $T = 30°C$; ○ = acid phosphatase (32 μg/ml); □ = invertase (50 μg/ml); △ = β-glucosidase (10 μg/ml); ● = urease (500 μg/ml). (From Cantarella, M. and Ragosta, G., *Boll. Soc. Ital. Biol. Sper.*, 51, 1092, 1975. With permission.)

It is possible to obtain the degree of conversion, X, from changes in either the feed stream or the permeate stream. These two values, X_1, and X_2, respectively, are plotted in Figure 9 as a function of axial flow rate, Q_A, measured in acid phosphatase experiments. In these experiments, the substrate solution was continuously recirculated. The degree of conversion increased with increasing Q_A, and while the two values X_1 and X_2 are not identical, the difference $X_1 - X_2$ was invariant with Q_A.

All the experimental results demonstrate that a gelled enzyme layer dynamically formed on a UF membrane retains its activity. Moreover, the stability is markedly increased relative to that of the free enzyme in solution. Some results comparing enzyme activity in free solution with that when tested as a dynamically formed gelled membrane are shown in Figure 10.[35] For all the enzymes tested (acid phosphatase, invertase, β-glucosidase, and urease) an increase in stability was observed.

CONCLUSIONS

In conclusion, the existence of a very thin secondary layer on primary membranes must be considered when studying hyperfiltration and ultrafiltration. Industrial waste waters or river

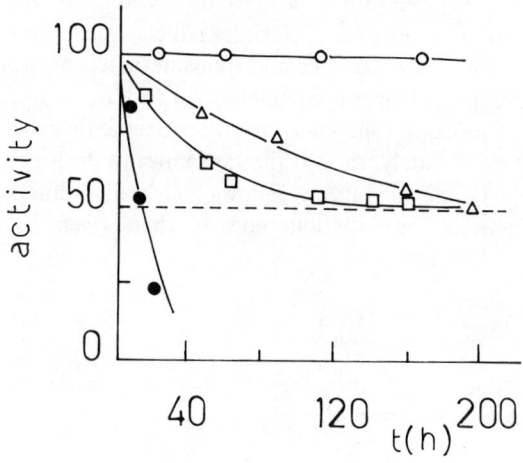

FIGURE 10B. Gelled enzyme stability at $T = 30°C$; ○ = acid phosphatase (16 μg); □ = invertase (17 μg); △ = β-glucosidase (5 μg); ● = urease (500 μg). (From Cantarella, M. and Ragosta, G., *Boll. Soc. Ital. Biol. Sper.*, 51, 1092, 1975. With permission.)

waters very often contain small amounts of species such as humic acid, surfactants, aluminum, etc., and these secondary membranes may be formed. A better understanding of their formation mechanism and of their properties can help to solve the problems such as fouling and rejection variation.

The effects observed when aluminum ions are present at low concentration in solution are probably related to the interaction of the linear polymers, resulting from the hydrolysis of the aluminum salt, with the negatively charged ionic groups present in cellulose acetate membranes.

A particular useful example of a dynamically formed membrane is one prepared from an enzyme. Such membranes have been shown to retain the activity of the enzyme while increasing its stability.

ACKNOWLEDGMENTS

The author would like to thank Dr. H. K. Lonsdale for helpful discussions during the preparation of the manuscript and Mrs. Lana Scott of Bend Research, Inc. for carefully typing the paper.

REFERENCES

1. Marcinkowsky, A. E., Kraus, K. A., Phillips, H. O., Johnson, J. S., and Shor, A. J., Hyperfiltration studies. IV. Salt rejection by dynamically formed hydrous oxide membranes, *J. Am. Chem. Soc.*, 88, 5744, 1966.
2. Kraus, D. A., Phillips, H. O., Marcinkowsky, A. E., Johnson, J. S., and Shor, A. J., Hyperfiltration studies. VI. Salt rejection by dynamically formed polyelectrolyte membranes, *Desalination*, 1, 225, 1966.
3. Kraus, K. A., Shor, A. J., and Johnson, J. S., Jr., Hyperfiltration studies. X. Hyperfiltration with dynamically formed membranes, *Desalination*, 2, 243, 1967.
4. Sachs, S. B., Baldwin, W. H., and Johnson, J. S., Jr., Hyperfiltration studies. XVI. Salt filtration by dynamically formed and cast poly(glutamic acid) and poly(acrylic acid) membranes, *Desalination*, 6, 215, 1969.
5. Jackson, J. M. and Landolt, D., About the mechanism of formation of iron hydroxide fouling layers on reverse osmosis membranes, *Desalination*, 12, 361, 1973.
6. Drioli, E., Gianfreda, L., Palescandolo, R., and Scardi, V., Activity of acid phosphatase as a gel-layer overlying an ultrafiltration cellulose acetate membrane, *Biotechnol. Bioeng.*, 17(9), 1365, 1975.
7. Shor, A. J., Kraus, K. A., Smith, W. T., Jr., and Johnson, J. S., Jr., Hyperfiltration studies. XI. Salt rejection properties of dynamically formed hydrous Zr(IV) oxide membranes, *J. Phys. Chem.*, 72, 2200, 1968.
8. Johnson, J. S., Jr., Minturn, R. E., and Wadia, P. H., Hyperfiltration. XXI. Dynamically formed hydrous Zr(IV) oxide-polyacrylate membranes, *J. Electroanal. Chem.*, 37, 267, 1972.
9. Van Heuven, J. W. and Bloebaum, R. K., Reverse osmosis by dynamically formed cation exchange membranes, *Desalination*, 14, 229, 1974.
10. Tanny, G. and Jagur-Grodzinski, J., Dynamically formed polyelectrolyte membranes on partially cured cellulose acetate, *Desalination*, 13, 53, 1973.
11. Minning, C. P. and Spiegler, K. S., Polarization at membrane-solution interfaces in reverse osmosis (hyperfiltration), in *Proc. NATO Advanced Study Inst. Polyelectrolytes II*, Selegny, E., Ed., Reidel Dordrecht, Holland, in press.
12. Pusch, W., Membrane potentials of asymmetric cellulose acetate membranes, in *Proc. NATO Advanced Study Inst. Polyelectrolytes II*, Selegny, E., Ed., Reidel Dordrecht, Holland, in press.
13. Sachs, S. B. and Lonsdale, H. K., Third Int. Symp. on Fresh Water From the Sea, 2, 561, 1970; Preparation and properties of polyelectrolyte membranes, *J. Appl. Polym. Sci.*, 15, 797, 1971.
14. Riley, R. L., Milstead, C. C., Lonsdale, H. K., and Mylseps, K. J., O.S.W. Res. Dev. Annual Progress Report, Gulf General Atomic, March 1971.
15. Grobl, R. and Pusch, W., Asymmetric behaviour of cellulose acetate membranes in hyperfiltration experiments as a result of concentration polarization, *Desalination*, 8, 277, 1970.
16. Manjikian, S., Loeb, S., McCutchan, J. W., Proc. First Int. Desalination Symp., Paper SWD/2, Washington, D.C., 1965.
17. Drioli, E., Lonsdale, H. K., and Pusch, W., Dynamically formed membranes prepared from aluminum ion, *J. Colloid Interface Sci.*, 51, 355, 1975.
18. Drioli, E., Lonsdale, H. K., and Pusch, W., Solute-Solute and Solute-Membrane Interactions in Hyperfiltration, paper presented at the DECHEMA Workshop, Industrielle Gevinnung von Susswasses aus dem Meer, Frankfurt am Main, January 1975.
19. Jacazio, G., Probstein, R. F., Sonin, A. A., and Yung, D., Porous Materials for Reverse Osmosis Membranes: Theory and Experiment, Fluid Mechanics Laboratory, Massachusetts Institute of Technology, Cambridge, 1973.
19a. Probstein, R. F., Sonin, A. A., and Yung, D., Brackish water salt rejection by porous hyperfiltration membranes, *Desalination*, 13, 303, 1973.
20. Committee report, State of the Art of Coagulation, presented at the Annual Conference of the *Journal of the American Water Works Association*, June 23, 1970.
21. *Gmelins Handbuch der Anorganischen Chemie*, 8th ed., System No. 35, Aluminum, Part A, Abtl.1, S.426, Verlag Chemie GmbH, Berlin, 1935.

22. Sullivan, J. H. and Singley, J. E., *Journal of the American Water Works Association,* 60, 1280, 1968.
23. Lonsdale, H. K., Riley, R. L., Milstead, C. E., LaGrange, L. D., Douglas, A. S., and Sachs, S. B., O.S.W. Res. Dev. Prog. Rep. No. 577, U.S. Government Printing Office, Washington, D.C., 1970, 118.
24. Lonsdale, H. K., Pusch, W., and Walch, A., Donnan-membrane effects in hyperfiltration of ternary systems, *Trans. Faraday Soc.,* 71, 501, 1975.
25. Perona, J. J., Butt, F. H., Fleming, S. M., Mayr, S. T., Spitz, R. A., Brown, M. K., Cochran, H. D., Kraus, K. A., and Johnson, J. S., Hyperfiltration processing of pulp mill sulfite wastes with a membrane dynamically formed from feed constituents, *Environ. Sci. Technol.,* 1, 991, 1967.
26. Collins, J. W., Boggs, L. A., Webb, A. W., and Wiley, A. A., *TAPPI,* 56, 121, 1973.
27. Moore, G. E., et al., Hyperfiltration and cross-flow filtration of kraft pulp mill wastes, ORNL, NSF, EP 14, (Contract No. W7405 eng-26), 1972.
28. Savage, H. C., Bolton, N. E., Phillips, H. O., Kraus, K. A., and Johnson, J. S., Jr., Hyperfiltration studies. XV. Hyperfiltration of plant effluents, *Water Sewage Works,* p. 102, March 1969.
29. Aurich, C., Brandon, C. A., Johnson, J. S., Jr., Minturn, R. E., Turner, K., and Wapia, R. H., Hyperfiltration. XX. Processing of textile dyeing wastes with dynamically formed membranes, *J. Water Pollut. Control Fed.,* 44, 1545, 1972.
30. Minturn, R. E., Johnson, J. S., Jr., Schofiled, W. M., and Todd, D. K., Hyperfiltration of laundry wastes, *Water Res.,* 8, 921, 1974.
31. Hester, J. C. and Brandon, C. A., *NASA Contract. Rep.,* 112127, 1972.
32. Drioli, E. and Scardi, V., Ultrafiltration processing with enzyme-gel composite membranes, *J. Membrane Sci.,* 1, 237, 1976.
33. Porter, M. C., Concentration polarization with membrane ultrafiltration, *Ind. Eng. Chem. Prod. Res. Dev.,* 11, 234, 1972.
34. Capobianco, G., Comportamento Cinetico di Enzimi Gelificati su Membrane Semipermeabili, Ph.D. thesis in Chemical Engineering, University of Naples, 1976.
35. Cantarella, M. and Ragosta, G., Stabilità di alcuni enzimi idrolitici in forma di geli, *Boll. Soc. Ital. Biol. Sper.,* 51, 1092, 1975.

ELECTROLYTIC DEACIDIFICATION

S. Suzuki and I. Karube

TABLE OF CONTENTS

Introduction 355

Electrolytic Purification and Deacidification of Sugar Solutions 355
   Recovery of Acids 356
   Removal of Heavy Metal Ions 357
   Diaphragm 357
   Preparation of Glucose 358
   Preparation of Xylose 359
   Preparation of Fructose 360
   Preparation of Galactose 360
   Electrolysis of an Acid-Leached Wood Solution 360
   Characteristics of Electrolytic Purification 360
   Conclusion 360

Electrolytic Preparation of Cysteine 361
   Electrodes 361
   Electrolysis under Acidic Conditions 362
   Electrolysis under Alkaline Conditions 362
   Conclusion 363

References 364

INTRODUCTION

Electrophoresis is the migration of charged particles suspended in a liquid under the influence of an applied electric field. This is the basic principle of electrolytic deacidification where acids are separated by migration in electrochemical cells. This process has been applied to the production of amino acids, sugars, and other compounds.[1-7]

Electrolytic production and deacidification of cysteine (an amino acid) is being employed in Japan. On the other hand, enzymatic methods are used for sugar production, and therefore the deacidification of sugar solutions is not required for sugar production at present. However, the deacidification of sugar solutions is a good example of the electrolytic purification process. In practice, neutralizers and ion exchange resins can be used for the deacidification; however, electrolytic reduction, codeposition, and purification can be simultaneously carried out with the electrolytic deacidification process. Therefore, the electrolytic deacidification process has a potential application in the food and pharmaceutical industries. A review of the electrolytic deacidification process is provided in this chapter.

ELECTROLYTIC PURIFICATION AND DEACIDIFICATION OF SUGAR SOLUTIONS

We have been studying fundamental problems of the electrolytic deacidification process for purification of sugar solutions in this laboratory since 1955.[8-10] Electrolysis of an acidified sugar solution placed in the cathode chamber of the electrolytic cell (Figure 1) causes the migration of acids to the anode chamber (Figure 2). Therefore, acids and impurities are removed from the catholyte. Thus, the sugar solution is neutralized and purified by electrolysis. A pure monosaccharide can be obtained from the catholyte. If sulfuric acid or perchloric acid is used as a catholyte, the acid can be completely recovered from the anolyte.

The electrolytic production of sugars comprises three operations: (1) hydrolysis of a polysaccharide with acid (sulfuric, hydrochloric, oxalic, or perchloric acid), (2) neutralization and purification of the solution by electrolysis, and (3) separation of a pure monosaccharide from the neutralized solution.

An example of electrolysis of the starch solution is shown in Figure 3. Sulfuric acid in the catholyte decreased with the quantity of current passed. On the other hand, the concentration of sulfuric acid in the anolyte gradually increased. This was caused by the migration of sulfate ions to the anode chamber.

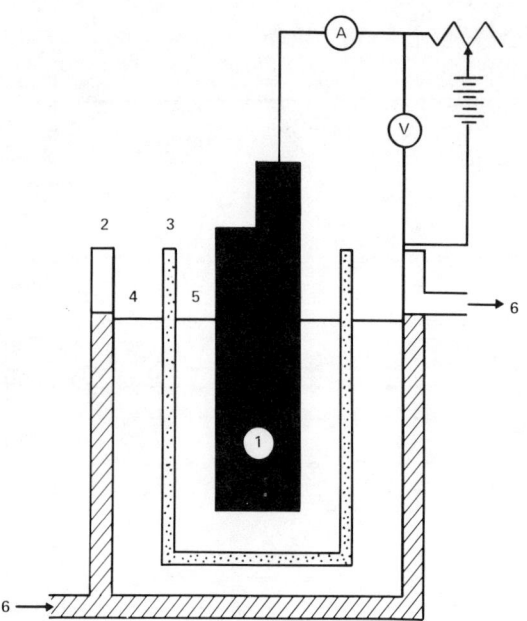

FIGURE 1. Electrolytic cell. 1. anode (Pb); 2. cathode (Pb vessel); 3. porcelain diaphragm (diameter 13 cm × 40 cm, thickness 0.3 cm); 4. cathode chamber (2.5 l); 5. anode chamber (3 l); 6. cooling water.

Recovery of Acids

Suitable acids for saccharification and electrolytic deacidification were examined. Acids such as sulfuric acid, hydrochloric acid, nitric acid, perchloric acid, and oxalic acid can be used for saccharification and electrolytic deacidification. However, nitric acid and oxalic acid undergo reduction during electrolysis. Sulfuric acid and perchloric acid are excellent for saccharification and electrolytic deacidification. Sulfate ions in the catholyte migrate to the anode chamber during electrolysis. The acid concentration of the anolyte increases with the quantity of current passed. Therefore, sulfuric acid is completely recovered from the anolyte and can be reused for saccharification. The optimum concentration of sulfuric acid for the catholyte is between 5% and 10%. The electrolytic removal of sulfate ions is not complete above 30% w/v.

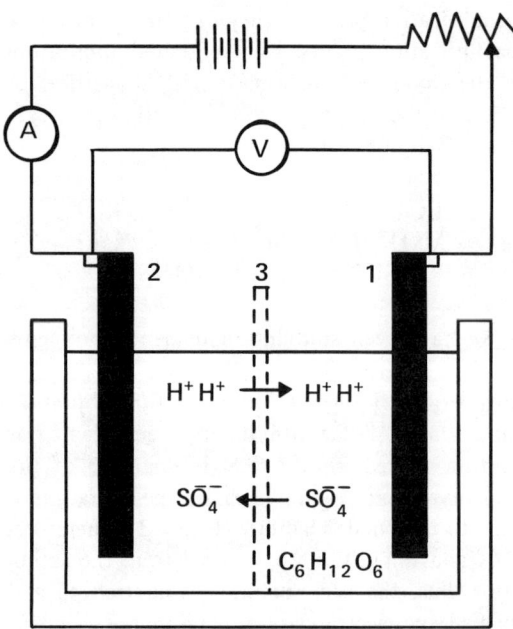

FIGURE 2. Electrolytic deacidification of the saccharified solution. 1. cathode; 2. anode; 3. diaphragm.

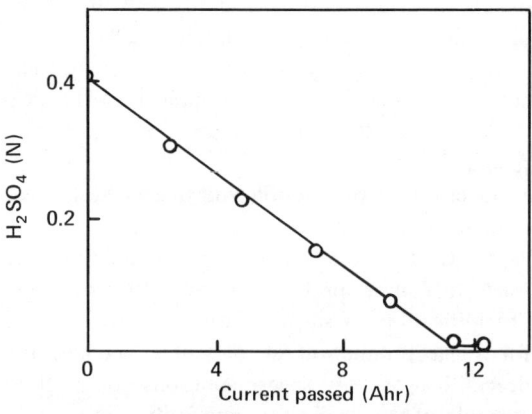

FIGURE 3. Electrolytic deacidification of the starch-saccharified solution. Catholyte: 1.5 l of 0.4 N sulfuric acid containing 414 g of glucose. Anolyte: 1.5 l of 1% of sulfuric acid. Voltage: 9.8 to 10 V. Current: 18 to 2.6 A. Porcelain was used for diaphragm.

Removal of Heavy Metal Ions

Heavy metal ions can be removed from the sugar solution during electrolysis. Ferrous ions were added to the starch-saccharified solution. Figure 4 shows the electrolytic removal of ferrous ions from the starch solution. The amount of ferrous ion in the catholyte decreased with the quantity of current passed. These heavy metal ions are removed from the monosaccharide solution by adsorption and reduction at the cathode. The electrolytic removal of various heavy metal ions such as Fe^{+3}, Al^{+3}, Cu^{++}, and Hg^{++} in the sugar solution is also shown in Table 1. These added heavy metal ions are removed from the monosaccharide solution. Furthermore, colored materials are removed from the solution during electrolysis by precipitation and coagulation.

Diaphragm

There are two kinds of porcelain diaphragms. The comparison of the performance of two porcelain diaphragms is shown in Table 2. For the gelatinized porcelain diaphragm, the loss of glucose was greater than in the case of the porcelain one. Therefore, no advantage of the gelatinized diaphragm was observed. An anion exchange membrane (Daia® ion exchange membrane, Mitsubishi Kasei Kogyo Co.) was also effective for the diaphragm of the electrolytic cell. As the migration of hydrogen ions from the anolyte was blocked by the anion exchange membrane, the deacidification of the catholyte occurred more rapidly.

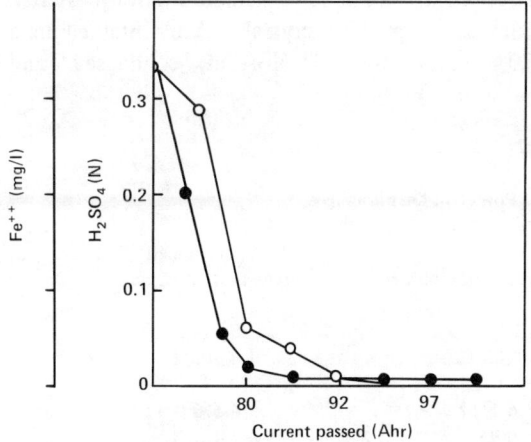

FIGURE 4. Electrolytic removal of ferrous ion from the starch saccharified solution. Catholyte: 1.5 l of 0.37 N sulfuric acid containing 360 g of glucose. Anolyte: 1.75 l of 5% sulfuric acid. Voltage: 11.7 V. Current: 38.2 A.
o———o Ferrous ion concentration.
●———● Sulfuric acid concentration in catholyte.

TABLE 1

Electrolytic Removal of Various Heavy Metal ions

	Metal ion				
	Fe^{+++}	Al^{+++}	Cu^{++}	Pb^{++}	Hg^{++}
Acidity of catholyte (M)	0.81	0.86	0.82	–	0.83
Duration (hr) of electrolysis					
Complete disappearance of the metal ions*	2 1/2	7 1/2	4	6 1/2	3 1/2
Complete electrolysis	9	10	6 5/6	8	7 1/2
Current passed (Ah)					
Complete disappearance of the metal ions*	21	26	23	26	26
Complete electrolysis	27	28	25	28	30
pH					
Complete disappearance of the metal ions*	2	3.6	3	3	2.2
Complete electrolysis	40–42	36–38	44–46	40–42	42–44

*Reagents for the detection of metal ions are as follows: Fe^{+++} [$K_3Fe(CN)_6$, Na_2S]; Al^{+++} [Alizarin S]; Cu^{++} [$K_4Fe(CN)_6$, Oxine, Na_2S]; Pb^{++} [K_2CrO_4, Oxine; Na_2S]; Hg^{++} [Na_2S].

A three-chamber cell (the acidified solution is placed in the middle chamber) can be also used for the electrolytic purification of sugars. As the resistance of two porcelain diaphragms is greater, the use of high voltage is required for deacidification of the solution.

Preparation of Glucose

A pure starch was saccharified with a 2% sulfuric acid solution by boiling at atmospheric pressure. This solution was then placed in the cathode chamber of the electrolytic cell and electrolyzed (Figure 5). The neutral and decolorized glucose solution obtained was condensed and crystallized. The crystalline glucose was found to meet all standards required by the Japanese pharmaceutical industry. The yield was 95% of the theoretical amount calculated from starch. Anhydrous glucose crystals were obtained in a 70% yield from decolorized, condensed, and

TABLE 2

Comparison of Performance of Porcelain Diaphragms

Electrolysis		With porcelain diaphragm	With gelatinized porcelain diaphragm
Voltage	(V)	13.1–113	23–142
Current	(A)	1.4–0.22	1.4–0.02
Temperature	(°C)	14–25	13–30
Duration	(hr)	6 1/12	4 5/6
Current passed	(Ah)	6.95	5.15
Deacidification		completed	completed
Volume change (cc)			
Catholyte		100–101	100–70
($C_2H_{12}O_6 + H_2SO_4$)			
Anolyte		100–95	100–123
(H_2SO_4 soln)			
Loss of glucose	(%)	3	26

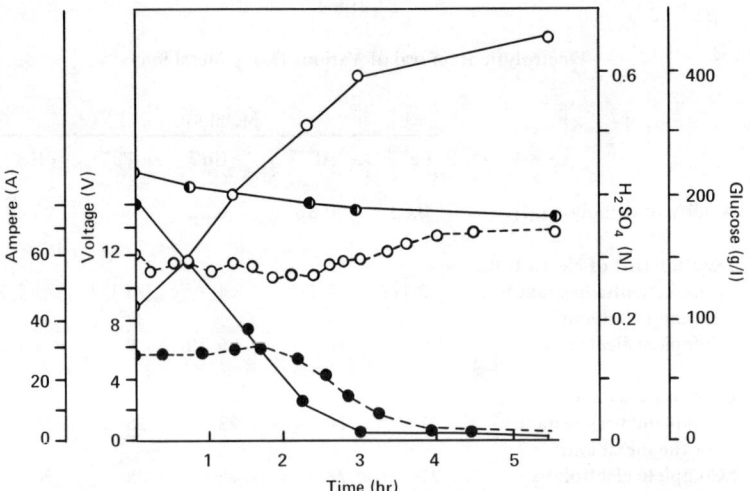

FIGURE 5. Electrolysis of the starch saccharified solution. Catholyte: 1.5 l of 0.4 N sulfuric acid containing 310 g of glucose. Anolyte: 1.5 l of 1% sulfuric acid.

o———o Concentration of sulfuric acid in anolyte.
•———• Concentration of sulfuric acid in catholyte.
⊙———⊙ Concentration of glucose in catholyte.
o--------o Terminal voltage.
•--------• Current.

deacidified solution by the addition of methanol. The loss of glucose occurred on the recrystallization process.

A flow-type electrolysis was performed for automatic, continuous neutralization of the starch-saccharified solution (Figure 6). Individual cells were connected in series and placed on different levels.

Preparation of Xylose

The electrolytic production of xylose from corn stalks was examined. Corn stalks were directly saccharified with the sulfuric acid solution at atmospheric pressure and electrolyzed by the method described above. Results obtained are shown in Figure 7. Heavy metal ions, color, and dissolved furfural were removed from the solution by electrolysis and a neutral xylose solution was obtained. Furfural produced by pentose hydrolysis was aggregated and precipitated by electrolytic treatment. Crude xylose (mp 108 to 117°C, 14% yield) was obtained from the decolorized and condensed syrup. Upon recrystallization, the pure product (mp 141 to 145°C, $[\alpha]_D^{10} = +19.4$) was obtained in a 6.5% yield. This low yield resulted from the low xylose content of the corn stalk.

A saccharified solution obtained from peanut shells (0.8 N $HClO_4$, 97 to 100°C for 6 hr) was placed in the cathode chamber of the diaphragm cell. Perchlorate ions and furfural in the saccharified solution were removed by electrolysis and perchloric acid was recovered from the anolyte. Crude xylose (mp 98 to 104°C, 13% yield) was obtained from electrolytically refined

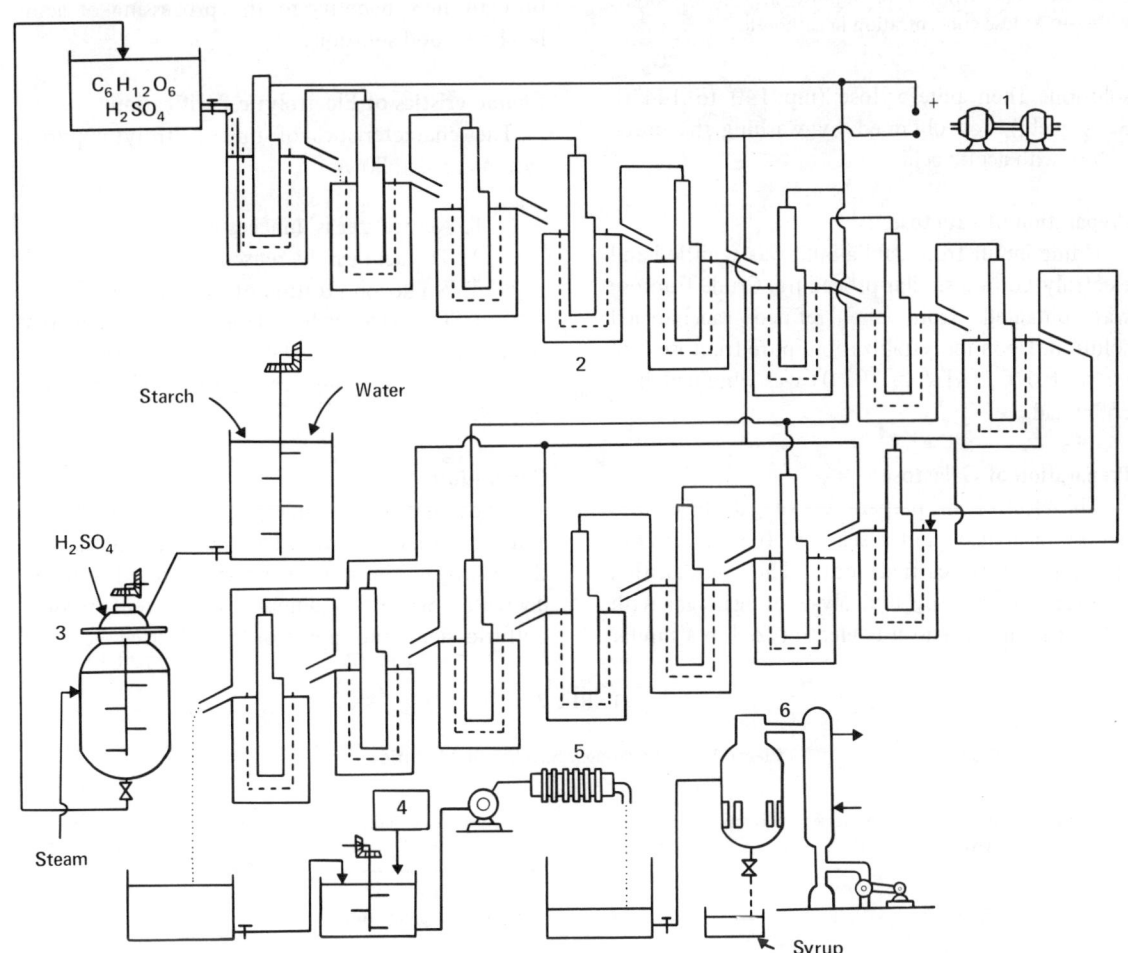

FIGURE 6. Diagram of the electrolytic preparation of glucose. 1. DC source; 2. electrolytic cell; 3. starch converter; 4. decolorization tank; 5. filter press; 6. evaporator.

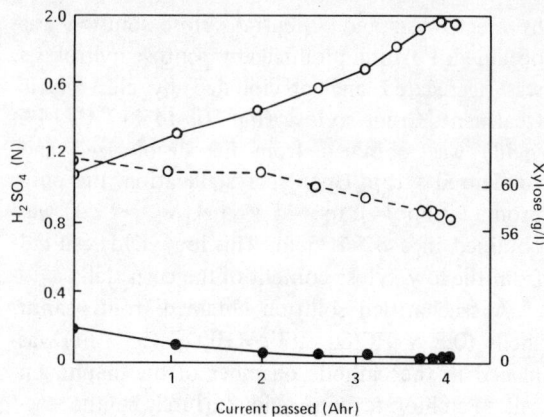

FIGURE 7. Electrolysis of the corn stalk saccharified solution. Catholyte: 100 ml (0.12 N) of sulfuric acid containing 6.42 g of xylose. Anolyte: 50 ml (1.1 N) of sulfuric acid. Voltage: 8 to 9.5 V. Current: 4.4 to 0.3 A. Temperature: 48 to 56°C.
○──────○ Concentration of sulfuric acid in anolyte.
●──────● Concentration of sulfuric acid in catholyte.
○--------○ Xylose concentration in catholyte.

solution. Then pure xylose (mp 140 to 144°C, 3.4% yield) was obtained by washing the crude xylose with acetic acid.

Preparation of Fructose

Crude inulin from dahlia bulb was acidified and electrolyzed in a similar procedure. Crude fructose was obtained from the refined saccharified solution. Upon recrystallization, pure fructose (mp 96 to 101°C, $[\alpha]_D^{16} = -93.0$) was obtained in a 35% yield.

Preparation of Galactose

The electrolytic preparation of galactose from lactose, agar-agar, and a seaweed (*Gracilaria confervoides*) was carried out. The saccharified solution obtained by hydrolysis of agar-agar with 0.1 N perchloric acid was electrolyzed in a similar process. The results are shown in Table 3. Crude galactose was obtained from the syrup of the refined solution in a 28% yield. In the case of *Gracilaria confervoides*, the pretreatment of the raw seaweed with an alkaline solution was effective for obtaining crystalline galactose.

Electrolysis of an Acid-leached Wood Solution

An acid-leached wood solution obtained by the usual method (sulfuric acid treatment at high pressure) was introduced into the cathode chamber of the diaphragm cell and electrolyzed by the method described above. Sulfate ions, furfural, and ferric ions in the solution were quantitatively removed by electrolysis. Thus, the saccharified solution was deacidified (37 kWh/kg H_2SO_4) and refined at the same time by this electrolytic process.

Application of the electrolytic process has brought new benefits to the processing of acid-leached wood solutions.

Characteristics of Electrolytic Purification

The characteristics of this electrolytic purification are as follows:

1. Neutralizer is unnecessary.
2. The acid can be reused.
3. No decomposition of sugar occurs.
4. The saccharified solution is refined (by removal of heavy metal ions, decolorization, decrease of ash, and coagulation of unknown impurities).

Conclusion

Acids that are suitable for the saccharification and have a large transport number can be used in this process. The acid in the saccharified solution decreases with the amount of current passed. Sulfuric acid and perchloric acid used for the

TABLE 3

Electrolysis of agar-agar Saccharified Solution

Cell voltage (V)	Current (A)	Current passed (Ah)	Temp. (°C)	Duration (hr)	pH	Sugar* in catholyte (g/l)		Apparent loss of sugar* (%)	Apparent recovery of $HClO_4$** (%)
						Before electrolysis	After electrolysis		
8.0–8.3	7.3–9.1	24	18–27	2 5/6	3.6	57	44	18	100

*Unclassified reducing sugar.
**Containing sulfuric acid from agar-agar.

saccharification are recovered quantitatively from the anolyte chamber. Pure glucose can be obtained in a 96 to 99% yield by electrolytic deacidification of the pure starch-saccharified solution. Monosaccharides such as glucose, xylose, fructose, and galactose are not electrolytically reduced.

ELECTROLYTIC PREPARATION OF CYSTEINE

Cysteine and its salts, which are used as food enrichers and as a component of some dermatological drugs, have been traditionally manufactured on a small scale by chemical reduction of cystine obtained from human hair. The common chemical method for the preparation of cysteine consists of many steps such as reduction of cystine with hydrochloric acid, neutralization of the acid with pyridine, removal of metal ions with hydrogen sulfide, and precipitation of the organic compounds. It was found in our laboratory that cystine can be readily reduced to cysteine and then the acid can be removed by the electrolytic deacidification method.

Various experiments have proved that pure cysteine or its salts can be obtained from cystine in a high yield by the electrolytic process. This electrolytic process is easier to carry out and there is no consumption of chemicals as described above. The electrolytic cell (for basic experiments) consisted of a 300-ml glass beaker in which an unglazed porcelain cylinder (diameter 3.7 cm, height 12.8 cm, and thickness 0.2 cm) was used for the preparation of cysteine. A platinum plate anode was placed inside the cylinder. The materials employed as a cathode were lead, silver, copper, zinc, and stainless steel. Hydrochloric acid, sulfuric acid, sodium hydroxide, sodium carbonate, and ammonium hydroxide were employed as electrolytes.

Electrodes

Various electrodes were employed at different temperatures and current densities. In all cases, cystine was readily reduced to cysteine. The lead electrode showed a higher current efficiency for reduction of cystine than other electrodes (Figure 8). No corrosion of the lead cathode was observed in the experiments, and lead was not detected in the product. Current efficiency was increased with increase in the hydrogen over voltage of the electrodes. The zinc electrode showed a higher current efficiency than the lead electrode, but the product was contaminated by zinc. In the case of a zinc electrode, nascent hydrogen evolved by zinc dissolution in acid also contributes to the reduction of cysteine. In industrial use, silver electrodes are used for the production of cysteine.

$$\begin{array}{c} NH_2 \\ | \\ CH_2-CH-COOH \\ | \\ S \\ | \\ S \\ | \\ CH_2-CH-COOH \\ | \\ NH_2 \end{array} \longrightarrow \begin{array}{c} CH_2-CH-COOH \\ | \quad\quad | \\ SH \quad NH_2 \end{array}$$

Cystine → Cysteine

Electrolysis Under Acidic Conditions

Electrolysis of the cystine solution was performed under acidic conditions. Reduction and deacidification occurred during electrolysis (Figure 9), and cystine was completely reduced in a high yield. Pure cysteine or its salt can be obtained very easily by evaporation of the deacidified or acidic catholyte. When hydrochloric acid was used as an electrolyte, chlorine gas was evolved from the anode. Therefore, a platinum electrode is required for the prevention of electrode corrosion. In the case of nonvolatile acids such as sulfuric acid, undesirable gases are not evolved from the anode. Pure cysteine sulfate is obtained from the concentrated catholyte. An anion exchange membrane (Daia ion exchange membrane, Mitsubishi Kasei Kogyo Co.) was also used for the diaphragm of the electrolytic cell (Figure 10). The catholyte could be completely deacidified by electrolysis and a neutral solution containing pure cysteine was obtained. An anion exchange membrane diaphragm was more effective for the deacidification than a porcelain diaphragm. As the migration of hydrogen ion from the anolyte was blocked by the anion exchange membrane, the deacidification of the catholyte occurred more rapidly. The results obtained from various experiments are summarized in Tables 4 and 5.

Electrolysis Under Alkaline Conditions

There was an appreciable loss of cystine and cysteine by electrolysis under alkaline conditions. As these substances have negative charges in alkaline solution, the loss was caused not only by the diffusion, but also by the migration of these substances from the cathode chamber to the anode chamber. The decomposed materials of cystine

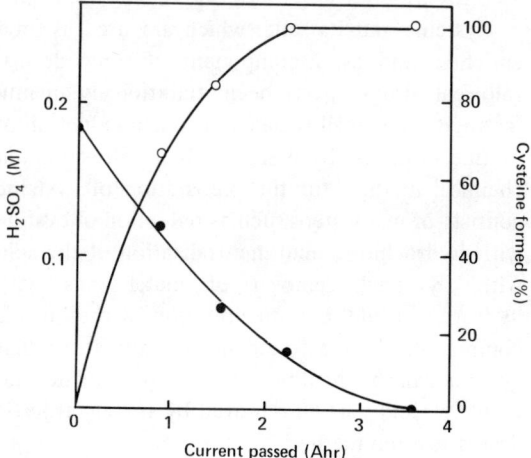

FIGURE 9. Electrolytic reduction of cystine and deacidification. Cathode: Pb 1.2 dm². Anode: Pb 0.6 dm². Catholyte: 200 ml of 2 N sulfuric acid containing 0.5 M cystine. Anolyte: 70 ml of 0.5 N sulfuric acid.
○────○ Cysteine formed.
●────● Concentration of sulfuric acid in catholyte.

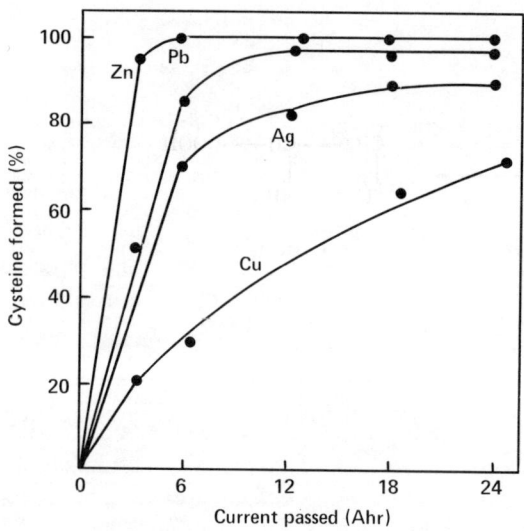

FIGURE 8. Electrolytic reduction of cystine with various electrodes. Cathode: 1.2 dm². Anode: Pt (0.5 dm²). Cathode current density: 5 A/dm². Catholyte: 200 ml of 2.7 N hydrochloric acid containing 0.5 M cystine. Anolyte: 70 ml of 1.4 N hydrochloric acid.

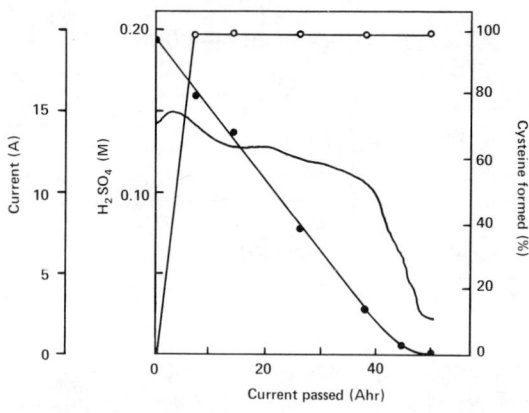

FIGURE 10. Electrolytic reduction of cystine and deacidification. Cathode: Pb 1.2 dm². Anode: Pb 0.6 dm². Catholyte: 100 ml of 0.5 N sulfuric acid containing 0.5 M cystine. Anolyte: 100 ml of 0.4 N sulfuric acid. Anion exchange membrane was used for diaphragm.
○────○ Cysteine formed.
●────● Concentration of sulfuric acid in catholyte.
──── Current.

TABLE 4

Electrolytic Preparation of Free Cysteine

Electrolyte	Cathode	Anode	Temp. (°C)	Current (A)	Duration of electrolysis (hr)	Current passed (Ah)	Yield (%)
HCl	Pb	Pt	21	7.1–0.2	10.5	33.6	96
	Ag		80	13.1–0.1	3.5	20.8	90
	Cu		80	13.1–0.3	3.5	20.0	90
H_2SO_4	Pb	Pb	20	4.0–0.4	22.5	67.2	93
			25	12.7–0.3	12.5	62.3	94
			80	14.8–2.0	5	49.5	91
			80	5.0–2.7	13	55.4	94

TABLE 5

Electrolytic Preparation of Cysteine Salts

Electrolyte	Cathode	Temp. (°C)	Yield (%)	Product Metal ion	Product Cystine ion	$[\alpha]_D$
HCl	Pb	20	97	not detected	not detected	+6.7
	Pb	80	82	not detected	not detected	+6.1
	Zn	25	98	detected	not detected	+0.7
	Cu	80	85	not detected	not detected	+6.2
	Ag	80	86	not detected	not detected	+6.2
H_2SO_4	Pb	27	97	not detected	not detected	+3.2
	Pb	80	92	not detected	not detected	+2.8

and cysteine are not detectable in the catholyte. After electrolysis, the catholyte was dried and the composition of the residual matter was examined.

Residual matter	9.0 (g) (74% of starting materials)
Cysteine	7.0
Cystine	2.0
Pb^{2+}	not detected
NH_4^+	not detected

In conclusion, electrolysis in alkaline condition gave a lower yield of pure cysteine or its salts than electrolysis in acidic condition.

Conclusion

Cystine was electrolytically reduced to cysteine when a lead, copper, silver, zinc, or stainless steel cathode was used. Moreover, in the case of the lead cathode, the product was obtained with a high current efficiency. The acidic catholyte was more favorable than the alkaline one. Pure cysteine hydrochloride and sulfate were readily obtained in high yields from the acidic catholyte. The catholyte could be completely deacidified by electrolysis and pure free cysteine was readily obtained in a high yield from the catholyte.

REFERENCES

1. **Mizuguchi, J., Karube, I., and Suzuki, S.,** Electrochemical deposition of collagen, *J. Chem. Soc. Japan Ind. Chem. Sect.* (translated from Japanese), 73, 2123, 1970.
2. **Karube, I., Suzuki, S., Kinoshita, S., and Mizuguchi, J.,** Electrochemical shaping of fibrous protein fibrils, *Ind. Eng. Chem. Prod. Res. Develop.*, 10, 160, 1971.
3. **Mizuguchi, J., Karube, I., Kobayashi, T., and Suzuki, S.,** Application of electrolysis to purification of fibrous protein, *J. Chem. Soc. Japan Ind. Chem. Sect.* (translated from Japanese), 74, 134, 1971.
4. **Karube, I., Mizuguchi, J., and Suzuki, S.,** Mechanism of electrochemical forming of collagen film, *J. Chem. Soc. Japan Ind. Chem. Sect.* (translated from Japanese), 74, 971, 1971.
5. **Karube, I. and Suzuki, S.,** Electrochemical aggregation of tropocollagen, *Biochem. Biophys. Res. Commun.*, 48, 320, 1972.
6. **Karube, I. and Suzuki, S.,** Studies on preparation and application of artificial protein membrane, *Res. Rep. Asahi Glass Found.* (translated from Japanese), 21, 53, 1972.
7. **Karube, I. and Suzuki, S.,** Effects of salts on electrochemical forming of collagen films, *J. Chem. Soc. Japan Ind. Chem. Sect.* (translated from Japanese), 74, 1267, 1971.
8. **Mizuguchi, J., Suzuki, S., Satoh, T., Aizawa, M.,** Electrolytic preparation of chemicals, *Chem. Ind.* (translated from Japanese), 23, 990, 1970.
9. **Suzuki, S.,** Utilization of organic waste waters, *Rep. Ind. Res. Lab. Shikoku* (translated from Japanese), 24, 15, 1973.
10. **Suzuki, S.,** Trends of recovery process: compounds from organic waste waters, *J. Chem. Eng. Soc. Japan* (translated from Japanese), 32, 532, 1968.

AUTHOR INDEX

A

Abdel-Alim, A. A., 125(ref. 8)
Abramowitz, M., 66(ref. 1)
Abrams, D. S., 66(ref. 2)
Acrivos, A., 78, 126(ref. 28), 127(ref. 80)
Aizawa, M., 364(ref. 8)
Albert, R. E., 191(ref. 33)
Al-Jariah, R. H., 292(ref. 19)
Allegra, G., 209(ref. 11)
Amano, H., 341 (ref. 29)
Amma, E. L., 209 (ref. 3, 4)
Anderson, J. R., 168(ref. 1)
Anderson, N. G., 66(ref. 2)
Anderson, R. E., 169(ref. 32)
Afinsen, C., 66(ref. 26)
Anon., 168(ref. 3)
Apelblatt, A., 75, 127(ref. 57)
Applebaum, S. B., 148(ref. 5)
Aramaki, G., 341(ref. 37, 38, 39)
Arden, T. V., 168(ref. 10), 169(ref. 41)
Aris, R., 66(ref. 2)
Arnold, G. B., 193(ref. 78)
Asada, M., 260(ref. 20)
Asher, W. J., 292(ref. 21)
Astill, K. N., 68(ref. 107)
A Study of Residential Water Use, 312(ref. 19)
Ataka, K., 341(ref. 30)
Atroshenko, L. S., 69(ref. 122)
Atwood, J. L., 209(ref. 6, 7, 8, 12, 15, 16, 18)
Aurich, C., 354(ref. 29)
Ayres, D. E. R., 169(ref. 19)

B

Bagley, E., 296, 312(ref. 8)
Bailey, M. F., 209(ref. 2)
Bakanov, S. P., 66(ref. 30)
Baker, B., 138
Baker, B., III, 148(ref. 10, 13), 149(ref. 46)
Baker, M., 340(ref. 11)
Bakker, C. A. P., 75, 127(ref. 54)
Baldwin, W. H., 353(ref. 4)
Ballard, D., 212, 223(ref. 5)
Banchero, J. T., 126(ref. 25)
Barduhn, A. J., 149(ref. 30)
Barker, P. E., 66(ref. 5)
Baron, T., 81, 109, 110, 118, 126(ref. 20)
Barthel, Y., 224(ref. 30)
Barr, J. T., 321(ref. 3)
Bassett, R. J., 241(ref. 41), 243, 259(ref. 3, 10)
Bassingthwaighte, J. B., 66(ref. 6), 67(ref. 72)
Batschelet, E., 75, 127(ref. 62)
Battaerd, H. A. J., 149(ref. 25, 42, 43, 44)
Bayadzhiev, L., 127(ref. 50)
Bdzil, J., 241(ref. 37)
Beavon, D. K., 224(ref. 25, 26, 27, 29)
Beck, W. J., 75

Beckmann, R. B., 116, 127(ref. 78)
Beek, W. J., 127(ref. 54)
Behr, J. P., 249, 260(ref. 19)
Benedict, M., 66(ref. 7, 8)
Bennett, L. C., 223(ref. 21)
Benson, H. F., 223(ref. 19)
Benson, R. C., 192(ref. 67)
Berens, A. R., 312(ref. 6, 9, 13, 14, 15)
Berg, J. G., 127(ref. 80)
Berman, A. S., 66(ref. 9)
Besserer, G. J., 192(ref. 68)
Bewley, J. L., 283(ref. 5), 284(ref. 35)
Bhattacharyya, D., 261, 267, 273, 274, 278, 283(ref. 1, 2, 3, 5, 6), 284(ref. 35)
Bidwell, R. M., 171, 190(ref. 6)
Bier, M., 66(ref. 10, 10a)
Bierlein, J. A., 192(ref. 62)
Bilal, B., 68(ref. 115)
Binder, M. J., 66(ref. 25)
Bird, R. B., 66(ref. 11), 67(ref. 54)
Birks, J. B., 127(ref. 65)
Blaschke, H. G., 100, 125(ref. 4)
Blesing, H. V., 149(ref. 25)
Blesing, N. V., 149(ref. 42, 43, 44)
Bloebaum, R. K., 353(ref. 9)
Bloomer, O. T., 192(ref. 55, 56)
Blum, D. E., 148(ref. 10, 14)
Blumenthal, R., 227, 241(ref. 38)
Bocard, J. P., 218, 224(ref. 36)
Boggs, L. A., 354(ref. 26)
Bojadjev, C., 75, 127(ref. 60)
Bolto, B. A., 148(ref. 8), 149(ref. 18, 19, 20, 21, 24, 25, 40, 42, 43, 44, 45)
Bolton, N. E., 354(ref. 28)
Bonnemayre, A., 312(ref. 11)
Booth, H. S., 171, 190(ref. 6)
Bough, W. A., 341(ref. 24)
Boussinesqu, J., 78, 126(ref. 29)
Bowles, L. K., 209(ref. 15)
Box, G. E. P., 66(ref. 12)
Bracewell, R., 69(ref. 120)
Brandon, C. A., 354(ref. 29, 31)
Bratzler, K., 223(ref. 18)
Brauer, H., 80, 84, 103, 106, 113, 114, 125(ref. 9), 128 (ref. 87).
Breiter, M. W., 241(ref. 37)
Brenner, H., 66(ref. 13, 14), 67(ref. 52)
Bright, P. B., 240(ref. 16)
Bring, J. L., 81
Brink, J. L., 82, 103, 109, 110, 118, 126(ref. 18)
Brooks, H., 126(ref. 34)
Brounshtein, B. I., 79, 126(ref. 13)
Brown, C. J., 169(ref. 16)
Brown, M. K., 354(ref. 25)
Brun, B., 100, 127(ref. 75)
Brunke, U., 72, 73, 75, 125(ref. 5), 126(ref. 33)
Buckingham, P. A., 211, 223(ref. 1)
Burrows, G., 66(ref. 15)
Butt, F. H., 354(ref. 25)

Byers, C. H., 75, 127(ref. 55)

C

Cahn, R. P., 286, 292(ref. 4, 11, 13)
Caldwell, K. D., 67(ref. 48)
Calmon, C., 168, 169(ref. 35, 38)
Cantarella, M., 354(ref. 35)
Capobianco, G., 354(ref. 34)
Carlier, C. C., 241(ref. 37)
Carslaw, H. S., 66(ref. 17)
Catsimpoolas, N., 66(ref. 18)
Cerullo, A., 68(ref. 107)
Chamberlain, D. G., 169(ref. 34)
Chandrasekhar, S., 126(ref. 37)
Chapman, S., 66(ref. 16)
Chen, S. S., 68(ref. 107)
Chludzinski, P. J., 283(ref. 16)
Chan, R. K. S., 312(ref. 3)
Chandrasekhar, S., 96, 99, 100
Channabasappa, K. C., 149(ref. 27)
Chao, K. C., 192(ref. 69)
Chappelear, D. C., 191(ref. 39)
Chu, C., 148(ref. 17)
Chueh, P. L., 176, 191(ref. 17, 18, 19)
Ciani, S., 243, 259(ref. 1)
Claytor, E. E., Jr., 224(ref. 28)
Clegg, J. W., 169(ref. 29)
Cline, G. B., 66(ref. 19)
Clinebell, P. W., 340(ref. 13)
Close, R. E., 192(ref. 59)
Cobble, J. W., 284(ref. 32)
Cochran, H. D., 254(ref. 25)
Coghlan, C. A., 193(ref. 78)
Cohen, J., 168(ref. 4)
Cohen, K., 23, 66(ref. 20)
Colburn, A. P., 85, 126(ref. 44)
Collins, J. W., 354(ref. 26)
Collins, R. E., 240(ref. 2)
Colton, C. K., 226
Committee report, State of the Art of Coagulation, 353 (ref. 20)
Condon, E. U., 69(ref. 123)
Connolly, J. F., 187, 193(ref. 81)
Conway, J. B., 126(ref. 45)
Conway, R. A., 340(ref. 4)
Cook, D., 192(ref. 64)
Cooney, D. O., 20, 66(ref. 21, 22, 23)
Cope, A. F., 149(ref. 25, 42, 43, 44)
Cordeiro, D. J., 191(ref. 25, 26, 27)
Cotton, F. A., 292(ref. 18)
Cowling, T. G., 66(ref. 16)
Cox, H. C., 66(ref. 24)
Coxon, M., 66(ref. 25)
Crane, J. S., 68(ref. 87)
Crank, A., 127(ref. 63)
Crank, J., 303, 304, 305, 306, 308, 312(ref. 7)
Crump, J. S., 191(ref. 29)
Cuatrecasas, P., 66(ref. 26)
Cundiff, D. W., 66(ref. 14)
Curl, R. F., 175, 191(ref. 13)
Curtiss, C. F., 67(ref. 54)

Cussler, E. J., Jr., 66(ref. 27), 68(ref. 91)
Cussler, E., 226, 238, 240(ref. 24, 25, 26), 243, 259(ref. 5), 292(ref. 22)

D

Dahl, L. F., 209(ref. 2)
Dailey, L. W., 212, 223(ref. 2)
Daniels, G. A., 306, 308, 311, 312(ref. 15, 16)
Danneil, A., 193(ref. 83)
Darryl, H., 68(ref. 112)
Dastur, S. P., 191(ref. 44)
Davenport, A. J., 191(ref. 22)
Davey, D., 169(ref. 16)
Davies, J. T., 120, 126(ref. 35), 127(ref. 82), 128(ref. 83, 86)
Davis, H. E., 149(ref. 31)
Davis, R. E., 78, 126(ref. 28)
Dazai, M., 323, 341(ref. 23)
DeCapita, E. G., 309, 312(ref. 17)
DeClerk, K., 2, 66(ref. 28)
Dehnicke, K., 197, 209(ref. 13, 14, 17)
de la Garza, A., 24, 67(ref. 40, 41)
de Groot, S. R., 66(ref. 29)
De Nie, L. H., 80, 100, 126(ref. 27)
Derjaguin, B. V., 66(ref. 30)
DeSimone, J. A., 228, 241(ref. 40)
Diamond Shamrock Chemical Co., 149(ref. 39)
Dibbs, H. P., 241(ref. 53)
Diepen, G. A. M., 173, 190(ref. 9), 191(ref. 10)
Dietz, J. C., 319, 340(ref. 13)
DiMarzio, E. Z., 66(ref. 31)
Dimian, A., 127(ref. 51)
Dingman, J. C., 223(ref. 4)
Doerges, A., 223(ref. 18)
Dohner, J., 67(ref. 68)
Donaldson, T. L., 228, 227, 233, 235, 240(ref. 8, 34), 241 241(ref. 52)
Done, J. N., 66(ref. 33)
Donnelly, H. G., 192(ref. 63)
Donnelly, S. T., 220, 224(ref. 44)
Dorfner, K., 168, 168(ref. 15), 169(ref. 36)
Douglas, A. S., 283(ref. 19), 354(ref. 23)
Dresner, L., 265, 283(ref. 26, 30)
Drew, T. B., 127(ref. 82)
Drickamer, H. G., 128(ref. 94)
Drioli, E., 353(ref. 6, 17, 18), 354(ref. 32)
Dunn, C. L., 223(ref. 16, 17)

E

Eakin, B. E., 192(ref. 56)
Eckenfelder, W. W., Jr., 340(ref. 2)
Edmister, W. C., 222, 224(ref. 48)
Edwards, D. O., 67(ref. 73)
Eguchi, W., 128(ref. 89)
Ehrlich, P., 192(ref. 58)
Eickmeyer, A. G., 223(ref. 6, 9)
Eisenberg, B., 220, 224(ref. 41, 42)
Eisenman, G., 259(ref. 1)
Elenkov, D., 127(ref. 50)

Elgin, J. C., 180, 191(ref. 37, 38, 39)
Ellington, R. T., 192(ref. 56)
Ellis, S. R. M., 171, 190(ref. 3)
Elzinga, E. R., Jr., 126(ref. 25)
Enns, T., 227, 240(ref. 27)
Eppinger, K., 149(ref. 40)
Erbar, J. H., 224(ref. 47)
Erk, S., 126(ref. 24)
Evans, D. F., 259(ref. 5), 292(ref. 22)
Evans, R. B., III, 66(ref. 34, 35)
Ewald, A. H., 175, 191(ref. 12)

F

Faltermayer, E., 193(ref. 97)
Farrant, L., 259(ref. 12)
Feynman, R. P., 66(ref. 36)
Fickett, A. P., 283(ref. 16)
Fields, J. W., 312(ref. 10)
Filey, R. L., 283(ref. 19)
Fishbein, G. A., 79, 126(ref. 13)
Fitzgerald, K. J., 217, 224(ref. 32)
Fleming, S. M., 354(ref. 25)
Foley, D. D., 169(ref. 29)
Folt, V. L., 312(ref. 9)
Forman, J. C., 192(ref. 52)
Forster, R. E., 226, 240(ref. 3)
Franck, E. U., 192(ref. 66), 193(ref. 83)
Frankenfield, J. W., 292(ref. 8, 21)
Fraser, H. J., 149(ref. 31)
Frazier, G. C., 240(ref. 10, 14)
Fredenslund, A., 66(ref. 38)
Freitas, F. R., 223(ref. 16, 17)
Fresch, H. L., 241(ref. 37)
Freyer, H., 69(ref. 115)
Friedberg, F., 66(ref. 39)
Friedlander, S. K., 241(ref. 48)
Froning, H. R., 224(ref. 28)
Fox, J. B., 66(ref. 37)
Fuller, J., 196, 209(ref. 1)
Furukawa, J., 260(ref. 20)
Fuyiyoshi, K., 128(ref. 89)

G

Galkin, N. P., 169(ref. 21)
Gal-or, B., 128(ref. 90, 91, 92)
Gami, D. C., 192(ref. 56)
Gardner, H. E., 112, 126(ref. 46), 127(ref. 77), 169 (ref. 23)
Garreau et al, 69(ref. 121)
Garrett, G. A., 67(ref. 41)
Garrison, K. A., 283(ref. 2, 6)
Gatsis, J. G., 189, 193(ref. 93)
Gaufres, R., 127(ref. 75)
General Mills Corp., 292(ref. 14)
George, D. R., 169(ref. 30)
George, J. H. B., 284(ref. 34)
Gerritsen, T., 67(ref. 50)
Gerstner, F., 168(ref. 14)
Gianfreda, L., 353(ref. 6)

Giddings, J. C., 2, 67(ref. 42, 43, 44, 48)
Gill, W. N., 49, 67(ref. 45), 68(ref. 98, 99)
Ginde, V. R., 148(ref. 17)
Glass, J. E., 312(ref. 10)
Glueckauf, E., 283(ref. 10)
Glueckauf, E., 67(ref. 46)
Gmelins Handbuch der Anorganischen Chemie 353 (ref. 21)
Gmehling, J., 67(ref. 47, 47a)
Goddard, J. D., 226, 228, 240(ref. 9, 13, 20), 241 (ref. 41), 259(ref. 9, 10)
Gold, H., 169(ref. 35)
Goodenbour, J. W., 223(ref. 17)
Goupil, D. W., 341(ref. 25)
Grabowski, E. F., 67(ref. 72)
Green, M. S., 190(ref. 8)
Greer, A. H., 169(ref. 25)
Greger, U., 128(ref. 93)
Gregor, H. P., 262, 283(ref. 4)
Grieves, R. B., 192(ref. 45, 46, 47, 48), 283(ref. 1, 2, 3, 5, 6), 284(ref. 35)
Griffith, B. L., 191(ref. 31)
Griffith, E. A. H., 209(ref. 3, 4)
Griffith, R. M., 78, 126(ref. 31)
Grigull, U., 126(ref. 24)
Grim, J. M., 312(ref. 2)
Griswold, J., 192(ref. 75)
Gröber, H., 108, 126(ref. 23, 24)
Groβl, R., 353(ref. 15)
Gros, G., 241(ref. 36)
Groseva, V. M., 68(ref. 85)
Grov, Ø., 128(ref. 95)
Grün, F., 75, 127(ref. 62)
Grushka, E., 67(ref. 48)
Gufres, R., 100
Gutknecht, J., 240(ref. 35)
Guttman, C. M., 66(ref. 31)
Guy, K. W. A., 178, 191(ref. 24)

H

Hadamard, I., 77, 81, 103, 125(ref. 7)
Hadamard, L., 120
Haglung, H., 67(ref. 51)
Hamielec, A. E., 79
Hamdi, A., 67(ref. 62)
Hamielec, A. E., 125(ref. 8), 126(ref. 47)
Handlos, A. E., 109, 110, 118, 126(ref. 20)
Handlow, A. E., 81
Hanji, K., 260(ref. 20)
Hannah, S. H., 168(ref. 4)
Hannig, K., 67(ref. 49, 50)
Hanson, C., 126(ref. 27), 127(ref. 69)
Happel, J., 67(ref. 52)
Harada, M., 96, 128(ref. 89), 341(ref. 41)
Harper, J. F., 80, 126(ref. 12)
Haselden, G. G., 192(ref. 60, 61)
Hayworth, K. E., 193(ref. 84)
Heertjes, P. M., 80, 100, 126(ref. 19, 27)
Heideger, W., 127(ref. 52)
Helfferich, F., 67(ref. 53)
Helfferich, F., 148(ref. 3), 168, 169(ref. 37)

Hemmingsen, E. A., 229, 240(ref. 4)
Henderson, D. R., 220, 224(ref. 44)
Henderson, H. T., 223(ref. 17)
Henkel, W. M., 126(ref. 49)
Herrin, J. P., 223(ref. 15)
Hess, M., 149(ref. 29)
Hessels, J. K. C., 66(ref. 24)
Hester, J. C., 354(ref. 31)
Hicks, C. P., 171, 190(ref. 4)
Higgins, I. R., 148(ref. 9)
Hill, F. S., 223(ref. 16)
Himmelblau, D. M., 127(ref. 56)
Hirano, K., 341(ref. 37, 38, 39)
Hirao, M., 341(ref. 32)
Hirschfelder, J. O., 67(ref. 54, 55)
Himmelblau, D. M., 75
Hinckley, R. B., 177, 191 (ref. 21)
Hinn, H. W., 321(ref. 18)
Hoché, G., 168(ref. 2)
Hochhauser, A., 292(ref. 22)
Hoelscher, H. E., 96, 99, 100, 126(ref. 37), 128(ref. 90)
Hoffer, E., 265, 283(ref. 28)
Hoffman, E. J., 193(ref. 94)
Hoglund, R. L., 68(ref. 100)
Hollrah, U. M., 191(ref. 31)
Holland, F. A., 192(ref. 61)
Holm, L. W., 191(ref. 30), 193(ref. 90)
Holmes, L. E., 149(ref. 23)
Holmes, E. L., 149(ref. 23)
Holve, W. A., 126(ref. 19)
Homan, H. R., 211, 223(ref. 1)
Hoofd, L. J., 232, 241(ref. 49)
Hooper, J. W., Jr., 127(ref. 82)
Hopfenberg, H. B., 293
Horii, K., 341(ref. 19, 30)
Horiguchi, K., 260(ref. 20)
Horvath, C., 67(ref. 59)
Hough, E. W., 192(ref. 54)
House, J. E., 292(ref. 7)
Howard, D. W., 67(ref. 55)
Hristescu, E., 127(ref. 51)
Hsi, C., 191(ref. 20)
Hsu, C., 176
Hubert, P., 193(ref. 85, 86, 87)
Huie, N. C., 192(ref. 72, 73)
Hull, P., 193(ref. 89)
Humphrey, M. B., 209(ref. 19)
Hussinger, R. C., 193(ref. 88)
Hwang, S-T., 67(ref. 56)
Hwang, S., 264, 283(ref. 25)

I

ICI Australia Ltd, 149(ref. 39)
Ida, S., 191(ref. 40)
Iitaka, Y., 209(ref. 5)
Ikeda, T., 191(ref. 40)
Imamura, T., 128(ref. 89)
Imhoff, K., 340(ref. 5)
Irani, C. A., 192(ref. 71)
Irving, H. M. N. H., 292(ref. 19)
Iwai, M., 341(ref. 40)

J

Jacazio, G., 353(ref. 19)
Jackson, J. M., 353(ref. 5)
Jacquez, J. A., 341(ref. 46)
Jaeger, A., 192(ref. 74)
Jaeger, J. C., 66(ref. 17)
Jagur-Grodzinski, J., 344, 348, 353(ref. 10)
James, B. R., 127(ref. 81)
Jepson, W. B., 191(ref. 12)
Johansson, M., 254, 260(ref. 17)
Johns, L. E., 116, 127(ref. 78)
Johnson, A. I., 126(ref. 47)
Johnson, C. R., 293, 312(ref. 1)
Johnson, J. S., 268, 354(ref. 29), 253(ref. 1, 2, 3, 4, 7, 8 25, 28, 30), 283(ref. 21, 22, 23, 26)
Jones, G. E., 149(ref. 30)
Jones, G. L., 318, 340(ref. 6)
Jones, J. H., 216, 217, 224(ref. 28)
Jones, R. L., 66(ref. 38)
Joyce, R. S., 341(ref. 26)
Jumawan, A. B., 283(ref. 6)
Jung, D., 241(ref. 44), 353(ref. 19a)

K

Kabat, E. A., 67(ref. 57)
Kammermeyer, K., 67(ref. 56), 264, 283(ref. 25)
Kanagawa, A., 67(ref. 58)
Kapelyushnikov, M. A., 191(ref. 42)
Karger, B. L., 2, 67(ref. 59)
Karube, I., 364 (ref. 1, 2, 3, 4, 5, 6, 7)
Kasch, J. E., 192(ref. 75)
Katchalsky, A., 38, 75, 67(ref. 60), 127(ref. 57), 227, 241(ref. 38)
Katz, D. L., 192(ref. 76)
Kay, W. B., 33, 34, 179, 191(ref. 32), 192(ref. 62), 193(ref. 82)
Kedem, O., 38, 67(ref. 60), 265, 283(ref. 24, 27, 28)
Keller, K. H., 241(ref. 48)
Kennedy, G. C., 192(ref. 65)
Kent, R. L., 220, 224(ref. 41, 42)
Kertes, A. S., 259(ref. 11)
Kerr-McGee, 189
Kim, J. C., 223, 224(ref. 49)
Kim, Y. J., 191(ref. 23)
Kimura, S. G., 241(ref. 54, 55, 56), 283(ref. 15)
King, A. D., 192(ref. 50)
King, C. J., 67(ref. 61), 68(ref. 110), 148(ref. 12)
King, F. W., 224(ref. 25, 26, 27)
King, J. C., 75, 127(ref. 55)
King, M. B., 192(ref. 61)
Kinoshita, S., 364(ref. 2)
Kirchoff, R. H., 67(ref. 62)
Kistaller, G., 127(ref. 66)
Kitagawa, T., 292(ref. 8, 12)
Kitamura, H., 341(ref. 18)
Klein, J. P., 223(ref. 3)
Knight, J. R., 47, 67(ref. 63)
Knox, J. H., 66(ref. 33)
Knox, W. G., 149(ref. 29)

Kobayashi, M., 341(ref. 18)
Kobayashi, R., 192(ref. 76)
Kobayashi, T., 364(ref. 3)
Kobayashi, Y., 209(ref. 5)
Kobe, K. A., 193(ref. 77)
Kobuke, Y., 252, 260(ref. 20)
Koch, H. A., Jr., 191(ref. 28)
Koelzer, W., 127(ref. 67)
Kohn, J. P., 178, 179, 184, 189, 191(ref. 23, 25, 26, 27), 192(ref. 73)
Koike, T., 340(ref. 1), 341(ref. 43)
Kolin, A., 67(ref. 64, 65, 66)
Kono, M., 341(ref. 21)
Kordosky, G. A., 292(ref. 9)
Koster, R., 209(ref. 9)
Kozinski, A. A., 67(ref. 67)
Kramer, G. M., 193(ref. 95)
Krause, K. A., 283(ref. 22, 23, 26), 353(ref. 1, 2, 3, 7, 25, 28)
Kressman, T. R. E., 169(ref. 38)
Kreter, E., 299, 312(ref. 12)
Kreuzer, F., 226, 232, 240(ref. 22), 241(ref. 49, 150)
Krylov, V. S., 128(ref. 88)
Krochta, J., 68(ref. 112)
Kronig, R., 81, 82, 103, 109, 110, 118, 126(ref. 18)
Kubota, Y., 341(ref. 40)
Kulipla, S., 340(ref. 9)
Kumabe, K., 341(ref. 32)
Kunin, R., 148(ref. 4), 168, 169(ref. 22, 23, 29, 39)
Kuribayashi, M., 341(ref. 34)
Kurihara, S., 341(ref. 40)
Kurita, A. O., 168(ref. 11), 169(ref. 33)
Kurpen, J. J., 192(ref. 58)
Kutchai, H., 231, 233, 241(ref. 46)
Kyuchukov, G., 127(ref. 50)

L

Lacey, R. E., 149(ref. 26)
LaConti, A. B., 262, 265, 266, 283(ref. 16)
LaForce, R. C., 231, 241(ref. 45)
LaGrange, L. D., 283(ref. 19), 354(ref. 23)
Lakshminarayanaiah, N., 283(ref. 29)
Lamaze, C. E., 262, 283(ref. 14)
Lander, R. J., 227, 235, 239, 241(ref. 51), 259(ref. 7)
Landolt, D., 353(ref. 5)
LaRicerca, 192(ref. 57)
Larsen, S. V., 190(ref. 8)
Latty, J. A., 133, 145, 148(ref. 16)
Leder, F., 192(ref. 71), 193(ref. 95), 218, 224(ref. 38)
Lee, H. L., 51, 67(ref. 68, 69), 68(ref. 90)
Lee, J. I., 217, 220, 224(ref. 33, 34, 35)
Lee, J. L., 224(ref. 43)
Lefemine, G., 312(ref. 5)
Lehmkuhl, H., 209(ref. 9, 10)
Lehn, J. M., 249, 260(ref. 19)
Leland, T. W., 193(ref. 77)
Lento, H. G., 341(ref. 45)
Leonard, R. P., 341(ref. 25)
Levich, V. A., 77
Levich, V. B., 80, 125(ref. 1)
Lewis, C. J., 168(ref. 6), 285, 292(ref. 3)

Li, N., 244, 260(ref. 16)
Li, N. N., 168(ref. 6), 285, 286, 287, 289, 290, 292(ref. 1, 2, 4, 5, 6, 8, 10, 11, 13, 20, 21)
Licht, W., 126(ref. 45, 45a)
Light, W. G., 133, 134, 149(ref. 22)
Lightfoot, E. N., 20, 32, 33, 51, 60, 66(ref. 11, 22), 67(ref. 55, 67, 68, 69, 70, 71, 72, 73, 74), 68(ref. 89, 90)
Lode, T., 127(ref. 52)
Loeb, S., 149(ref. 26), 346, 353(ref. 16)
Long, F. A., 296, 312(ref. 8)
Long, R. B., 191(ref. 41)
Lonsdale, H. K., 345, 348, 353(ref. 13, 14, 17, 18), 354(ref. 23, 24), 268, 283(ref. 8, 19), 284(ref. 33)
Lu, C. Y., 176, 191(ref. 20)
Ludviksson, V., 67(ref. 74)
Luks, K. D., 191(ref. 25, 26, 27), 192(ref. 73)
Luner, S. J., 67(ref. 66)
Luttinger, L. B., 168(ref. 2)

M

Mack, W. A., 312(ref. 4)
MacKay, K. D., 292(ref. 9)
Macpherson, A. S., 148(ref. 8), 149(ref. 18, 19, 20, 21, 40, 45)
Maddox, R. N., 216, 217, 222, 224(ref. 40, 46, 47)
Maeda, M., 341(ref. 42)
Malanowski, S. K., 191(ref. 24)
Maltby, P., 169(ref. 28)
Maltoni, C., 312(ref. 5)
Manjikian, S., 346, 353(ref. 16)
Mantell, G. J., 312(ref. 3)
Mapstone, G. E., 218, 224(ref. 37)
Marcinkowsky, A. E., 353(ref. 1, 2)
Marcus, Y., 247, 254, 257, 259(ref. 11)
Maruyama, T., 168(ref. 4)
Mason, E. A., 66(ref. 34)
Mason, E. A., 66(ref. 35)
Masouka, H., 191(ref. 40)
Matesich, M., 259(ref. 5)
Mather, A. E., 224(ref. 33, 34, 35, 43)
Mather, F. J., 241(ref. 46)
Matschke, K., 127(ref. 73)
Matson, S. L., 241(ref. 54, 55, 56)
Matsuda, K., 341(ref. 41)
Matsuura, T., 283(ref. 12)
Matulevicius, E. S., 292(ref. 10)
Mayers, G. R. A., 128(ref. 83, 83a)
Mayland, B. J., 218, 224(ref. 36)
Mayr, S. T., 354(ref. 25)
Mazur, P., 66(ref. 29)
McBain, J. W., 241(ref. 42)
McCabe, W., 67(ref. 75)
McCarthy, J. M., 283(ref. 3)
McCaully, J. J., 189, 193(ref. 92)
McCoy, D. D., 216, 224(ref. 31)
McCutchan, J. W., 346, 353(ref. 16)
McDonald, C. W., 292(ref. 17)
McKelvey, J. G., 262, 283(ref. 13)
McKetta, J. J., Jr., 193(ref. 77)
McNeil, R., 149(ref. 18, 19, 20, 21)

McNeill, R., 148(ref. 8), 149(ref. 45)
Meiboom, F. W., 120
Meldon, J. H., 226, 227, 231, 233, 240(ref. 15, 33)
Mensing, W., 126(ref. 38)
Merigold, C. R., 292(ref. 7), 159, 169(ref. 18)
Merten, U., 148(ref. 6), 293(ref. 9)
Mewes, D., 125(ref. 9)
Meyboom, F. W., 128(ref. 84, 85)
Meyhack, U., 127(ref. 70)
Michaels, A. S., 68(ref. 88), 266, 284(ref. 284)
Michaels, S., 190(ref. 8)
Miller, W. S., 168(ref. 12), 169(ref. 27)
Milstead, C. C., 353(ref. 14)
Milstead, C. E., 283(ref. 18, 19), 354(ref. 23)
Milton, P. A., 209(ref. 12)
Mindler, A. B., 168(ref. 7), 169(ref. 20, 25)
Mining, C. P., 353(ref. 11)
Minturn, R. E., 353(ref. 8), 354(ref. 29, 30)
Misono, M., 340(ref. 10)
Miyashita, T., 341(ref. 34)
Miyazawa, S., 341(ref. 27)
Mizuguchi, J., 364(ref. 1, 2, 3, 4, 8)
Mochizuiki, M., 226
Mochizuki, K., 341(ref. 32)
Mochizuki, M., 240(ref. 3)
Modin, R., 254, 256, 260(ref. 14, 15, 17)
Mohan, R. R., 289, 290, 292(ref. 6, 20)
Molof, A. H., 341(ref. 28)
Moll, W., 241(ref. 36)
Moller, W. J. H. M., 68(ref. 76)
Molstad, M. C., 223, 224(ref. 49)
Moore, D. W., 80, 126(ref. 12)
Moore, G. E., 354(ref. 27)
Moore, F. L., 289, 292(ref. 15, 16, 17)
Moore, J. H., 243, 259(ref. 8)
Moore, T. F., 223(ref. 5)
Morrison, W. S., 168(ref. 9)
Muirbrook, N. K., 191(ref. 19)
Müller, H., 127(ref. 71)
Muller, W. J., 340(ref. 5)
Muntean, O., 127(ref. 51)
Murphy, J. E., 67(ref. 41)
Murray, J. D., 240(ref. 11, 12)
Myers, N. M., 67(ref. 48)
Myers, M. N., 67(ref. 44)
Mylseps, K. J., 353(ref. 14)

N

Nachod, F., 168, 169(ref. 40)
Nachod, F. C., 159, 168(ref. 6, 9, 14), 169(ref. 17)
Naden, D., 169(ref. 24)
Nady, L., 68(ref. 112)
Nagase, T., 341(ref. 33)
Nakayama, Y., 260(ref. 20)
Neilson, H. B., 213, 223(ref. 10)
Neilson, J. E., 75, 127(ref. 59)
Nelsen, L., 262, 266, 283(ref. 7), 284(ref. 31)
Nemerow, N. L., 340(ref. 3)
Newberry, W. R., 209(ref. 6, 7, 8)
Newitt, D. M., 192(ref. 60)
Newman, A. B., 103, 106, 109, 110, 118, 126(ref. 16)

Newman, J. S., 68(ref. 77)
Nishikawa, N., 292(ref. 12)
Nishikawa, Y., 292(ref. 8)
Nitsch, W., 99, 126(ref. 48), 127(ref. 73, 74)
Nivens, T. D., 191(ref. 34)
Noguchi, S., 341(ref. 40)

O

O'Brien, L. J., 193(ref. 90)
O'Conner, D. J., 340(ref. 2)
Odishaw, H., 69(ref. 123)
Ogawa, M., 341(ref. 23)
O'Grady, T. M., 186, 193(ref. 81)
Ohara, H., 341(ref. 30)
Okubo, H., 341(ref. 17, 33)
Olander, D. R., 68(ref. 78), 259, 260(ref. 21)
O'Mara, M. M., 309, 312(ref. 17)
Onken, V., 67(ref. 47, 47a)
Ono, H., 340(ref. 7, 8, 9), 341(ref. 23, 34, 35)
Orlander, D. R., 126(ref. 22)
Ornstein, L., 68(ref. 79)
Otocka, E. P., 68(ref. 80)
Otto, F. D., 224(ref. 33, 34, 35, 43)
Otto, N. C., 227, 240(ref. 29, 30), 243, 259(ref. 4)
Otto, W., 126(ref. 40, 41, 42), 128(ref. 95)
Otto, W. M., 293, 312(ref. 1)
Overbeek, J. T. G., 68(ref. 76)
Owens, W. R., 222, 224(ref. 46)
Ozaki, A., 341(ref. 19)

P

Pageau, L., 283(ref. 12)
Palescandolo, R., 353(ref. 6)
Pan, Y. C., 191(ref. 23)
Papadopoulos, M. N., 223(ref. 17)
Pansing, F., 126(ref. 45a)
Parent, J. D., 192(ref. 55, 56)
Parrish, R. W., 213, 223(ref. 10, 19)
Patel, J. M., 84, 126(ref. 26)
Paul, D. R., 68(ref. 81)
Paul, P. F. M., 171, 190(ref. 1)
Pauschmann, H., 2, 24, 28, 68(ref. 82)
Perego, G., 209(ref. 11)
Perona, J. J., 349
Perry, E. S., 66(ref. 5, 19), 67(ref. 66), 68(ref. 108)
Persson, B., 256, 260(ref. 14, 18)
Pfann, W. G., 68(ref. 83)
Pfeffer, R., 68(ref. 84)
Phillips, C., 68(ref. 85)
Phillips, H. O., 353(ref. 1, 2, 28)
Phillips, S. C., 69(ref. 125)
Philpot, J. S. L., 68(ref. 86)
Pietsch, G., 75, 125(ref. 2)
Pigford, A. L., 130
Pigford, R. L., 148(ref. 10, 11), 149(ref. 32, 46)
Pigford, T. H., 66(ref. 8)
Pilz, V., 171, 190(ref. 5)
Pitzer, K. S., 175, 191(ref. 13)
Podall, H. E., 149(ref. 28)

Pohl, H. A., 68(ref. 87)
Porter, M. C., 68(ref. 88), 262, 266, 283(ref. 7), 284(ref. 31), 354(ref. 33)
Potter, W. D., 284(ref. 34)
Powers, G. J., 68(ref. 96)
Prausnitz, J. M., 66(ref. 2, 38), 174, 176, 191(ref. 11, 17, 18, 19), 192(ref. 67)
Prescott, W. G., 149(ref. 23)
Probstein, R. F., 229, 241(ref. 44), 353(ref. 19, 19a)
Proctor, D. E., 306, 308, 311, 312(ref. 16)
Pruess, A. F., 169(ref. 29)
Pusch, W., 268, 284(ref. 33), 345, 353(ref. 12, 15, 17, 18), 354(ref. 24)

Q

Quinn, J. A., 227, 228, 233, 235, 240(ref. 8, 30, 34), 243, 259(ref. 4, 7)

R

Ragosta, G., 354(ref. 35)
Rathmell, J. J., 193(ref. 88)
Rathore, R. N. S., 68(ref. 105)
Ravoo, E., 68(ref. 112)
Rebert, C. J., 193(ref. 82)
Reid, R. C., 175, 177, 191(ref. 14, 21)
Reinert, K., 209(ref. 9)
Reis, J. F. G., 67(ref. 68), 68(ref. 89, 90)
Reusch, C. F., 68(ref. 91)
Rice, A., 69(ref. 119)
Rice, A. W., 149(ref. 35)
Richardson, J. A., 217, 224(ref. 33)
Richardson, M. I., 175
Richardson, M. J., 171, 174, 190(ref. 7)
Rideal, E. K., 128(ref. 86)
Rieke, R. D., 148(ref. 15)
Rietema, K., 68(ref. 92)
Riley, R. L., 353(ref. 14), 354(ref. 23)
Robb, W. L., 227, 240(ref. 28), 259(ref. 2)
Robert, C. J., 193(ref. 84)
Robertson, W. W., 192(ref. 50)
Robin, S., 192(ref. 49)
Robinson, D. B., 192(ref. 68)
Robinson, R. L., Jr., 192(ref. 69)
Rodesiler, P. F., 209(ref. 3)
Rolke, R. W., 149(ref. 35)
Rolks, R., 69(ref. 119)
Rombach, R., 68(ref. 107)
Rony, P. R., 2, 6, 68(ref. 93)
Roselius, W., 193(ref. 85)
Rosen, J. B., 68(ref. 94)
Rosenberg, N. W., 270, 284(ref. 34)
Ross, I. A., 223(ref. 21)
Rotem, Z., 75, 127(ref. 59)
Rowlinson, J. S., 174, 191(ref. 24), 193(ref. 79)
Rowlins, J. S., 191(ref. 22)
Rowlinson, J. S., 171, 175, 190(ref. 7), 191(ref. 12)
Rozen, A. M., 68(ref. 95)
Ruckenstein, E., 126(ref. 15)
Rudd, D. F., 68(ref. 96, 97)
Rybczynski, W., 77, 81, 103, 120, 125(ref. 6)

S

Sammon, D. C., 283(ref. 19), 341(ref. 47), 345, 353(ref. 4, 13), 354(ref. 23)
Sanchez-Palma, R. J., 67(ref. 73)
Sankarasubramanian, R., 49, 67(ref. 45), 68(ref. 98, 99)
Satoh, T., 364(ref. 8)
Savage, H. C., 354(ref. 28)
Savic, P., 78, 126(ref. 30)
Saville, D. A., 78, 119, 126(ref. 32)
Sawistowski, H., 119, 127(ref. 68, 81)
Scardi, V., 353(ref. 6), 354(ref. 32)
Scammell, G. W., 319, 340(ref. 15)
Schaaf, D. P., 283(ref. 1)
Schadow, E., 72, 73, 75
Schaeffer, F. E. C., 190(ref. 9), 191(ref. 10)
Schechter, R. S., 259(ref. 8)
Schecter, R. S., 243
Scheffer, F. E. C., 173
Schiffer, D. K., 292(ref. 22)
Schill, G., 249, 259(ref. 13, 14)
Schindler, A., 262, 266, 283(ref. 14, 17)
Schmidt, F. P., 67(ref. 67)
Schmidt-Traub, H., 103, 104, 113, 117, 125(ref. 10), 127(ref. 79)
Schneider, G., 192(ref. 70)
Schofiled, W. M., 354(ref. 30)
Scholander, P. F., 225, 226, 240(ref. 1)
Schomp, W. G., 284(ref. 35)
Schroder-Nielsen, M., 256, 260(ref. 15)
Schropfer, C. J., 319, 340(ref. 12)
Schubert, J., 168(ref. 6, 9, 14)
Schügerl, K., 126(ref. 39, 41, 42), 128(ref. 95, 96)
Schultz, J. S., 241(ref. 41), 253, 244, 259(ref. 3, 9, 10)
Scriven, L. E., 93, 127(ref. 69)
Seale, S. K., 209(ref. 12, 18)
Segima, T., 341(ref. 37, 38)
Seiko, Y., 340(ref. 9)
Shachter, J., 23, 68(ref. 100)
Shadow, E., 125(ref. 3)
Shah, S. M., 192(ref. 60)
Sheeler, J. F. R., 223(ref. 16)
Sherwood, T. K., 148(ref. 11), 149(ref. 32), 175, 191(ref. 14)
Shiba, T., 127(ref. 72)
Shih, T-S. T., 149(ref. 33)
Shimbashi, T., 99, 127(ref. 72)
Shor, A. J., 264, 265, 283(ref. 22, 23), 353(ref. 1, 2, 3, 7)
Shrier, A. L., 286, 287, 292(ref. 4, 5, 13)
Shubert, J., 169(ref. 40)
Shumate, S. E., II, 240(ref. 14)
Shultz, J. S., 226, 227, 232, 240(ref. 9, 13, 20, 32)
Siirola, J. J., 68(ref. 96)
Simmons, P. J., 169(ref. 16)
Simons, R., 283(ref. 27)
Sinfelt, J. H., 99, 128(ref. 94)
Singley, J. E., 348, 354(ref. 22)
Siudak, R., 148(ref. 8), 149(ref. 18, 19, 20, 21, 40, 45)
Skelland, A. H., 112, 126(ref. 21), 127(ref. 76)
Skelland, A. H. P., 126(ref. 14, 46)
Skelland, R., 81, 83, 84
Slattery, J. C., 68(ref. 101)

Slobod, R. L., 191(ref. 28)
Smith, D. R., 241(ref. 43), 243, 259(ref. 7)
Smith, H. B., Jr., 149(ref. 29)
Smith, J. C., 67(ref. 75)
Smith, K. A., 231, 240(ref. 33)
Smith, S. B., 168(ref. 5)
Smith, W. T., Jr., 353(ref. 7)
Smith, W. T., 283(ref. 22, 23)
Smolders, C. A., 283(ref. 20)
Snedeker, R. A., 191(ref. 36)
Snyder, A. E., 148(ref. 7)
Snyder, L. R., 67(ref. 59)
Sobocinski, D. P., 192(ref. 53)
Soda, A., 341(ref. 39)
Solomon, H. J., 191(ref. 43)
Sonin, A. A., 353(ref. 19a)
Sonoda, Y., 319, 340(ref. 14)
Sourirajan, S., 283(ref. 11, 12)
Spaan, J. A., 231, 241(ref. 47)
Spear, R. R., 224(ref. 47)
Spiegler, K. S., 148(ref. 1, 2), 283(ref. 131), 345, 353 (ref. 11)
Spitz, R. A., 354(ref. 25)
Sørensen, J. P., 68(ref. 104)
Stalkup, F. I., 193(ref. 88)
Staverman, J. A., 38, 68(ref. 102)
Steffen, A. J., 319, 340(ref. 11)
Stegemeier, G. L., 192(ref. 54)
Stegun, I., 66(ref. 1)
Stein, W. A., 226
Stein, W. D., 240(ref. 23)
Stephens, G. K., 149(ref. 25, 38, 42, 43, 44)
Sternling, C. V., 93, 127(ref. 69)
Stewart, W. E., 66(ref. 11), 68(ref. 103, 104, 114)
Stone, H. L., 191(ref. 29)
Streicher, R., 119, 126(ref. 42, 43), 128(ref. 96)
Strickland, R. D., 69(ref. 124)
Stringer, B., 341(ref. 47)
Stroeve, P., 228, 240(ref. 7), 241(ref. 39)
Strub, A. L., 340(ref. 13)
Suchdeo, S. R., 226, 227, 231, 232, 240(ref. 9, 13, 20, 31, 32), 259(ref. 9)
Sudarikov, D. N., 169(ref. 21)
Sullivan, J. H., 354(ref. 22)
Sussman, M. V., 68(ref. 105, 106, 107)
Sussman, S., 159, 169(ref. 17)
Suzuki, S., 364(ref. 1–10)
Sweed, N., 69(ref. 119)
Sweed, N. H., 68(ref. 108), 149(ref. 34, 35, 36, 37)
Swinton, E. A., 148(ref. 8), 149(ref. 18, 19, 20, 21, 45)
Szabo, G., 259(ref. 1)

T

Tagami, M., 283(ref. 18)
Taitel, Y., 74, 75, 127(ref. 53, 61)
Takao, K., 341(ref. 17)
Takenouchi, S., 192(ref. 65)
Takiguchi, Y., 341(ref. 20)
Talsma, H., 126(ref. 19)
Tamir, A., 127(ref. 61)
Tan, F. O., 179, 191(ref. 27)
Tanaka, M., 319, 340(ref. 8, 9, 14), 341(ref. 34)

Tang, Y. P., 75, 127(ref. 56)
Tanny, G., 344, 348, 353(ref. 10)
Tayeban, M., 112
Taylor, G. I., 68(ref. 109)
Taylor, G. T., 319, 341(ref. 16)
Tenen, J. M. G., 66(ref. 24)
Termini, J., 169(ref. 20)
Termini, J. P., 169(ref. 25)
Testing Method for Industrial Waste Water, 341(ref. 44)
Tezuka, T., 341(ref. 17, 33)
Thirkell, H., 223(ref. 20)
Thistlethwayte, D. K. B., 340(ref. 5)
Thodos, G., 191(ref. 44), 192(ref. 45, 46, 47, 48, 52)
Thompson, R. W., 68(ref. 110)
Todd, D. B., 191(ref. 37)
Todd, D. K., 354(ref. 30)
Todheide, K., 193(ref. 83), 192(ref. 66)
Tomiyama, M., 341(ref. 32)
Tosteson, D. C., 240(ref. 35)
Trayeban, M., 127(ref. 77)
Tregan, R., 312(ref. 11)
Treybal, R. E., 124(ref. 45), 126(ref. 17), 221, 222
Tsonopoulos, C., 175, 191(ref. 15, 16)
Tsuchiya, Y., 341(ref. 36)
Tsuru, S., 341(ref. 31)
Turner, K., 354(ref. 29)
Tuthill, E. J., 68(ref. 111)

U

Underwood, J. C., 335, 341(ref. 45)
Ulanowicz, R. E., 240(ref. 10)

V

Valteris, R. L., 171, 190(ref. 2)
Valteris, V. R. L., 187
van der Velden, P. M., 283(ref. 20)
Van Heuven, J. W., 353(ref. 9)
Van Oss, C. J., 66(ref. 5), 67(ref. 66), 68(ref. 76, 108)
Van Oss, J. C., 66(ref. 19)
Vauck, W., 127(ref. 71)
Vei, D., 219, 224(ref. 39)
Vermeulen, T., 68(ref. 112), 127(ref. 82)
Villadsen, J. V., 68(ref. 113, 114)
Vinograd, J. R., 241(ref. 42)
Virnig, M. J., 292(ref. 9)
Vitzthum, O., 187, 193(ref. 85, 86, 87)
Vodar, B., 192(ref. 49)
Vofsi, D., 283(ref. 24)
Voigtländer, J. G., 120, 128(ref. 84, 85)
von Berg, R. I., 126(ref. 49)
Von Halle, E., 68(ref. 100)
Voronina, S. M., 69(ref. 122)

W

Wagener, K., 68(ref. 115)
Walch, A., 354(ref. 24)
Wadia, P. H., 353(ref. 8)
Wang, L. K., 324, 341(ref. 25)

Wang, M. H., 341(ref. 25)
Wankat, P. C., 68(ref. 116, 117)
Wapia, R. H., 354(ref. 29)
Ward, A. F. H., 126(ref. 34)
Ward, W. J., 226, 227, 231, 233, 243, 240(ref. 18, 19, 21, 28), 241(ref. 37), 259(ref. 2, 6)
Ward, W. J., III, 241(ref. 55)
Watanabe, K., 341(ref. 35)
Watson, C. C., 68(ref. 97)
Watson, G. M., 66(ref. 34)
Webb, A. W., 354(ref. 26)
Weber, M. E., 80, 125(ref. 11)
Weinstock, J. J., 180, 191(ref. 38)
Weiss, C. O., 168(ref. 1)
Weiss, D. E., 148(ref. 8), 149(ref. 18, 19, 20, 21, 24, 25, 40–45), 132, 134
Weisz, P. B., 68(ref. 118)
Wellek, A. H., 81, 83, 84
Wellek, R., 112, 126(ref. 21)
Wellek, R. M., 126(ref. 26), 127(ref. 76)
Weller, F., 197, 209(ref. 13, 14, 17)
Welsh, D. G., 85, 126(ref. 44)
Wendt, C. J., 212
Wendt, C. J., Jr., 223(ref. 2)
Westwood, R. J., 169(ref. 19)
Whitehead, J. C., 193(ref. 91)
Wiggill, J. B., 126(ref. 35)
Wiley, A. A., 354(ref. 26)
Wilhelm, R. H., 69(ref. 119), 149(ref. 34, 35, 36)
Wilke, C. R., 149(ref. 32)
Wilkinson, G., 292(ref. 18)
Williams, D. F., 193(ref. 91)
Willis, D., 148(ref. 8), 149(ref. 18, 19, 20, 21, 24, 25, 40, 42, 43, 44, 45)
Willits, C. O., 341(ref. 45)
Wilson, I. L., 209(ref. 13)
Wise, W. S., 171, 190(ref. 1)
Wittenberg, J. B., 226, 240(ref. 5, 17)
Woertz, B. B., 219, 223(ref. 14)
Wolf, F., 312(ref. 12)
Wood, W., 159, 169(ref. 17)
Worboys, J. C., 149(ref. 25, 42, 43, 44)
Wyllie, M. R. J., 283(ref. 13)
Wyman, J., 240(ref. 6, 12)

Y

Yabe, K., 168(ref. 11), 169(ref. 33)
Yamamoto, D., 168(ref. 11), 169(ref. 33)
Yamazaki, H., 209(ref. 5)
Yamir, A., 74, 75, 127(ref. 53)
Yang, F. J., 67(ref. 44)
Yano, T., 341(ref. 19, 30)
Yaron, I., 128(ref. 92)
Yasuda, H., 266, 283(ref. 14, 17)
Yasuda, Y., 262
Yokouchi, H., 341(ref. 22)
Yorizane, M., 180, 191(ref. 40)
Yoshino, Y., 341(ref. 23, 42)
Young, C. L., 171, 190(ref. 4)
Yudovich, A., 192(ref. 69)
Yung, P., 229

Z

Zaalishvili, D., 192(ref. 51)
Zahka, J. G., 240(ref. 7)
Zhuze, T. P., 179, 191(ref. 35, 42)
Ziegler, K., 196, 209(ref. 9)
Ziemke, N. R., 340(ref. 12)
Zogg, M., 75, 127(ref. 58)
Zuckerman, M. M., 324, 341(ref. 28)
Zurmühl, R., 127(ref. 64)

SUBJECT INDEX

A

Absolute permeant flux
  in carrier-mediated transport in synthetic membranes, 235
Achromobacter sp.
  isolated from the aeration tanks, 334
Acid
  recovery of, 356
Acid-leached wood solution
  electrolysis of, 360
Acid stripping
  in the cleanup of waste waters containing chromates, 287
Activated carbon adsorption
  as a chemical treatment for waste waters
    batchwise treatment, 324
    countercurrent continuous, 324
    treatment using granular carbon columns, 324
    using pulverized carbon, 324
Activated sludge process
  as the primary treatment of fermentation waste waters, 337
Activated sludge treatment
  as a method of aerobic biological waste treatment, 316–318
Adair reaction sequences
  in analyzing the O_2-hemoglobin system, 226
Adduct-forming agent, 257
Adduct forming liquid-liquid extraction system, 245
Adip® process
  used for acid gas removal, 212
Adsorption, as a process that separates salt from the solution, 129
Aerobic type process
  as a biological treatment of waste waters, 316
Affinity chromatography, 2, 12
Agar-agar, 360
Alcohol distillation waste, utilization of feed stuffs, 326
Aliquat 336®
  as a liquid ion exchange agent, 289
Alkali metal pseudohalides
  inclusion of, 196
Alkaline leaching systems, 161
Amberlite LA-2®, 289
Amisol® process
  which combines two solvents in gas sweetening, 214
Ammonia stripping
  as a chemical treatment for waste waters, 325
  in technology for treatment of fermentation wastewaters, 315
Anaerobic digestion (methane fermentation)
  as a method of anaerobic biological waste treatment, 318
Anaerobic type process
  as a biological treatment of waste waters, 316
Analogs of Gravitational Settling, 15
Analytical solutions
  in carrier-mediated transport in synthetic membranes, 231

Analytical ultracentrifuge, 85
Anhydrous glucose crystals, 358
Anions
  summary of, 197
Annular electrophoresis, 19
Antisymmetric enrichment, 7
Aromatic molecules
  interaction with, 196
Aspects of morphology peculiar to pseudo-continuum descriptions, 28

B

Basidiomycetes, 336
Beavon sulfur removal process
  in gas sweetening, 216
Benfield process
  as an activated carbonate process, 213
Benzene
  reaction with, 196
B.F. Goodrich Company
  in the removal of solvent and monomers residuals from glossy polymers, 294
Bicarbonate
  carrier role of, 227
Bicarbonate-carbonate system
  buffering action of in facilitating the transport of hydrogen sulfide, 227
Biochemical oxygen demand, 314, 318, 324, 331–332, 337
Biochemical oxygen demand and chemical oxygen demand in biological treatment, behavior of, 330–332
Biological enzyme systems as trapping agents, 289–291
Biological transport, 225
Biological treatment of waste waters, 316
Biological treatment with physicochemical treatments for fermentation waste waters, 336–337
Bisulfide ions
  in coupled gas counter diffusion, 238
Blowdown feed, 156
Blue water, 157
BOD, see Biochemical oxygen demand
Brewery wastes, ultilization of as feed stuffs, 326
Brownian dispersion, 19, 64
Brownian effects, 35, 54
Brownian motion, 30, 38, 40, 42, 53, 54, 64
Brownian particles, 34
Bucket brigade mechanism
  in facilitated oxygen transport, 226
Bureau of mines, 164

C

Carbon adsorption, 153
Carbon dioxide
  in separations using supercritical gases, 184–186
Carbon dioxide transport

facilitated, 226–227
Carbonate ions
  in coupled gas counter diffusion, 238
Carboxylic weak acid exchanger in the hydrogen form, 156
Carrier-containing liquid membranes
  industrial applications of, 226
Carrier-mediated transport
  in synthetic membranes, 227
Cascades, 23, 31, 59
Catacarb® process
  as an activated carbonate process, 213
Cationic species, 154
Centrifugation, 54
Cerenkov radiation, 97
Channel ratio method, 87
Characteristic magnitudes in pressure dissuation, 38
Charged membrane ultrafiltration
  with actual and synthetic systems, 274–280
Chelating agent, 252, 257
Chemical coagulation and sedimentation
  as a chemical treatment for waste waters, 322
Chemical oxidation
  as a chemical treatment for waste waters, 324
Chemical oxygen demand
  in biological waste treatments, 319, 331–332
  in chemical and physical treatments for waste waters, 322
  in chemical oxidation as a chemical treatment for waste waters, 324
  in the activated sludge treatments for waste waters, 318
  legislative limits, 314
  removal of, 325, 334, 337–339
Chemical and physical treatments of waste waters, 319–326
Chitosan as a natural polyelectrolyte
  in chemical coagulation and sedimentation, 324
Chlorella
  as a biological waste treatment, 319
  in the alcohol fermentation waste, 327
Chromatography in a fixed bed of uniform spheres, 45
Clathrate compounds
  definition of, 195–196
  range of, 195
Claus sulfur unit, 215
Cleanup of waste waters containing chromates, 287–288
Cloete-streat contactor, 165
Closed system
  in technology for treatment of fermentation waste waters, 316
Clusius-Dickel columsn, 42
Coagulants, inorganic
  in chemical coagulation and sedimentation, 322
Coagulation and precipitation, 152–153
Coal Research Council in England, 189
COD, see Chemical oxygen demand
Coleman-Drew mass-transfer coefficients, 221
Colored substances, 334
Combination processes
  in gas sweetening, 214
Complex formation
  as a mechanism for the extraction of ionic species, 247
Components and their removal by microorganisms, 334–336

Compton electrons, 88, 97
Concentration polarization
  in charged membrane ultrafiltration, 266
Concentration profile, 100
Concentration profiles of polyisopotylen, 93
Concentration wave
  in the cycling zone adsorption process, 131
Conductivity
  rejection of, 275
Continuous electrophoresis, 2
Continuous chromatography, 2
Continuous column operation
  as part of the removal of salt from water by thermal cycling of ion-exchange resins, 141–146
Continuous countercurrent ultracentrifugation, 42
Continuum level of organization, 10
Continuum migration, 64
Convective augmentation of diffusional separations, 41
Cooling tower blowdown, 155, 156, 166, 167
Copper, 156–157
Correlation of vapor-liquid equilibrium data for the absorption of acidic gases in amines, 220
Countercurrent regeneration, 152
Countercurrent steam-stripping columns, 302
Counting rate method, 87
Coupled gas counterdiffusion
  in carrier-mediated transport in synthetic membranes, 237–239
Crank-Nicholson method, 75
Crank's solution
  as a model to approximate the sorption of vinyl chloride monomers by poly vinyl chloride resins, 295
Crude inulin, 360
Crystallization, as a process that separates water from the solution, 129
Cyanides, 157
Cycle time
  as part of the removal of salt from water by thermal cycling of ion-exchange resins, 140–141
Cycling Zone Adsorption
  description of, 130–132, 143, 145, 147
Cysteine
  electrolytic preparation of, 361–363
CZA, see Cycling Zone Adsorption

D

Daia® ion exchange membrane, 357, 362
Damkohler number
  in carrier-mediated transport in synthetic membranes, 229
DAO, see Deasphalted oil
Darcy flow, 11
DEA, see Diethanolamine
Deacidification
  of sugar solutions, 355–361
Deasphalted oil
  in the overhead phase for subcritical and supercritical (115°C) propane, 182
Decaffeination of coffee
  in development of dense-gas separations, 187–188
Demineralized water, 154
Dense-gas extraction, 172

as a separation process, 171, 173
distillation, 172
of low volatility liquids, 176
theory of, 173–177
Desalination of sea water
in charged membrane ultrafiltration, 281
Determination of species concentration distributions, 9
Diaphragms
definition of, 357–358
Dielectrophoresis, 2, 34
Diethanolamine
as a natural gas sweetening agent, 212
used for refinery gases, 212
Differential settling, 15
Diffusion
in decreasing thermal and concentration wave amplitudes, 132
of vinyl chloride monomer through poly vinyl chloride, 304
Diffusion coefficient
concentration dependence of, 300
determined for emulsion poly vinyl chloride, 297–299
determined for suspension poly vinyl chloride, 299
temperature dependence of, 300
Diffusion in the polymeric phase
in the removal of low molecular weight materials from polymers, 294
Dimensional analysis
in carrier-mediated transport in synthetic membrane, 229–231
Dimensionless groupings
in carrier-mediated transport in synthetic membranes, 229
Dirac, 6
Discrete level of organization, 12, 23
Dissolved solids (hardness) reduction
in charged membrane ultrafiltration, 278
Distillation
as a separation process, 129, 173
Distributed parameter description
in carrier-mediated transport in synthetic membranes, 227
Donnan equilibrium effects
in charged membrane ultrafiltration, 263–264
Donnan exclusion effects, 347
Donnan exclusion mechanism
for the rejection of ionic solutes, 261, 262, 347
Donnan ion exclusion theory, 344
Donnan membrane effect, 348
Donnan potential
in charged membrane ultrafiltration, 273
Donnan rejection trend
in charged membrane ultrafiltration, 268
Drag out, 154
Droplets
hydrohynamics of, 77
Dual-layer zirconium oxide-polyacrylic acid (Zr-PAA) membranes
in charged membrane ultrafiltration, 262
Doulite® A-368 (weak base) resin, 133, 134
Doulite® CC-3 (weak acid) resin, 133, 134
Dynamically formed membranes
industrial potential, 343, 348, 349

E

Eddy dispersion coefficient, 46
EDTA, 157
Effluent concentration profile
in the cycling zone adsorption process, 131
Electrodecantation, 42
Electrodes
in the electrolytic preparation of cysteine, 361
Electrodialysis, as a process that separates salt from the solution, 129
Electrolysis
alkaline conditions, 362–363
as treatment for waste waters, 325
flow-type, 359
under acidic conditions, 362
Electrolytic deacidification
definition of, 355
Electrolytic purification
characteristics of, 360
of sugar solutions, 355–361
Electrophoresis
definition of, 355
Electropolarization chromatography, 43
Electrostatic repulsion of ions
in charged membrane ultrafiltration, 262
Eluex process, 161
Elution, 160–163
Enrichment ratios, 7
Equilibrium breakthrough experiments
as part of the removal of salt from water by thermal cycling of ion-exchange resins, 139–140
ES, see External standard
Estimation of separation potential, 54–55
Ethane
in separations using supercritical gases, 179
Ethylene
in separations using superciritical gases, 179
Evaporation
as a process that separates water from the solution, 129
as treatment for waste waters, 325
Excess carrier
in carrier-mediated transport in synthetic membranes, 233
Experimental apparatus
in the removal of salt from water by thermal cycling of ion-exchange resins, 136
External standard, 88
Extraction of lanolin from wool
in development of dense-gas separations, 187
Extraction of residual monomers with alcohols, 293

F

Facilitated carbon dioxide transport
buffering species, 227
Facilitated transport
in analogous liquid phase systems, 225, 226
oxygen
hydration reaction, 227
nitric oxide-gerrous chloride system, 227
through liquid membranes, 286

Facilitation ratio
  in carrier-mediated transport in systhetic membranes, 229
Feasibility of available separations techniques, 53
Feed concentration, 130
Feed patterns and species velocities at the microscopic continuum level, 25
Feed preparation and ingenuity, 40–41
Feed schedule, 27
Fermentation
  high-temperature
    as a kind of methane fermentation, 318
  industries
    in Japan, 314
  medium-temperature
    as a kind of methane fermentation, 318–319
  waste
    as fertilizers, 327–329
    problems, 330–339
    treatment, 330–339
Fermentation waste water
  characteristics of, 314–315
  microflora investigation, 332
  technology for treatment of, 315
Fick's law, 11, 55, 296
  as applied to experimental situations involving vinyl chloride monomers and poly vinyl chloride, 295
  in carrier-mediated transport in synthetic membranes, 228–229
Field flow, 42
Field-induced migrations in pseudobinary systems, 32
Filtration, 15
Finite mass transfer resistance
  in the cycling zone adsorption process, 131
Fixed bed, 163
Fixed bed co-flow ion exchange, 168
Fixed bed downflow contacting method, 163
Fixed charge model
  in charged membrane ultrafiltration, 265
Flavobacterium sp.
  isolated from the aeration tanks, 334
Flocculation aids, 160
Flory-Huggins theory, 179
Fluid extraction, see Dense-gas extraction
Fluor Econamine process
  using ethanol as a sweetening agent, 212
Fluor solvent process
  as a physical solvent in gas sweetening, 214
Foam fractionation, 42
Fourier analysis, 73
Fourier number, 80
Fractional water recovery
  in charged membrane ultrafiltration
    definition of, 273
Fractionation, 42
Frasch process for sulfur production, 211
Free-flow electrophoresis, 41
Freezing, as a process that separates water from the solution, 129
Froth flotation techniques, 40
Fructose
  preparation, 360
Frumkin-Levich theory, 78
Fugacity coefficient, 174
Fungi
  as biological waste treatment, 319

G

Galactose
  preparation of, 360
Gas absorption and/or stripping in adiabatic packed towers, 221
Gas chromatography, 84
Gas extraction, see Dense-gas extraction
Gasification
  as a process in methane fermentation, 318
Gas-solid systems
  in the cycling zone adsorption process, 132
Gas sweetening, 211
Gaussian probability function, 21
Gel model theory, 351
Giammarco-Vetrocoke process
  activated carbonate process, 213
Gibbs relation, 95
Glucose
  crystalline, 358
  preparation of, 358–359
Glueckauf's correlation, 46
Glutaraldehyde solution
  as part of an ion-exchange resin, 135
Gracilaria confervoides, 360
Gravitational-hydrodynamic dipole, 35

H

Hadamard-Rybczynski, 121
Halibide-bridged complexes
  structure of, 197
Handlos-Baron model, 84
Headspace GC analytical method, 309
Heavy metal ions
  removal of, 357
Hemoglobin
  as carriers for oxygen and carbon monoxide, 226
Henry's constant, 177
Henry's law, 176, 220
  in carrier-mediated transport in synthetic membranes, 228
HF, see Hyperfiltration
Higgins contactor, 165
High-pressure distillation, 172
High-speed, 2
Himsley contactor, 165
Hindered settling, 15
Hi-Pure process
  which combines two solvents in gas sweetening, 214
Hot carbonate processes
  for the removal of carbon dioxide from synthesis gas streams, 213
sp Hybridization
  of the fluorine atom, 196

Hydration reaction
  in facilitated carbon dioxide transport, 227
Hydrochloric acid
  used as an electrolyte, 362
Hydrodynamic diffusion theory, 64
Hydrodynamics and mass transfer
  mathematical models of, 72–84
Hydrogen bonding, 345
Hydrolysis of a polysaccharide with acid, 356
Hydrous zirconium (IV) oxide
  incharged membrane ultrafiltration, 262
Hydroxyl ions
  in coupled gas counter diffusion, 238
Hyperfiltration, 343, 352
Hyperfiltration membranes, 350

I

Ideal cascades, 23
Immunodiffusion, 19
Inclusion compounds
  host molecules, 195
  macromolecular, 195
  thiourea, 195
  with a host lattice constructed from small molecules, 195
  urea, 195
Infrared studies, 197
In *situ* regeneration, 343
Instantaneous Sherwood numbers, 115
Interfacial resistance, 100
Intermediate droplet Reynolds numbers, 117
Institut Francais du Pétrole process
  in gas sweetening, 216
Ion association model
  in charged membrane ultrafiltration, 265
Ionic considerations in facilitated transport, 228
Ion exchange, 152, 167
  application
    chromium, 154–167
  as a process that separates salt from the solution, 129
  equipment, 163
  resins, 154
    thermally sensitive, 132
  separation of metal ions from water, 152
  system, 132–136
Ion-pair formation
  as a mechanism for the extraction of ionic species, 247
Ion-pair forming liquid-liquid extraction system, 245
Ion-retardation resin
  as part of an ion-exchange system, 134
Isoelectric focusing electrophoresis, 15
Isotachophoresis, 2

J

Jigging, 15

K

Klebsiella sp.
  isolated from the aeration tanks, 334
Knudsen flow range, 10
Kraft pulping and bleaching processes
  treatment of waste effluents generated in, 349
Kronig-Brink model, 104
Kyowa Hakko Kogyo Co., Ltd.
  in classification of wastewaters, 316

L

Lactose, 360
Langmuir isotherms
  as part of the removal of salt from water by thermal cycling of ion-exchange resins, 137
Laplace operator method, 75
Laplace transformation, 47
Lattice energy, 196, 207
Leach liquor, 159
Lennard-Jones 6-12, 175
Lewis acids, 189
Limiting solution
  in carrier-mediated transport in synthetic membranes, 232–235
Linear irreversible thermodynamic analysis, 33
Liquefaction
  as a process in methane fermentation, 318
Liquid chromatography, 2
Liquid clathrate behavior
  definition, 198–199
  origin of, 207
  solid state clues to, 205–207
Liquid clathrates
  applications to, 207–208
  aromatic: anion ratio in, 200–202
  aromatic separation, 207–208
  behavior of, 196
  composition of, 200–202
  factors affecting the constitution of, 202–205
  in NMR spectra, 203
  preparation of, 199–202
  stability of, 207
  synthesization of, 199
Liquid extraction, as a process that separates salt from the solution, 129
Liquid ion exchange, 153
Liquid ion exchange reagent, 285
Liquid membrane – liquid ion exchange process
  in removal of anions from wastewater streams, 286
Liquid membranes
  characteristics of, 285–286
Liquid membrane system
  a water-immiscible emulsion dispersed in water, 285
  an oil-immiscible emulsion dispersed in oil, 285
Liquid-phase activity coefficients
  description of, 176

Liquid scintillation technique, 86
Liquid-solid systems
  in the cycling zone adsorption process, 132
Loading, 160, 162
Lo-Cat process
  in gas sweetening, 216
Loeb-Sourirajan cellulose acetate, 344
Long-chain fatty acid, 252
Longitudinal mixing
  in decreasing thermal and concentration wave amplitudes, 132
Low droplet Reynolds number, 103, 116–117
Luminescence, 86
Lumped parameter description
  in carrier-mediated transport in synthetic membranes, 227–228

M

$M[Al_2 R_6 X]$
  anionic structure of, 197–198
Magneto-electrophoresis, 34
Manhattan Project, 6, 55
Manufacturing processes, changing of
  in technology for treatment of fermentation waste waters, 316
Marangoni instabilities, 90
Mass spectrometry, 34
Mass transfer, 120
  across the interface of two liquids, 72
  between the droplets and the continuous phase, 79
  from the toluene to the water phase, 100
  from the water to the toluene phase, 90–93, 97
  on single droplets, 100–120
  surface-active agents, 120
  stationary two-phase laminar flow, 74
  with a swarm of droplets
    theory, 121
  with droplets
    experimental methods, 121–123
Mass-transport modeling in well-defined flow situations, 44
Mathematical models
  in carrier-mediated transport in liquid-liquid membrane systems, 245–247
Max Planck Institute for Biophysics, 345
MEA, see Monoethanolamine
Mechanisms of separations, 30–53
Medium droplet Reynolds numbers, 108
MEK, 180
Membrane characteristics
  in charged membrane ultrafiltration, 266
Membrane processes
  as treatment for waste waters, 325
  in treatment of fermentation waste waters, 339
Membrane rejection and water flux models, 263–266
Membranes
  carrier-mediated, 258
  with specific transport properties, 243
Membranes, dynamically formed gelled enzyme, composite, 350
Membrane separations, 2

Membrane swelling
  in charged membrane ultrafiltration, 269, 270, 272
Membrane transport, 37
Merry-go-round system of adsorption, 160
Methane
  in separations using supercritical gases, 177–179
Methanobacterium
  as a methane-producing bacteria, 318
Methanococcus
  as a methane-producing bacteria, 318
Methanosarcina
  as a methane-producing bacteria, 318
Michaelis constant, 350
Michaelis-Menten kinetics, 351
Microbial cells grown on waste waters, utilization of, 326
Microbial denitrification
  as a biological waste treatment, 319
Microbial flocs
  formation of, 332
Micrococcus denitrificans, 289
Microflora characteristics in the biological treatment, 332–334
Microresin mixtures
  as part of an ion exchange system, 134
Microscopic continuum formulation, 30
Migration of vinyl chloride monomer residuals from poly vinyl chloride pipes and bottles into noninteracting fluid contents, 303–311
Millipore® filter membranes
  in carrier-mediated transport in synthetic membranes, 235
Millipore molecular filter, 344
Mobile carrier
  in oxygen transport, 226
Molecular asymmetry, 35
Molecular level of organization, 9, 22
Molecular sieves, 214
Molecules of solvation (inclusion compounds), 196
Molybdenum, 162
Monoethanolamine
  for use as a sweetening agent, 212
Morphological classes of separations, 24
Morphology, 61–64
Morphology of separations processes, 12
Moving bed column systems, 163
Myoglobin, 231
Myoglobin
  as carriers for oxygen and carbon monoxide, 226

N

Navier-Stokes equations, 72, 88
Near-diffusion regime
  in carrier-mediated transport in synthetic membranes, 231
Nernst-Planck equation, 38
  in charged membrane ultrafiltration, 265
Nernst-Planck model
  in charged membrane ultrafiltration, 265
Nernst-Planck term, 228
Neutralization and purification of a solution by electrolysis, 356

Newman model, 104
Nickel, 157–159
NIM contactor, 165
Nitric oxide-ferrous chloride system
  in facilitated oxygen transport, 227
Nonisotropic systems, 33
Nonlinear effects, 33
Nonstoichiometric liquid enclosure compounds, see
  Liquid clathrate behavior

O

Oak Ridge National Laboratory, 262, 344
Occupational Safety and Health Administration
  for minimizing exposure of production workers to solvent and monomer vapors, 294
Oil recovery
  in development of dense-gas separations, 188–189
One-dimensional migration with nonuniform species velocity, 21
Operating characteristics of uranium, 160
Optical density
  rejection of, 275
d Orbital participation
  by the aluminum atom, 196
Organic carbon
  rejection of, 275
Organic scintillator molecules, 86
Organophosphorous compound, 252
ORNL, see Oak Ridge National Laboratory
Oxygen consumption rate during the biological treatment and treatment of high loads of waste, 332
Oxygen transport, 225
  facilitated, 226
  through films of immobilized and free hemoglobin, 226

P

PAA, see Polyacrilic acid
Parametric pumping, 2
Parent compounds $M[Al_2 R_6 X]$
  preparation of, 196–197
Particulate adsorption, 42
Peclet number, 46, 51
Penetration theory, 99
Perchloric acid, 356
Permselective membranes
  in carrier-mediated transport in synthetic membranes, 239
Perpendicular shifts, 15
Pfeffer's correlation, 46
pH involving a selenium solution
  in ultrafiltration of tonic metal constituents from metal manufacturing wastewater, 276–277
Phase equilibrium curve
  in the cycling zone adsorption process, 131
Photosynthetic bacteria
  as a biological waste treatment, 319
Physical characteristics of uranium, 160
Physical solvent processes
  in gas sweetening, 213

Plating rinse waters, 154–155
Platinum electrode, 362
Polarization chromatography, 2, 42, 54
Polarization chromatography and taylor diffusion, 48
Pollution control processes
  in gas sweetening, 215
Polyacrylic acid, 344
Porter contactor, 165
Poynting correction, 177
Poynting factor, 174
Precipitation, 163
  of uranium, 161
Prediction of equipment performance, 59
Preferential sorption-capillary flow
  in charged membrane ultrafiltration, 262
Pressure diffusion, 54
  of skew particles, 34
Pretreatment, 162
Process developments
  in removing hydrogen sulfide from natural gas, 212
Propane
  in separations using supercritical gases, 180–183
Protein separation by SDS-complexing, 40
Pseudo-binary diffusion, 38
Pseudo-continuum level of organization, 11, 12, 22, 25
Pseudohalide complexes, 197
Pseudomonas sp.
  isolated from the aeration tanks, 334
Pseudo-ternary approximations, 38
Purisol process
  as a physical solvent in gas sweetening, 214
PVC, see Poly vinyl chloride

R

Raman studies, 197
Raoult's law, 176
Rate of vinyl chloride monomer loss, 309
Rate of vinyl chloride monomer migration, 309
Reaction equilibrium limit
  in carrier-mediated transport in synthetic membranes, 229
Reclaimer
  definition of, 212
Recovered nickel concentration, 158
Rectisol® process
  as a physical solvent in gas sweetening, 214
Redlich-Kwong equation of state, 176
Redundancy, 28
Rejection and solute flux
  in charged membrane ultrafiltration, 264–266
Rejection behavior
  of chloride salts
    incharged membrane ultrafiltration, 268
  of oxyanions
    in charged membrane ultrafiltration, 267–268
  of sulfate salts
    in charged membrane ultrafiltration, 268–272
Rejection of hydroquinone as a function of pH
  in ultrafiltration of photographic processing water constituents, 278
Removal of anions from waste water streams, 286–287

Reserve osmosis, 153
Residual chemical oxygen demand, 334
Resin characteristics of uranium, 159
Resin-in-pulp process, 154, 160, 164
Resin ion exchange, 153–154
Resin poisons, 161
Reverse osmosis
  as a process that separates water from the solution, 129
  as a tool of fermentation waste treatment, 337–339
  hyperfiltration process
    in charged membrane ultrafiltration, 262
Reynolds number, 35, 46, 64, 111
RIP, see Resin-in-pulp
R.O., see Reverse osmosis
Rochelle, 157

S

System
  carrier-mediated transport, 253

T

Takahax process
  in gas sweetening, 216
Takuma Co. of Japan
  in liquid membrane systems, 291
  in waste water treatment by liquid ion exchange, 291
Tapered cascades, 23
Taylor dispersion, 11, 28, 41, 46, 48, 50, 54
Taylor series, 75
Temporal gradients in velocity, 64
Thermal diffusion processes, 57
Thermodynamic polarization diffusional polarization, 42
Thick film theory
  in carrier-mediated transport in synthetic membranes, 232
Thin film theory
  in carrier-mediated transport in synthetic membranes, 231
Thiourea, see Inclusion compounds
TOD, see Total oxygen demand
Toluene
  reaction with, 196
Toluene phase, 97, 100
Total oxygen demand
  in the measurement of organic matter in a waste water, 330
Tracer diffusion
  in carrier-mediated transport in synthetic membranes, 235–237
Traveling wave mode
  in the cycling zone adsorption process, 132
Tricarb 3003, 102
True phase boundaries, 37
Two-phase film flow, 84
  experimental methods, 121–122
  experimental setup and measurement, 88–100
  hydrodynamics of, 72
  theory, 121
  velocity profiles, 84
Two-film theory, 72

U

UF, see Ultrafiltration, 343
Ultracentrifuges, 57
Ultrafiltration
  dynamically formed membrane, 343
  for the separation and concentration of various inorganic salts present in aqueous solution, 261
  secondary layer, 352
Ultrafiltration-induced polarization chromatography, 42, 48
Ultrafiltration membranes, 350
Ultrafiltration of photographic processing water constituents
  in charged membrane ultrafiltration, 278–280
Ultrafiltration of plating rinse waters
  in charged membrane ultrafiltration, 274–276
Ultrafiltration of toxic metal constituents from metal manufacturing wastewater
  in charged membrane ultrafiltration, 276–277
Ultrafiltration results with various inorganic salts
  in charged membrane ultrafiltration, 267–273
Uniform velocity, 64
United States Atomic Energy Commission, 159
UPC, see Ultrafiltration-induced polarization chromatography
Uranium, 159–167
Urea, see also Inclusion compounds, 195
Utilization of useful waste components as feed stuffs, 326–327
Utilization of useful waste components as fertilizers, 327

V

Vanadium, 162
Vanadium salts
  in gas sweetening, 216
van Laar model, 176
Variants on gravitational settling, 13
VCM, see Vinyl chloride monomer
Velocity profiles, 84, 90
Vinyl chloride monomer
  removal of, 294
Vinyl chloride monomer loss during storage, 304–305
Vinyl chloride monomer migration from freshly formed poly vinyl chloride products into finite fluid contents, 305–308
Vinyl chloride monomer migration from thin-walled bottles into finite closed contents, 311
Vinyl chloride monomer migration in closed-system service from previously aged poly vinyl chloride products, 308–311
Vinyl chloride monomer transport and equilibria in poly vinyl chloride
  experimental characterization of, 294–296
Vinyl chloride monomer transport in poly vinyl chloride powders, 295–296

W

Waste components, useful

technology for recovery and utilization of, 326
Waste feed
 equalization of, 316
Waste flows
 surveys of seasonal and daily changes of, 316
Waste waters
 classification of, 316
 high-strength, 316, 318
 lower-strength, 316
Water
 in separations using supercritical gases, 186–187
Water bulk phase, 100
Water flux
 with rinse waters, 275
Water flux and concentration polarization
 in charged membrane ultrafiltration, 266
 high-strength, 318
Water flux behavior with chloride salts
 in charged membrane ultrafiltration, 272–273
Water flux behavior with sulfate salts
 in charged membrane ultrafiltration, 272–273
Water phase, 97, 100
Water-toluene-pyridine system, 97
Weak base resins, 156
Weiss parameter, 60

West Rand Consolidated Mines, 159
Woertz process
 as a physical solvent in gas sweetening, 214

X

Xylose
 electrolytic production of, 359–360
 preparation of, 359–360

Y

Yeast
 as a biological waste treatment, 319
 in the alcohol fermentation waste, 327
Yellow cake, 161

Z

Zinc Removal By Ion Exchange, 166
Zonal Centrifugation, 2
Zone Melting, 15

CRC PUBLICATIONS OF RELATED INTEREST

CRC HANDBOOKS

CRC HANDBOOK OF TABLES FOR APPLIED ENGINEERING SCIENCE, 2nd Edition
Edited by **Ray E. Bolz, D.Eng.,** Case Western Reserve University and **George L. Tuve, Sc.D.,** Case Institute of Technology.

This Handbook is designed to provide a wide spectrum of data covering many fields of modern engineering with reference to more complete sources and includes presentation of data in metric as well as in conventional units.

CRC HANDBOOK OF CHEMISTRY AND PHYSICS, 58th Edition
Edited by **Robert C. Weast, Ph.D.,** Consolidated Natural Gas Co., Inc.

This Handbook is the definitive reference for chemistry and physics and maintains the tradition that has earned it the reputation as the best scientific reference in the world.

CRC HANDBOOK OF CHROMATOGRAPHY, Vols. I and II
Edited by **Gunter Zweig, Ph.D.,** Chief, Chemistry Branch, EPA and **Joseph Sherma,** Lafayette College.

This two-volume set provides comprehensive information concerning chromatographic data, methods and literature. It also contains a Compound Index that lists the more than 12,000 compounds referenced in this data collection.

CRC HANDBOOK OF LABORATORY SAFETY, 2nd Edition
Edited by **Norman V. Steere.**

This Handbook is a complete treatise on personal hazards and safety for laboratory personnel and includes eight color pages on accidents and safety hazards in the laboratory.

CRC UNISCIENCE PUBLICATIONS

CHEMISTRY AND PHYSICS OF SOLID SURFACES
Edited by **Ralf Vanselow, Ph.D.,** University of Wisconsin, and **S. Y. Tong, Ph.D.,** Naval Research Laboratory.

This book is based on the proceedings of the 1975 International Summer Institute on Surface Science held at the University of Wisconsin, Milwaukee.

FUNDAMENTAL MEASURES AND CONSTANTS FOR SCIENCE AND TECHNOLOGY
By **Frederick D. Rossini,** Rice University.

Invaluable to working scientists, as well as students in science or engineering, who need to know the basis and current status of the measurements involved in their respective disciplines.

CRC CRITICAL REVIEW JOURNALS

CRC CRITICAL REVIEWS™ IN ANALYTICAL CHEMISTRY
Edited by **Bruce H. Campbell, Ph.D.,** J. T. Baker Chemical Co.